Methods of
BEHAVIOR ANALYSIS
in NEUROSCIENCE

Edited by

Jerry J. Buccafusco

CRC Press
Boca Raton   London   New York   Washington, D.C.

Publisher/Acquiring Editor: Barbara Norwitz

Library of Congress Cataloging-in-Publication Data

Methods of behavior analysis in neuroscience / edited by Jerry J. Buccafusco.
      p.   cm. — (Methods and new frontiers of neuroscience)
   Includes bibliographical references and index.
   ISBN 0-8493-0704-X (alk. paper)
   1. Neurosciences—Handbooks, manuals, etc. 2. Animal behavior—Handbooks,
   manuals, etc. 3. Laboratory animals—Psychology—Handbooks, manuals, etc. 4.
   Behavioral assessment—Handbooks, manuals, etc. I. Buccafusco, Jerry J. II. Series.
[DNLM: 1. Behavior, Animal. 2. Neurosciences—methods. 3. Animals,
Laboratory—psychology. WL 100 M5925 2000]
RC343.M45 2000
616.8′027—dc21

00-039762

Visit the CRC Press Web site at www.crcpress.com

© 2001 by CRC Press LLC

No claim to original U.S. Government works
International Standard Book Number 0-8493-0704-X
Library of Congress Card Number 00-039762
Printed in the United States of America   2 3 4 5 6 7 8 9 0
Printed on acid-free paper

Series Preface

Our goal in creating the Methods & New Frontiers in Neuroscience Series is to present the insights of experts on emerging experimental techniques and theoretical concepts that are, or will be at the vanguard of Neuroscience. Books in the series will cover topics ranging from methods to investigate apoptosis, to modern techniques for neural ensemble recordings in behaving animals. The series will also cover new and exciting multidisciplinary areas of brain research, such as computational neuroscience and neuroengineering, and will describe breakthroughs in classical fields like behavioral neuroscience. We want these to be the books every neuroscientist will use in order to get acquainted with new methodologies in brain research. These books can be given to graduate students and postdoctoral fellows when they are looking for guidance to start a new line of research.

The series will consist of case-bound books of approximately 250 pages. Each book will be edited by an expert and will consist of chapters written by the leaders in a particular field. Books will be richly illustrated and contain comprehensive bibliographies. Each chapter will provide substantial background material relevant to the particular subject. Hence, these are not going to be only " methods books." They will contain detailed "tricks of the trade" and information as to where these methods can be safely applied. In addition, they will include information about where to buy equipment, web sites that will be helpful in solving both practical and theoretical problems, and special boxes in each chapter that will highlight topics that need to be emphasized along with relevant references.

We are working with these goals in mind and hope that as the volumes become available the effort put in by us, the publisher, the book editors, and individual authors will contribute to the further development of brain research. The extent that we achieve this goal will be determined by the utility of these books.

Preface

Behavioral techniques used in animals to model human diseases and to predict the clinical actions of novel drugs are as varied as the numbers of scientists who use them. For behaviors as simple as locomotor performance in an open field, there are dozens of experimental approaches and various components of movement that contribute to overall motor activity. As behavioral models become more complex, there is often a bewildering array of perturbations for a given task extant in the literature. Yet behavioral analysis, as a tool for the basic neuroscientist, is becoming indispensable as information gained at the molecular and cellular level is put into practice in fully behaving animal subjects.

Since the neuroscientist trained in methodologies directed toward the molecular and cellular level does not often have experience in the intricacies of animal behavioral analyses, there is often much time devoted to assessing a complex literature, or to developing an approach *de novo*. Specialists who are recognized experts in several fields of cognitive and behavioral neuroscience have provided chapters that focus on a particular behavioral model. Each author has analyzed the literature to describe the most frequently used and accepted version of the model. Each chapter includes: (1) a well-referenced introduction that covers the theory behind, and the utility of the model; (2) a detailed and step-wise methodology; and (3) approach to data interpretation. Many chapters also provide examples of actual experiments that use the method.

The primary objective of the book is to provide a reference manual for use by practicing scientists having various levels of experience who wish to use the most well-studied behavioral approaches in animal subjects to better understand the effects of disease, and to predict the effects of new therapeutic treatments on the human cognition. In view of the large numbers of transgenic animals produced on an almost daily basis, special attention is given to procedures designed for testing mice. While there has been no attempt to cover all areas of animal behavior and sensory processing, this text will help take the guesswork out of designing the methodology for many of the most widely used animal behavioral approaches developed for the study of brain disorders, drug abuse, toxicology, and cognitive drug development.

One cautionary note to the newcomer to the behavioral field is to not be deceived by the manner in which this book presents its topics. As a matter of convenience, the topics have been arranged in chapter form, and there may be the false sense of security that each method described is the last word on the subject. However, it is often not sufficient to employ one of these methods to assess the cognitive status of an animal. For example, when studying memory or recall, it is prudent to use a test battery that can better provide a comfortable level of interpretation of the effect of the perturbation applied to the subject. Spatial and non-spatial tasks should be considered. If a negative reinforcer is involved, such as electrical shock, the animal should be tested for his response to pain. Drugs or other manipulations that might alter pain sensitivity could give false impressions in a shock-motivated memory task. Drugs that affect motor activity may alter maze activity or swimming behavior, and drugs that alter taste, appetite, or that induce GI disturbances could affect food-motivated behaviors.

Whenever possible, the animal should be observed (at least initially) while performing the task. It is often surprising to some investigators (this one included) to find that the animal is using a technique to solve the problem posed to it that was not considered in designing the task. A good example is the mediating or non-mnemonic strategies that rats use to solve matching problems in various operant paradigms. Most animals would rather use such strategies (such as orientating to a proffered lever) to obtain food rewards than to use memory. Whenever possible, our authors have provided some of these pitfalls in their chapters, although every possible contingency could not be anticipated. Thus, it is in the best interest of the investigator to use this book to help develop several strategies to understand the complex behaviors of animals as they respond to drugs, new diets, surgical interventions, or to additional or fewer genes. While danger in anthropomorphizing the behavior of animals always exists, the investigator should feel some level of confidence that much of the behavioral literature is replete with instances of high predictive value for similar perturbations in humans. Of course, species and strain differences can limit such interpretations. Mice are clearly not little rats, and rats are not non-human primates. Each species has a specific level of predictive value that should be assessed.

A final cautionary note is that investigators make every attempt to be as reproducible as possible when studying animal behavior. Handlers, experimenters, food, water, bedding, noise, surrounding visual cues, are just a few of the factors that should be held constant when performing behavioral studies. Inconsistency contributes mightily to response variability in a population, and may even lead to a completely opposite behavior to the one expected.

At this point I would like to express my sincere thanks to the many authors who contributed these chapters. Their difficult task in preparing this information will make easier the tasks of our readers in their own efforts to assess animal behavior. I would also like to acknowledge the support (moral and technical) of the CRC staff, Publisher Barbara Norwitz, and Editorial Assistant Amy Ward, and the Methods in Neuroscience Series Editors, Sidney Simon and Miguel Nicolelis. Finally, I would like to thank my office administrator Vanessa Cherry for her many contributions in getting this book together for publication, and to my wife Regina Buccafusco, who is an education design specialist, for helping in the proofreading of the chapters.

The Editor

Jerry J. Buccafusco, Ph.D., is director of the Alzheimer's Research Center, in the Department of Pharmacology and Toxicology of the Medical College of Georgia. He holds a joint appointment as Research Pharmacologist at the Department of Veterans Affairs Medical Center. He holds the rank of Professor of Pharmacology and Toxicology and Professor of Psychiatry and Health Behavior. Dr. Buccafusco was trained classically as a chemist, receiving an MS degree in inorganic chemistry from Canisius College in 1973. His pharmacological training was initiated at the University of Medicine and Dentistry of New Jersey where he received a Ph.D. degree in 1978. His doctoral thesis concerned the role of central cholinergic neurons in mediating a hypertensive state in rats. Part of this work included the measurement of several behavioral components of hypothalamically mediated escape behavior in this model. His postdoctoral experience included two years at the Roche Institute of Molecular Biology under the direction of Dr. Sydney Spector. In 1979 he joined the Department of Pharmacology and Toxicology of the Medical College of Georgia. In 1989 Dr. Buccafusco helped found and became the director of the Medical College of Georgia, Alzheimer's Research Center. The Center hosts several core facilities, including the Animal Behavior Center, which houses over 30 young and aged rhesus monkeys who participate in cognitive research studies.

Awards and honors resulting from Dr. Buccafusco's research include the New Investigator Award, National Institute on Drug Abuse, 1980; Sandoz Distinguished Lecturer, 1983; Distinguished Faculty Award for the Basic Sciences, School of Medicine, Medical College of Georgia, 1988; Callaway Foundation of Georgia, Center Grant recipient, 1989; and the Distinguished Alumnus Award, University of Medicine and Dentistry of New Jersey, 1998. Dr. Buccafusco also served as a member of the Pharmacology II Study Section of the National Institute on Drug abuse from 1989–1991, and he is a member the Scientific Advisory Board of the Institute for the Study of Aging in New York.

Dr. Buccafusco holds memberships in several scientific societies. In his professional society, the American Society for Pharmacology and Experimental Therapeutics, he serves as Chairman of the Graduate Student Convocation subcommittee, and member of the Education Committee. He also serves as Associate Editor (Neuro-Behavioral Pharmacology section) of the *Journal of Pharmacology and*

Experimental Therapeutics. Finally, Dr. Buccafusco was an invited speaker and discussant at the National Institutes on Aging symposium on Age-Related Neurobehavioral Research: An Integrative Cognitive Neuroscience Agenda for the 21st Century, held in 1999.

Dr. Buccafusco's research area includes the development of novel treatment modalities for Alzheimer's disease and related disorders. In 1988 his laboratory was the first to report the cognitive enhancing action of low doses of nicotine in nonhuman primates. Since that time he has studied numerous novel memory-enhancing agents from several pharmacological classes in this model. His most recent work is directed at the development of single molecular entities that act on multiple CNS targets, not only to enhance cognitive function, but also to provide neuroprotection, or to alter the disposition and metabolism of amyloid precursor protein. Dr. Buccafusco also has studied the toxic effects of organophosphorus anticholinesterases used as insecticides and as chemical warfare agents. In particular, he has studied the behavioral/cognitive alterations associated with low level, chronic exposure to such agents. Finally, his work in the area of drug abuse has centered around the role of central cholinergic neurons in the development of physical dependence on opiates, and in the expression of withdrawal symptoms. These studies have been supported by continuous federally sponsored grants awarded by the National Institutes of Health, the Department of Defense, and the Veterans Administration.

Contributors

Adam C. Bartolomeo
Wyeth Neuroscience
Wyeth-Ayerst Research
Princeton, NJ

Carl A. Boast, Ph.D.
Behavioral Neuroscience
Wyeth-Ayerst Research
Princeton, NJ

Bruno Bontempi, Ph.D.
Laboratoire de Neurociences
  Comportementales et Cognitives
Talence, France

Peter J. Brasted, Ph.D.
Department of Experimental
  Psychology
University of Cambridge
UK

Jerry J. Buccafusco, Ph.D.
Alzheimer's Research Center
Department of Pharmacology and
  Toxicology
Medical College of Georgia and
  Department of Veterans Affairs
  Medical Center
Augusta, GA

Philip J. Bushnell, Ph.D.
Neurotoxicology Division
U.S. Environmental Protection
  Agency
Research Triangle Park, NC

Peter Curzon
Abbott Laboratories
Abbott Park, IL

Michael W. Decker, Ph.D.
Abbott Laboratories
Abbott Park, IL

Stephen B. Dunnett, Ph.D.
MRC Cambridge Center for
  Brain Repair
University Forvie Site
Cambridge, UK

Ralph L. Elkins, Ph.D.
Department of Psychiatry and
  Health Behavior
Medical College of Georgia and
  Medical Research Service
Department of Veterans Affairs
  Medical Center
Augusta, GA

Gerard B. Fox, Ph.D.
Abbott Laboratories
Abbott Park, IL

John H. Graham, Ph.D.
Biology Department
King College
Bristol, TN

Leonard L. Howell, Ph.D.
Department of Psychiatry and
  Behavioral Sciences
Yerkes Regional Primate Center
Emory University
Atlanta, GA

William J. Jackson, Ph.D.
Department of Physiology and
  Endocrinology
Medical College of Georgia
Augusta, GA

Robert Jaffard, Ph.D.
Laboratoire de Neurociences
  Comportementales et Cognitives
Talence, France

John R. James, Ph.D.
Department of Pharmacy and
  Pharmaceutics
School of Pharmacy
Virginia Commonwealth University
Richmond, VA

Edward D. Levin, Ph.D.
Department of Psychiatry and
  Behavioral Sciences
Duke University Medical Center
Durham, NC

Frédérique Menzaghi, Ph.D
Arena Pharmaceuticals, Inc.
San Diego, CA

T. Edward Orr, Ph.D.
Department of Psychiatry and
  Health Behavior
Medical College of Georgia and
  Medical Research Service
Department of Veterans Affairs
  Medical Center
Augusta, GA

Merle G. Paule, Ph.D.
Behavioral Toxicology Laboratory
National Center for Toxicology
  Research
Jefferson, AR

Mark A. Prendergast, Ph.D.
Tobacco and Health Research
  Institute
University of Kentucky
Lexington, KY

John A. Rosecrans, Ph.D.
Department of Pharmacology
Virginia Commonwealth University
Richmond, VA

Jay S. Schneider, Ph.D.
Department of Pathology
Anatomy and Cellular Biology
Thomas Jefferson University
Philadelphia, PA

Sheldon B. Sparber, Ph.D.
Department of Pharmacology
University of Minnesota
  Medical School
Minneapolis, MN

Alvin V. Terry, Jr., Ph.D.
University of Georgia Clinical
  Pharmacy Program Medical
  College of Georgia
Augusta, GA

Thomas J. Walsh, Ph.D.
Department of Psychology
Rutgers University
New Brunswick, NJ

Paul A. Walters
Department of Psychiatry and Health
  Behavior
Medical College of Georgia and Medical
  Research Service
Department of Veterans Affairs Medical
  Center
Augusta, GA

Kristin M. Wilcox, Ph.D.
Division of Neuroscience
Yerkes Regional Primate
  Research Center
Emory University
Atlanta, GA

Richard Young, Ph.D.
Department of Medicinal
  Chemistry
School of Pharmacy
Virginia Commonwealth University
  Richmond, VA

In Memoriam

Sadly, Dr. Thomas J. Walsh, born in 1952, died suddenly on May 12, 2000. His many contributions to the understanding of basic concepts related to learning and memory and to improving animal models of cognitive impairment is very much appreciated. As a teacher, mentor, friend, researcher and collaborator, he is greatly missed.

Contents

Chapter 1

Choice of Animal Subjects in Behavioral Analysis

William J. Jackson

Contents

I. Introduction

Many researchers using behavioral techniques are not primarily interested in animal behavior, as such. Typically, behavioral animal research in physiology and pharmacology is designed to provide a model for human processes, and great effort is given toward the development of animal models that reflect behavioral processes shared by animals and humans.[1] Whenever using animals as research subjects, behavioral physiologists, pharmacologists, and geneticists commonly describe their work as animal models of human characteristics, or justify their work on the basis of relevance to human pathology. The search for treatment and cure of illness with behavioral implications will continue to lean heavily upon animal models. Animal subjects will assist in evaluation of the effectiveness of putative treatments, and in providing further insight into underlying physiologic mechanisms of human pathology. Subsequent chapters of this book focus on many of these models. Anxiety, addiction, taste-aversion, attention deficit, and disorders of learning and memory are examples of behavioral disorders that are often studied *via* animal models.

Behavioral researchers usually plan most aspects of their research projects to fine detail, but may fail to give the same level of planning toward selection of the species that is to be used as a model. The goal of this chapter is to provide a glance at some of the most popular species used for behavioral neuroscientific research. Most of the discussion is given to rats and non-human primates, but there is a section on mice and notes about other species as well.

II. Origin of the Albino Laboratory Rat

Contemporary strains of albino laboratory rats were bred from captured wild Norway rats. The wild Norway rat is believed to have originated from temperate regions of

Asia and southern Russia. As civilization developed, these animals found a suitable ecological niche in the castoffs and trash heaps of man and, as economic pests, spread rapidly over the world. Norway rats were common throughout Europe and the British Isles by the early 18th century. By 1775, Norway rats were common in the northeastern U.S.[2] Because the large number of rats represented an economic hazard, special breeds of rat-catching terrier dogs were bred. In a roundabout way, the breeding of these dogs was responsible for the beginnings of the albino laboratory rat.

The recorded speculation of early breeders of albino strains was that they were products of the "sport" of rat baiting, which was outlawed by decree.[3] Rat baiting involved the release of 100 or 200 newly trapped rats into a fighting pit. A trained terrier dog was then put into the pit, and a measurement of the time until the last rat was killed was taken. Wagers were placed on the speed of the various dogs. Rat baiting required that many rats be trapped and retained in pounds. Historical records relate that albinos were removed from these pounds and kept for show purposes and breeding. Many albino show rats were tamed and offspring were selected for docility and color. Because the captured wild Norway rats were fierce and difficult to handle, the more docile albinos were the stock from which early European laboratory researchers selected their first animals.

III. The Laboratory Rat in Behavioral Research

After serving for a hundred years as a subject in behavioral and physiological research experiments, the albino rat is a generally accepted model. However, early researchers had to justify their selection of animals, as opposed to humans, and it would be valuable for us to remember their rationale. Memories of the initial battles to gain acceptance for animal models of human cognition have dimmed over time; today, studies using animals are commonly accepted.

Behavioral investigation using rats as subjects in the U.S. arose from work at Clark University Biological Laboratory during the 1890s. According to Miles,[4] Stewart was working with wild gray rats to determine the effects of alcohol, diet, and barometric change on the activity of the animals as early as 1894. The feral-captured gray rats were fierce and difficult to handle, and Stewart was forced to switch to the more docile white rats by 1895. The rats were trained to run through the maze to earn food reward. Kline[5] invented several problem boxes that served as prototypes for contemporary devices still used to measure cognition and learning/memory in rodents. Small used a maze patterned after the garden maze found at Hampton Court Palace in England to measure observable behavior that would indicate learning by the rats. Use of the white rat as a research subject in behavioral research was given a great impetus by the investigations of Watson[6] at Chicago University. Watson's ideas firmly cemented the white rat as a fixture of experimental studies in behavior.[7]

At the same time of Stewart's behavioral work in the 1890s, Donaldson was using the white rat at the University of Chicago for anatomical and physiological research. Donaldson's rat colony became the parent stock at the Wistar Institute of

Anatomy and Biology in Philadelphia. In 1924, a book entitled *The Rat*[14] by H. H. Donaldson gave impetus to the white rat as an acceptable research subject and provided wisdom about the use of animal subjects in biomedical research. In one comment from *The Rat*, Donaldson stated that "in enumerating the qualifications of the rat as a laboratory animal, and in pointing out some of its similarities to man, it is not intended to convey the notion that the rat is a bewitched prince or that man is an overgrown rat, but merely to emphasize the accepted view that the similarities between mammals having the same food habits tend to be close, and that in some instances, at least, by the use of equivalent ages, the results obtained with one form can be very precisely transferred to the other."

IV.   Advantages of Rat Models

Rats are commonly employed as animal subjects in contemporary medical research, and general acceptance of the white rat as an animal research subject has increased in synchrony with an increased appreciation of the value of behavioral research. Since rats are small, clean, relatively inexpensive, easily handled and maintained, widely available, have short twenty-one day gestational periods, and a short two to three year lifespan, their use as research subjects offers many advantages. These advantages are amplified in application to research problems that require large numbers of animals. Likewise, the relatively short twenty-one day gestation and approximate three year lifespan of the rat provide a practical opportunity to study the stages of development and aging.

V.   Disadvantages of Rat Models

Despite the numerous advantages offered by the laboratory rat model, there are difficulties that must be considered. First, it is more difficult to draw parallels between rodent and human behavior and physiology, than to compare non-human primates with humans. The behavioral characteristics of the rodent subject are more primitive, making behavioral comparison more complicated. Second, it is more difficult to establish stimulus control of the rodent's behavior in training paradigms. Often, it is necessary to use aversive electrical shock or drastic food deprivation to motivate the rodent subjects. These severe control procedures further complicate the comparison with humans, since such control measures are unacceptable for human research.

VI.   Strain Selection

Once it has been decided to use rats as subjects in a given research endeavor, the question of strain selection becomes important. There are many outbred and inbred strains of rats available on the commercial market, and the effects of many common treatments differ according to the strain selected. Early in my scientific training I received a lesson about this important fact by finding that hippocampectomy in Sprague-Dawley rats increased several forms of activity, and increased one type of error in a Lashley III maze. However, when the same hippocampectomy was effected in Fischer strain rats, the animals consistently became less active, and did not make the same Lashley III error. Normally, Fisher rats are much less active than normal Sprague-Dawley rats, and the hippocampectomy may have removed cortical inhibition of behavioral tendencies that were already present in the normal unoperated controls. There are many published papers that show differences among strains of laboratory rats regarding the effects of various treatments, shown in the following sections.

A.   The Wistar Rat Colony

The first rats brought from Chicago to the Wistar Institute in Philadelphia by Donaldson in 1906 became the parent stock of a rat colony, whose offspring were sold to research facilities throughout the U.S. and many other countries until 1960. The Wistar Institute was a leader in determining laboratory animal husbandry practices necessary to support a large rat colony. By 1922, the colony had a total population of about 6000 rats. The commercial rights to the sale of the Wistar Rat were sold in 1960.

From the beginning, the Wistar Institute maintained a random bred, heterogeneous colony. Therefore, there was considerable variability in the commercial colony maintained by the Institute. It is unknown whether albino lines other than those provided by Donaldson were introduced into the Wistar Colony. According to Lindsey,[2] it is documented that outside breeders were brought into the colony to boost breeding production.

Most of the albino laboratory rats used in the U.S. are linked to the colony of the Wistar Institute. In previous discussion, the role of Donaldson in the development of the albino laboratory rat model was mentioned. Donaldson's rat colony became the parent stock at the Wistar Institute of Anatomy and Biology. Even Donaldson himself did not know whether stock from the European labs found its way to the U.S., or whether the first albinos in the U.S. were derived from wild rats captured in the U.S.[8,9]

B.   The Long-Evans Strain

Prior to 1920, two members of the faculty at the University of California at Berkley, J. A. Long and H. M. Evans, were interested in the estrous cycle of the rat. To support their research interests, Long and Evans established one of the leading strains of rats that continues to bear their names. The origins of the Long-Evans rat colony were described in a monograph entitled *The Oestrous Cycle in the Rat and Its Associated Phenomena*. In that monograph, it was stated that the colony descended from a cross made around 1915 "between several white females and a wild gray male" that was trapped along the bank of Strawberry Creek, which ran through the university campus. Dr. Leslie Bennett, a colleague of Dr. Long, is quoted as saying that the white females were supplied by the Wistar Institute.[2]

The Long-Evans rats exhibited varied fur color. The coats represented in the colony are black, gray, and hooded. The hooded animals are characterized by pigmented fur on their heads and often along the spine. These animals have a pigmented iris, and their visual acuity far exceeds that of the albino strains.

C.   Strains from Columbia University

The Crocker Institute of Cancer Research at Columbia University began in 1913 to inbreed six major bloodlines of rats to be used in cancer research. Researchers at the institute had noticed that certain rats were more susceptible to sarcomas induced by feeding tapeworm eggs. The inbreeding program was designed to determine if the susceptibility to sarcoma was genetic.[2] Dr. Maynie R. Curtis was the chief developer of the inbred rat colony. Dr. Curtis purchased a few breeding pairs of rats from each of four local breeders whose names were August, Fischer, Marshall, and Zimmerman. The obtained rats had different coloring, and these external characteristics were used as markers to help identify the various strains. The Marshall rats were albinos. The Fischer and Zimmerman rats were non-agouti piebalds, but did carry the albino gene. The August rats were the most varied, and included some with pink eyes. In 1941, a group of rats with red eyes were obtained from a breed in Connecticut. These animals were seed stock for brother-sister mating and are progenitors of several popular inbred strains. The first litter of pedigreed rats at Columbia were from mating number 344, and were the first representatives of the Fischer 344 strain.

D.   Sprague-Dawley Rats

There appears to be little record of the origin of the Sprague-Dawley strain. The primary stock is believed to have been established by Robert W. Dawley, who was a physical chemist at the University of Wisconsin. Mr. Dawley included his wife's maiden name, which was Sprague, to name the rats. Mr. Dawley later established Sprague-Dawley, Inc. to advance the commercial sale of his rats. Lindsey[2] cites a letter from Mr. Dawley to the National Institutes of Health (NIH), dated July of 1946, in which he states that the original parents were a hybrid hooded male rat of exceptional size, and vigor, that was genetically half albino. He was mated to a white female, and subsequently to his white female offspring for seven generations. The origin of the hooded male is unknown, but the first white female is believed to be from the Wistar colony. Selection was on the basis of many factors, including high lactation, rapid growth, vigor, good temperament, and high resistance to arsenic trioxide. The original company continues today under the name Harlan Sprague-Dawley.

There are many sub-lines of the Sprague-Dawley animals, and they are often used in behavioral research. They are randomly bred strains, but differ from one another. The Sprague-Dawley strain is also sold by the Charles River Co. There is variability in the stock marketed by the different suppliers. For example, Pollock and Rekito[10] found that the Sprague-Dawley rats marketed by Harlan differed from Sprague-Dawley rats marketed by the Charles River Co. in regard to hypertensive response to chronic L-NAME-induced nitric oxide synthase inhibition.

E.   Holtzman Rats

A major sub-line of the Sprague-Dawley line is the Holtzman rat. These rats are provided by a company established by E. C. Holtzman, who was a former employee of the Sprague-Dawley Co. Sprague-Dawley animals were the original seed stock of the Holtzman line.

F.   N/Nih Rats

The National Institutes of Health (NIH), through systematic interbreeding of eight inbred rat strains, created a heterogeneous stock of rats.[11] This strain, called N/Nih, has been maintained by a strict breeding policy to ensure that mating pairs of subsequent generations are distantly related. Predictably, the large genetic variability among the N/Nih rats results in varied phenotypes. Several selective breeding programs that originated with N/Nih rats have developed strains of inbred rats with characteristics of great interest to behavioral research. These inbred strains include rats chosen for high and low alcohol consumption and sensitivity. The genealogy of the animals maintained by the NIH can be found on their web site (http://www.nih.gov/od/ors/dirs/vrp/s&slst.htm#ratinbred), or by correspondence with them.

G.   Wild Norway Rats

Not many researchers are hardy enough to explore the use of wild rats as animal subjects, although some have.[12,13] Discussion of the laboratory rat would be incomplete without some mention of how the typical albino laboratory rat differs from the original wild stock. Albino laboratory rats were originally selected for docility, i.e., a reduced tendency to flee from humans or to struggle and bite when handled. There are implications of this selection for docility. For example, novel objects that induce avoidance or fear in wild rats often elicit approach or apparent curiosity in the albino laboratory strains. Albino males do not attack other male rats with the intensity typical of wild male rats. The food preferences of the wild rats also differ from the albino strains.[12,14] Differences in behavior are paralleled by changes in growth and in the relative weight of the adrenal glands. Wild rats are genetically heterogeneous, and thus there is considerable behavioral variation among wild rats. Barnett[15] used the techniques of ethology to observe the innate behavioral characteristics of wild Norway rats in groups. Under these conditions, it is obvious that rats have a complex social structure that is not generally measured in typical laboratory testing. Barnett also describes rat behavior that most of us have casually noticed, but generally have not understood. Barnett explained and illustrated the role of many body postures and gestures in rat society. Barnett has also reviewed techniques for testing wild Norway rats.[15]

VII. Inbred Rat Strains Selected for Various Behavioral Traits

With the explosion of interest in the genetic basis of physiology and behavior, there are many papers in which various inbred strains of rats are compared to outbred strains according to the criteria of interest. Populations of outbred rats such as the Wistar, and hetereogenous strains such as the N/Nih rats, manifest considerable variability along almost any behavioral or physiological attribute. Selective breeding programs for high or low manifestations of various phenotypes have resulted in inbred strains that are useful for many areas of behavioral research.

The names of these strains do not follow a systematic nomenclature. One attempt to standardize the rat strains[16] recommended that the rat nomenclature system follow that of inbred mice. The following quote from Festing and Staats presents the problem: "In many cases a strain name has been changed whenever a strain has been transferred to a new laboratory, and in other cases strains which have only a distant relationship have been given the same name. This is particularly the case with strains descended from Wistar outbred stock, which tend to be named 'WIS' or some other name beginning with W." The nomenclature system recommended by Festing and Staats has not been universally followed. Many strains are named after the university or some other prominent, but not obvious, feature of their development. The following sections describe basic aspects of some common inbred strains used in behavioral research.[16]

A.   Rat Strains Selected for Preference of — and Sensitivity to — Alcohol

Normal out-bred (e.g., Wistar) and heterogeneous rats (e.g., N/Nih) do not typically drink much alcohol, but there are large individual differences in alcohol drinking over a large population of these rats. By selectively breeding for high or low alcohol intake and high and low neurosensitivity, lines of rats that exhibit these characteristics regarding alcohol have been derived. The inbred alcohol-preferring rats typically consume up to ten times the amount of alcohol taken by normal out-bred or heter-ogeneous rats. The selectively bred rat lines include the alcohol-preferring (P), alcohol-accepting (Alko Alcohol — AA), Sardinian alcohol-preferring, and high alcohol drinking (HAD) rats.[17] These lines do not share all other behavioral traits.

Higher than normal alcohol drinking has also been observed in several groups of rats that were selectively bred for other specific behavioral traits. These include the Tryon Maze-Bright and Tryon Maze-Dull rats,[18] the Roman High- and Low-Avoidance rats,[19] and the Fawn-Hooded rats that have serotonin receptor abnormality.[20,21] Typically, the ACI strain of rats will not voluntarily drink alcohol.

B.   ACI Strain

The ACI line was originated by Curtis and Dunning at the Columbia University Institute for Cancer Research. Initially, the primary phenotype of interest was susceptibility to estrogen-induced tumors; however, there are also a number of behavioral differences associated with this strain. For example, the Brh sub-line shows low defecation response and high activity response in the open field test.

C.   Strains Bred for Various Serotonin Receptors

Selective inbreeding of N/Nih rats has resulted in strains that vary in sensitivity to 5-HT_{1A} receptor stimulation. Overstreet[21] has established a selective breeding program for high (HDS) and low (LDS) sensitivity to the hypothermic response of the 5-HT_{1A} agonist 8-OH-DPAT. These two rat lines are believed to differ in behavioral tests of depression, but not of anxiety. The lines also differ in post-synaptic 5-HT_{1A} receptors. Pre-synaptic mechanisms are not affected.

D.   Roman Strain

The Roman strain of rats was selectively bred from Wistar stock for high and low performance in two-way active avoidance learning.[22] RHA/Verh rats acquire active avoidance (shuttle box) performance quickly, because they are less emotionally reactive, but more active in regard to locomotion. RLA/Verh rats cope with the active avoidance problem more passively, and become immobile when faced with the avoidance task. They show increased defecation in the open field and increased activity in the hypothalamic-pituitary-adrenal axis. The two lines of the Roman strain differ in many respects at the behavioral and neurochemical level (for review).[19]

E.   Maudsley Strains

The Maudsley Reactive (MR) rats were selectively bred for high defecation in the open field test.[23] Maudsley Non-Reactive rats (MNRA) were selectively bred for low rates of defecation in the open field test. The strains were genetically selected from outbred Wistar progenitors.

F.   Tryon's Maze-Bright and Maze-Dull Rats

The genetics of maze-learning ability was investigated by Tryon and his group of researchers from the late 1920s through the 1940s. Tryon[24,25] was successful in segregating animals that were adept at learning a variety of mazes (Tryon maze-bright) from those rats that learned less well (Tryon maze-dull). Tryon conducted many experiments to identify the factors that were involved in maze learning. Some of the differences between the two groups can be attributed to emotionality and others seem to be a function of the ability of rats to perceive helpful guiding cues outside of the maze.

G.   Spontaneously Hypertensive Rats

The adult spontaneously hypertensive (SHR) rat includes behavioral changes in its phenotype, in addition to changes in blood pressure. SHR rats are also impaired in ability to perform learning and memory-related tasks and exhibit a decrease in the expression and nicotine-stimulated function of brain nicotinic acetylcholine receptors. These cholinergic factors are known to be important to learning performance.[26,27]

H.   Flinders Sensitive Line and Flinders Resistant Line

The Flinders Sensitive Line (FSL) and the Flinders Resistant Line (FRL) were selectively bred at Flinders University, in Australia, by selective breeding for differences in effects of the anticholinesterase, di-isopropylfluorophosphate (DFP) on temperature, drinking, and body weight. The FSL rats are more sensitive to DFP, as well as cholinergic agonists, and have more brain muscarinic receptors than the FRL rats (for review).[28] Because of the known relationship between cholinergic activity and many functions in humans and animals, these strains are useful for a wide variety of research topics, including learning, memory, and depression.

I.   Dahl Salt Sensitive Rats

Another hypertensive rat strain is the Dahl salt sensitive rat. Correspondingly, there is a Dahl salt resistant strain. The salt sensitive rats exhibit reduced learning ability, and also reduction in nAChR in brain regions known to be important for learning and memory. These include the hippocampus and the amygdala.

VIII. Comparison of Various Rat Strains for Behavioral Characteristics

From the preceding sections, it is clear that the most common outbred, and certainly the inbred rat strains, differ in regard to important baseline behavioral characteristics. Some years ago, Harringon[29,30-37] and others presented behavioral standardization data for several outbred and inbred rat strains. The behavioral comparisons included activity in both stabilimeters,[30,31] open-field[38] and rotating wheels,[39] free operant lever press levels,[37] cued lever press levels,[32,36] passive avoidance conditioning,[33] shuttle-box avoidance conditioning,[35] home cage behavior,[31] and runway learning.[34] Other comparisons measured anatomic and physiologic variables, including basal metabolism Harrington and Hellwig,[40] organ weights,[41] and cholinesterase.

The value of these baseline studies is that they compare the building blocks of behavior and some important physiologic substrates for a number of popular rat strains. Each of these studies of elementary behavioral characteristics presents data obtained from the behavior of approximately 500 rats. Though the studies were conducted some years ago, they remain valuable as a catalog of standards by which strains of rats may be specifically selected for particular behavioral traits.

IX. Mice in Behavioral Research

The major reason for the use of the mouse in behavioral research resides in the abundance of knowledge about the mouse genome and in the large number of genetically defined strains that display characteristics valuable for research. There is considerable excitement about the possibilities of measuring data from genetically engineered strains. The opportunities afforded by "knockout" of specific genes have garnered considerable interest. Likewise, transgenic mice have been genetically engineered and altered by injection of one or more genes, such as the human gene for apolipoprotein E (ApoE), which has been linked with the pathogenesis of Alzheimer's disease.

The small size of the mouse poses technical problems; however, there is commercially available, specialized behavioral equipment designed for their use. It is possible to obtain behavioral data from mice through many of the traditional paradigms designed for use with the rat. Mice are often tested by mazes and simple behavioral measures, such as drinking behavior.

One of the first things that will be noticed when reading research literature pertaining to mice is the complicated system of nomenclature for the many strains. A review of this system has been written by Lyon,[42] and is necessary for examination of various strains. The vast number of mouse strains developed for biomedical research exceed any other species. The Jackson Lab in Bar Harbor, Maine is a leader in the development of the laboratory mouse.

One of the first steps toward learning more about the relationship of behavior to the mouse genome will be behavioral comparisons of various strains. The NIH has funded a multicenter effort toward that goal, and the Department of Psychology at the State University of New York at Albany hosts a website that updates the results of the multi-center effort. At the time of this writing, the website is available through www.albany.edu.

X. Pigeons and Other Species Performing Traditional Non-human Primate Tasks

Pigeons are able to perform visually complex discriminated operant problems, such as non-spatial delayed matching-to-sample; however, there are important differences between the strategies that the pigeons use in comparison with primate species. For example, Cumming and Berryman[43] trained pigeons to match red, blue, and green disks to a high level of accuracy. To test for transfer of the matching concept, a novel yellow disk was then substituted for the blue disk. The pigeons regressed to chance levels of accuracy on trials in which the sample was the yellow stimulus. The birds resorted to a more concrete position strategy when the novel stimulus appeared. Jackson[44] tested a series of similar transfers in rhesus monkeys and found that the monkeys continued to match at 100% accuracy when presented with a novel yellow stimulus. However, if the novel stimulus was not a disk (e.g., cross, triangle), then the monkeys did not transfer the matching behavior either. These species differences relate very much to the Transfer Index (TI) of Rumbaugh,[45] a useful tool for comparing cognitive skills between species. Pigeons, in fact, do not form concepts of matching-to-sameness, but instead form individual stimulus-response links that are not mediated by a matching concept.

There are numerous papers in the recent literature in which the titles indicate results of experiments using rodents and other sub-primate species while performing tasks traditionally associated with non-human primates. Delayed response tasks, such as delayed matching-to-sample are examples of this. In most instances, these behavioral tasks are actually quite different from the typical non-spatial delayed response tasks traditionally administered to non-human primates. In the experience of this author, rodents are surprisingly proficient at delayed tasks that involve a spatial dimension, such as position or location; however, rodents are very much less proficient in non-spatial tasks. Pigeons and other birds have highly developed visual skills and can perform non-spatial visual tasks with high levels of accuracy.

XI. Non-human Primates

A. Advantages

As research subjects, non-human Primates offer several potential advantages over other species. The behavioral capabilities of non-human primate species resemble human capability along many criteria, and the near phylogenetic position of non-human primate species to humans offers opportunity for modeling many aspects of human anatomy and both normal and abnormal function.[46] Some behavioral tests can be adapted to allow testing of both humans and non-human primates by the same paradigm.[47,48] This offers a distinct advantage for the modeling of human physiology and behavior.

B.   Disadvantages

Despite the attractive advantages of using non-human primates as research subjects, there are numerous disadvantages to their use. Non-human primates are expensive to obtain and to maintain. Non-human primates are large, powerful, and potentially dangerous animals. The similarity to man that makes them attractive as research subjects also carries increased risk of animal-to-human disease transmission. Available animals, whether feral captured or colony-reared, are from diverse populations; thus there is great inter-subject variability among animals. The problems of such variability are magnified by generally small subject populations, which are a consequence of cost factors that often limit the number of subjects to be obtained and maintained for a given project.

C.   Commonly Used Non-human Primates in Biomedical Research

Table 1.1, adapted from Whitney,[49] lists some of the commonly employed primate species, and serves as a guide for the different family groups of non-human primates.

D.   Basic Behavioral Differences Among Monkey Species

Despite the large number of research studies that have been conducted upon various species of non-human primates, it is difficult to evaluate which species to select for a given project. Often, economic factors or availability determine the final choice; however, there is considerable difference in basic behavior among the non-human primates, and these differences have implication for the results of many treatments. These special species differences should be considered, if the results of the selected treatment are to be fully appreciated.

One comparison of basic behaviors among several common non-human primate species was conducted by Davis.[50] The authors observed and measured lemurs, Old World monkeys, and New World monkeys in a runway that measured 9 ft long by 18 in. wide by 24 in. high. Located on both ends of the runway was an "observation" box (18 by 18 by 24 in.) into which an animal of the same species was introduced. The monkey to be observed was introduced into the long central runway area. The

TABLE 1.1
Some Common Species of Non-human Primates

Prosimians

Tree Shrews	*Tupaia glis*
Galagos	*Galago spp.*

New World Monkeys

Capuchins	*Cebus spp.*
Marmosets	*Callithris spp.*
Owl Monkeys	*Aotus t rivirgatus*
Spider Monkeys	*Ateles spp.*
Squirrel monkeys	*Saimiri sciureus*
Tamarins	*Saguinus spp.*
Wooly monkeys	*Lagothrix spp.*

African Monkeys

Guenons	*Ceropithecus spp.*
Green monkeys	*Ceropithecus aethiops*
Mangabeys	*Cerocebus spp.*
Patas monkeys	*Erythrocebus patas*
Talapoins	*Miopitecus talapoin*

Baboons

Savanna baboons	*Papio cynocephalus*
Hamadryad baboons	*Papio hamadryas*

Asian Macaques

Celebes black ape	*Macaca nigra*
Cynomolgus (Crab eating macaque)	*Macaca fasicularis*
Japanese macaque	*Macaca fuscata*
Pig-tailed macaque	*Macaca nemistrina*
Rhesus	*Macaca mulatta*
Stump-tailed macaque	*Macaca speciosa* (a.k.a. *arctoides*)

Apes

Chimpanzees	*Pan troglotydes*
Gibbons	*Hylobates spp.*

categories of behavior that were measured were: visual surveillance, cage manipulation, social behavior, rapid energy expenditure, self-involved behavior, and vocalization. The monkey species surveyed were: rhesus macaques, pig-tailed macaques, stump-tailed macaques, Cebus (Apella) monkeys, Squirrel monkeys, Wooly monkeys, and Lemurs. The following summarizes some of the characterizations of basic behavioral tendencies for each species.

The rhesus monkeys (*Macaca mulatta*) were described as curious and manipulative. Rhesus spent more time in visual survey than either the stump-tailed or the pig-tailed macaques. Like the other macaques, rhesus monkeys also spent more time in presenting and grooming than was observed in the New World monkeys. Rhesus monkeys are also highly manipulative. They spend considerable time licking and examining objects with their hands.

Aside from the findings of Davis et al., rhesus monkeys are popular research subjects due to their relatively low cost and high cognitive skill; however, these are large, powerful, and potentially dangerous animals, and require special caution in their handling. In the wild, the defensive strategy of the rhesus troupe is to form a line of aggressive male defenders, who menace and threaten intruders. This defensive line allows the females and young to evacuate to safer territory. Everyone who visits a zoo or non-human primate housing area is confronted with the strong defensive and attack behavior that is elicited by eye contact with rhesus monkeys. This aggressive cage-rattling behavior in captive rhesus monkeys is a manifestation of the wild defensive posture.

In the Davis et al. studies, the stump-tailed macaques (*Macaca speciosa*) spent most of their time presenting and grooming. This preoccupation with presenting and grooming by the stump-tailed monkeys resulted in lower scores in the remaining behavioral categories.

The reports of Kling and Orbach[51] and Orbach and Kling[52] regarding the docility of stump-tailed macaques initiated great interest in this species for laboratory use. Juvenile stump-tails are considerably more compliant than are juvenile rhesus monkeys (*Macaca mulatta*). Larger stump-tailed macaques are noticeably more aggressive than the stump-tailed juveniles, but still considerably less difficult to handle than are rhesus monkeys. The defensive strategy of wild stump-tailed monkey troupes is in contrast to that of rhesus monkeys. When threatened by intruders, wild stump-tailed monkeys remain quiet and still and thus blend into the heavy foliage cover that is characteristic of their normal habitat. This stillness in the face of potential threat is the basis of the once prevalent idea that these monkeys are extremely docile; however, quietude and docility are not the same. Adult stump-tails become quite large and are known to become gradually more aggressive toward human handlers.[53] They are particularly dangerous if perched at eye level or above. One characteristic of stump-tailed macaques that frequently elicits comment is their musty body odor, but it is not so strong as to eliminate their use. An ethological study of stump-tailed macaque gestures and facial expression has been presented by Jones and Trollope.[54]

Pigtailed macaques (*Macaca nemestrina*) were intermediate to rhesus monkeys and stump-tailed macaques in the Davis et al. tests. The only exception to this was vocalization, which was the same as the other macaque species. Pig-tails become very large and, as with stump-tailed macaques, may gradually become quite aggressive toward their handlers. Our aged non-human primate project at the Medical College of Georgia has been forced to abandon several well-trained pig-tails because of extreme aggressiveness toward human handlers.

Cynomolgus or Crab-eating macaques (*Macaca fasicularis*) have become more popular as biomedical research subjects, as imported rhesus monkeys are difficult to obtain. These animals have very long tails. Their behavior and general cognitive skill resembles that of rhesus monkeys.

Apella monkeys (*Cebus Apella*) are the mechanics of the monkey world. They spend a large amount of time manipulating inanimate objects with their hands and tongues. Like stump-tailed and pig-tailed macaques, the Davis et al. studies found that they spent little time in visual survey, and, as other New World monkeys, they spent large amounts of time pacing, swinging, and bouncing on all four legs. They spent little time grooming and presenting.

Squirrel monkeys (*Saimiri sciurea*) were characterized by high levels of self-involved behavior (licking, biting, manipulation of their own body) in the Davis et al. tests. Also, many of the squirrel monkeys exhibited high levels of bouncing and pacing, but this was an individual characteristic that was not exhibited by all squirrel monkeys. Squirrel monkeys have a unique form of vocalization, characterized by frequent high-pitched shrieks and less frequent cooing.

Wooly monkeys (*Lagothrix humbolti*) are known for frequent vocalizations, but in the Davis et al. tests, this characterization varied considerably among individuals. Wooly monkeys will frequently threaten and scold, but are not known for their aggression toward handlers. Only rarely do they scratch or bite.

Chimpanzees (*Pan troglotydes*) are rare and special animals. The endangerment of these animals as a species, in conjunction with the great cost of housing and caring for them, results in few opportunities to work with these great apes. Most research personnel that work with these animals are profoundly impressed by their similarity to man, and many become ardent opponents of invasive research techniques to be applied to them. One source of information about the anatomical, physiological, and behavioral study of chimpanzees is provided by a volume edited by H. H. Reynolds.[55] Chimpanzees and other great apes, such as gorillas and orangutans, are generally superior to both Old and New World monkeys in learning skills, if the learning task is sufficiently difficult to distinguish the species.[56] The fact that chimpanzees can master rudimentary language skills has aroused great interest.[57]

E.   Primate Cognitive Skills

An important assumption that underlies most non-human primate research is that it is possible to align primate species in a graduated series of cognitive skill, so as to increasingly approximate humans.[58,45] Without entering into argument about the theory of linear evolution, most would agree that the behaviors of great apes, such as gorillas, chimpanzees, and orangutans, resemble human behavior more than do the behaviors of other primates, such as lemurs or marmosets. One factor that correlates highly with human similarity is development of the cerebral cortex. Among the primates, cortical development is greatest in humans, followed in order by the great apes, the lesser apes (e.g., gibbons) Old World monkeys, New World monkeys, and finally the pro-simians.[59,45]

F.   Transfer of Training

There is considerable difference among various non-human primate species in cognitive capability. One cognitive skill that has been extensively studied across species is transfer of training. Humans and animals have the ability to generalize aspects of what has been learned in one situation to a different situation. Likewise, animals and humans improve in generalization skills as a result of experience. One relevant body of literature began with the classic series of experiments by Harlow.[60] This work was continued and expanded by his many colleagues and students. Harlow demonstrated that, after repeated experience with discrimination problems, animals learned strategies for solving problems that allowed them to become more proficient at solving new problems that were similar, but not exactly the same. Harlow called this phenomenon "learning set" or "learning-how-to-learn." There is a distinct correlation between cerebral development and the ability to establish learning sets; thus, the great apes are more proficient at forming learning sets than are the prosimians. Figure 1.1 compares learning set formation in three species of primates: Old World rhesus monkeys, New World squirrel monkeys, and marmosets. Chimpanzees are even more adept at learning set formation than are rhesus monkeys. These data and the collective body of information[45] clearly support the hypothesis that, at least within the order Primates, there is a relationship between cortical development and complex learning skill. The generalizations exemplified by learning set formation are foundational to the conceptualizations shown by human children and adults. Learning set formation is only one example of cognitive processing that can discriminate between species of primates. The ability to abandon previously learned, but no longer appropriate behavior (extinction), is another behavioral characteristic that varies naturally throughout the order Primates.

It is normal and adaptive to extinguish a response pattern when it is no longer rewarded. The ability to perform this basic cognitive skill also distinguishes the various primates. Since cortical development correlates with the cognitive skills involved with learning sets, it would be expected that more developed animals would be able to more rapidly extinguish non-rewarded response patterns as well. This was confirmed in a study comparing highly developed *Cercopithecus* monkeys (includes African green monkeys and mangabeys) with less developed *Lemuridae* (lemurs) (Arnold and Rumbaugh, 1971). As seen in Figure 1.2, *Cercopithecus* monkeys abandoned the non-rewarded choice patterns more rapidly than did the *Lemuridae*, although the overall pattern of extinction was the same for both groups. This is only one example of the correlation between cognitive skill and cortical development with the order Primates.

Other behavioral test techniques also show this positive correlation between cortical development and cognitive skill among primate species. Rumbaugh[61,57] has developed a productive technique termed the Transfer Index (TI), which reflects transfer of training skill and the ability to abandon previously learned object choices. A typical procedure for the TI is to first train animals at two different criterion levels of accuracy on a discrimination task. Then, the cue valences of the stimulus objects

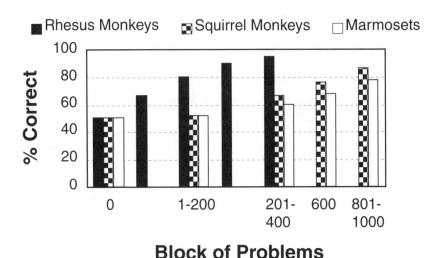

Block of Problems

FIGURE 1.1
Learning set curves for marmosets, squirrel monkeys, and rhesus monkeys.

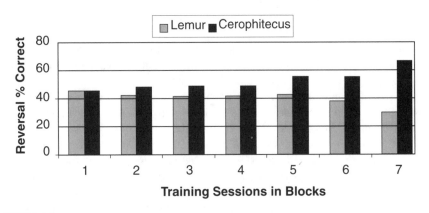

FIGURE 1.2
Reversal performance plotted as a function of pre-reversal performance.

are reversed. Thus, selection of the object that previously was rewarded with food resulted only in an empty food well and the reward was shifted to the previously incorrect object. The TI measures the percent accuracy on the reversal task (transfer). Transfer skills are measured in terms relative to amount of learning accuracy on the first discrimination task — not in terms of absolute percent accuracy. The TI was designed to be a species-fair measurement of cognitive functioning. The implication of the TI is that it allows a measurement of behavioral and cognitive flexibility. The ability to transfer small amounts of knowledge to a new situation can be an important

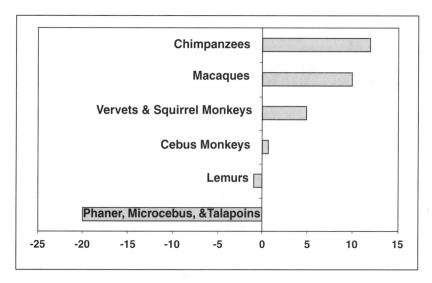

FIGURE 1.3
Transfer index of several non-human primate species.

advantage. Conversely, a negative transfer indicates an inflexibility that represents a disadvantage. Figure 1.3 compares transfer index of several non-human primates. There is a high correlation between the TI and cortical development. Primitive primates with less developed brains, such as lemurs, micro-cebus, and phaner, show negative transfer. Negative transfer indicates a behavioral inflexibility that is a handicap for the adaptation of these species. Macaques are intermediate to the New World monkeys and chimpanzees as measured by the TI. This cognitive flexibility of the Old World macaques facilitates adaptation to their environment, and the overall cognitive skill of the macaques accounts for their popularity as behavioral research subjects.

XII. Discussion

There are many variables to consider when selecting animals for behavioral experiments. Not only are there important species differences, but there is considerable variance among different individuals — even within the same lines. Although variance requires large subject populations for behavioral studies, these individual variances are not altogether without benefit. For example, the range of response to some treatments may be manifest most broadly in outbred groups. One example of this point is that there are numerous strains of laboratory rats available.[64,65,67] These strains arose from different stocks and have different behavioral characteristics. Importantly, these strains react differently to many common physiologic and pharmacologic treatments. Thus, many important research strategies should involve careful selection of the animal strain or species that is to be selected, and perhaps more consideration should be given to the replication of research findings across multiple species or strains. The classic standardization studies of Harrington[29,30-39] and Harrington and Hellwig[40,41] remain as immensely valuable standards by which to select rats for behavioral studies.

One practical implication of the variation among species and strains is that it is important for experimenters to personally spend some time observing the behavior of various species, both in the home cage and in the experimental test situation. Another practical implication of the variation among species and strains is that various populations may exhibit high or low values of a behavioral factor of interest. This has continued to be a valuable behavioral tool. There is important interaction between many treatments and species-specific behavior. Because the treatment is the usual focus of interest, it is tempting to accept experimental data values blindly as they are presented by an automated device or research technician. In such situations, the behavior is often viewed only as a dependent variable, and thus of lesser interest. Much valuable information can be lost by such lack of attention to the actual behavior of the animal. When the experimenter in charge spends some time observing the animals as they perform their particular task, the effort is often rewarded by increased insight into the behavioral intent of the animal. If we pay attention, they will teach us much and, as a bonus, we will learn much more about our treatments. Remember that animals can perform certain tasks for many different reasons — not only the ones originally construed by human planning.

Various animal species are uniquely endowed with characteristics that distinguish them from other plausible animal subjects. Animal species or strains often have specialized capabilities that allow them to cope with a narrow environment. Thus, so-called lower animals may have certain capabilities that exceed the ability of higher animals. A classic example of this can be seen in the olfactory functions of the amygdala and other limbic structures of macrosmatic species, such as rats, vs. the emotional/learning specialization of the same structures in microsmatic animals, such as primates. There are important differences among the behaviors of the species and different reactions to certain treatments are to be expected.

An interesting question that arises in this connection is whether phylogeny recapitulates the ontogeny of human development with respect to higher cognitive function.[61] Certainly, we are familiar with the idea that animals with greater cortical development have capacities that allow new functions to emerge.[62,63] Some highly developed primate species may have capabilities that are not represented in animals with less developed brains. Rumbaugh et al.[61] have presented convincing arguments that "emergents," which are new capabilities that were never directly rewarded by past experimental experience, are cognitive products of highly developed cortical structure. Transfer of training is an example of emergent behavior.[66] Emergents resemble human concept formation, and are present in many species, but are striking in primates, such as macaques, and especially striking in the great apes.

The selection of animal subjects is an immense topic, and additional information is available from many sources. Keying a search on any of the major search engines of the world wide web will identify animal suppliers and additional clearinghouses of information. Such key words as laboratory rat, laboratory mouse, primate, or mouse behavior will prove productive. The NIH-funded multicenter effort for comparison of mouse strains through the Psychology Department at SUNY at Albany should be a good source of up-to-date mouse behavioral data. Two publications of interest are *Laboratory Primate Newsletter* and *Rat News Letter.* The Oregon Regional Primate Center provides a valuable collection of primate-related journal citations, which are available for a modest fee. Many of these citations have been organized into databases of specific behavioral topics. The reference section of this chapter contains citations of papers written during the infancy of animal behavior research. These papers are rich in the justification of animals as models for human cognition and are worthy of rediscovery.

There are many reasons for conducting research upon animal subjects, but the decision to do so should be given some thought. Species selection is among the first problems to confront the researcher who hopes to pursue animal modeling of human behavior. Often, new researchers may select a model that has been commonly utilized, simply because it has become established and accepted. Although there is utilitarian value to this approach, it is valuable to recall that there is considerable difference between and among species and strains. Many opportunities are presented by these differences.

References

1. Bartus, R. T., Dean, R. L., and Beer, B. An evaluation of drugs for improving memory in aged monkeys: Implication for clinical trials in humans. *Psychopharmacology Bulletin* 19, 168–184, 1983.
2. Lindsey, J. R. Origin of the laboratory rat. In: Baker, H. J., Lindsey, J. R., and Weisbroth, S. H. (Eds.) *The Laboratory Rat, Vol. I, Biology and Diseases.* New York, Academic Press, 1979.
3. Richter, C. P. The effects of domestication and selection on the behavior of the Norway rat. *Journal of the National Cancer Institute* 15, 727–738, 1954.
4. Miles, W. R. On the history of research with rats and mazes. *Journal of General Psychology* 3, 324–337, 1930.
5. Kline, L. W. Methods in Animal Psychology. *American Journal of Psychology* 10, 256–279, 1899.
6. Watson, J. B. *Behavior: An introduction to comparative psychology.* New York, Holt Publishing, 1914.
7. Munn, N. L. *Handbook of Psychological Research on the Rat.* Boston, Houghton Mifflin, 1950.
8. Donaldson, H. H. A comparison of the European Norway and albino rats *Mus norvegicus albinus* with those of North America in respect to the weight of the central nervous system and to cranial capacity. *Journal of Comparative Neurology* 22, 71–77, 1912.

9. Donaldson, H. H. The history and zoological position of the albino rat. *Journal of the Academy of National Sciences Philadelphia* 15, 365–369, 1912.

10. Pollock, D. M. and Rekito, A. Hypertensive response to chronic NO synthase inhibition is different in Sprague-Dawley rats from two suppliers. *American Journal of Physiology,* 275(44), R1719–R1723, 1998.

11. Hansen, C. and Spuhler, K. Development of the National Institutes of Health genetically heterogeneous rat stock. *Alcoholism: Experimental and Clinical Research* 8, 477–479, 1984.

12. Barnett, S. A. Laboratory methods for behaviour studies of wild rats. *Journal of Animal Techs. Assistants* 9, 6–14, 1958.

13. Barnett, S. A. Social behavior among tame rats and among wild-white hybrids. *Proceedings of the Zoological Society of London* 134, 611–621, 1960.

14. Donaldson, H. H. *The Rat: Data and reference tables* (2nd ed). Philadelphia: Wistar Institute of Anatomy, 1924.

15. Barnett, S. A. *The Rat: A Study in Behavior.* Chicago: University of Chicago Press, 1975.

16. Festing, M. and Staats, J. Standardized nomenclature for inbred strains of rats. *Transplantation* 16, 221–245, 1973.

17. Sinclair, J. D., Le, A. D., and Kiianmaa, K. The AA and ANA rat lines, selected for differences in voluntary alcohol consumption. *Experientia* 45, 798–805, 1989.

18. Amit, Z. and Smith, B. R. Differential ethanol intake in Tryon maze-bright and Tryon maze-dull rats: Implications for the validity of the animal model of selectively bred rats for high ethanol consumption. *Psychopharmacology,* 108, 136–140, 1992.

19. Driscoll, P. L., Escorihuela, R. M., Fernandez-Teruel, A., Giorgi, O., Schwegler, H., Steimer, T., Wiersma, A., Corda, M. G., Flint, F., Koolhaas, J. M., Langhans, W., Schulz, P. E., Siegel, J., and Tobena, A. Genetic Selection and Differential Stress Responses: The Roman Lines/Strains of Rats. *Annals of the New York Academy of Sciences* 851, 501–510, 1998.

20. File, S. E., Ouagazzal, A.-M., Gonzalez, L. E., and Overstreet, D. H. Chronic fluoxetine in tests of anxiety in rat lines selectively bred for differential 5-HT$_{1A}$ receptor function. *Pharmacology Biochemistry and Behavior* 62, 695–701, 1999.

21. Overstreet, D. H. and Steiner, M. Genetic and environmental models of stress-induced depression in rats. *Stress Medicine* 14, 261–268, 1998.

22. Bignami, G. Selection for high rates and low rates of avoidance conditioning in the rat. *Animal Behavior* 13, 221–227, 1965.

23. Blizard, D. A. The Maudsley reactive and nonreactive strains: a North American perspective. *Behavioral Genetics* 11, 469–489, 1981.

24. Tryon, R. C. Genetics of learning ability in rats. *University of California Publications in Psychology* 4, 71–89, 1929.

25. Tryon, R. C. Genetic Differences in maze-learning ability in rats. *Yearbook of the National Society for Studies in Education* 39, 111–119, 1940.

26. Gattu, M., Pauly, J. R., Boss, K. L., Summers, J. B., and Buccafusco, J. J. Cognitive impairment in spontaneously hypertensive rats: role of central nicotinic receptors. I. *Brain Research* 771, 89–103, 1997.

27. Gattu, M., Terry, A. V., Pauly, J. R., and Buccafusco, J. J. Cognitive impairment in spontaneously hypertensive rats: role of central nicotinic receptors. Part II. *Brain Research* 771, 104–114, 1997.
28. Overstreet, D. H. The Flinders Sensitive Line rats: a genetic animal model of depression. *Neuroscience and Biobehavioral Review* 17, 51–68, 1993.
29. Harrington, G. M. Strain differences among rats initiating exploration of differing environments. *Psychonomic Science* 23, 348–349, 1971.
30. Harrington, G. M. Strain differences in activity of the rat in a shuttle stabilimeter. *Bulletin of the Psychonomic Society* 13, 149–150, 1979.
31. Harrington, G. M. Strain differences in activity of the rat using a home cage stabilimeter. *Bulletin of the Psychonomic Society* 13, 151–152, 1979.
32. Harrington, G. M. Strain differences in light-contingent barpress behavior of the rat. *Bulletin of the Psychonomic Society* 13, 155–156, 1979.
33. Harrington, G. M. Strain differences in passive avoidance conditioning in the rat. *Bulletin of the Psychonomic Society* 13, 157–158, 1979.
34. Harrington, G. M. Strain differences in runway learning in the rat. *Bulletin of the Psychonomic Society* 13, 159–160, 1979.
35. Harrington, G. M. Strain differences in shuttle avoidance conditioning in the rat. *Bulletin of the Psychonomic Society* 13, 161–162, 1979.
36. Harrington, G. M. Strain differences in simple operant barpress acquisition to an auditory stimulus by rats. *Bulletin of the Psychonomic Society* 13, 163–164, 1979.
37. Harrington, G. M. Strain differences in free operant leverpress levels in the rat. *Bulletin of the Psychonomic Society* 13, 153–154, 1979.
38. Harrington, G. M. Strain differences in open-field behavior of the rat. *Psychonomic Science* 27, 51–53, 1972.
39. Harrington, G. M. Strain differences in rotating wheel activity of the rat. *Psychonomic Science* 23, 363–364, 1971.
40. Harrington, G. M. and Hellwig, L. R. Strain differences in basal metabolism of behaviorally defined rats. *Bulletin of the Psychonomic Society* 13, 165–166, 1979.
41. Harrington, G. M. and Hellwig, L. R. Strain in organ weights of behaviorally defined rats. *Bulletin of the Psychonomic Society* 13, 165–166, 1979.
42. Lyon, M. F., Nomenclature. In: Foster, H. L., Small, J. D., and Fox, J. G. (Eds.) *The Mouse in Biomedical Research* (Vol. 1). New York, Academic Press, 1981.
43. Cumming, W. W. and Berryman, R. Some data on matching behavior in the pigeon. *Journal of the Experimental Analysis of Behavior*, 4, 281–284, 1961.
44. Jackson, W. J. and Pegram, G. V. Comparison of intra- vs. extra-dimensional transfer of matching by rhesus monkeys. *Psychonomic Science* 19, 162–163, 1970.
45. Rumbaugh, D. M. The importance of nonhuman primate studies of learning and related phenomena for understanding human cognitive development. In: Bourne, G. H. (Ed.) *Nonhuman primates and medical research*. New York, Academic Press, 1973.
46. Voytko, M. L. Nonhuman primates as models for aging and Alzheimer's disease. *Laboratory Animal Science* 48, 611–617, 1998.
47. Squire, L. R., Zola-Morgan, S., and Chen, K. S. Human amnesia and animal models of amnesia: Performance of amnesic patients on tests designed for the monkey. *Behavioral Neuroscience* 102, 210, 1988.

48. Irle, E., Kessler, J., and Markowitsch, H. J. Primate learning tasks reveal strong impairments in patients with presenile or senile dementia of the Alzheimer type. *Brain and Cognition* 6, 449, 1987.

49. Whitney, R. A., Johnson, D. J., and Cole, W. C. *Laboratory Primate Handbook*. New York, Academic Press, 1973.

50. Davis, R. T., Leary, R. W., Smith, M. D. C., and Thompson, R. F. Species differences in the gross behaviour of nonhuman primates. *Behaviour* 31, 326–338, 1968.

51. Kling, A. and Orbach, J. The stump-tailed macaque: A promising laboratory primate. *Science* 139, 45–46, 1963.

52. Orbach, J. and Kling, A. The stumped-tailed macaque: A docile asiatic monkey. *Animal Behaviour* 12, 343–347, 1964.

53. Bernstein, I. S. and Guilloud, N. B. The stumptail macaque as a laboratory subject. *Science* 147, 824, 1965.

54. Jones, N. G. B. and Trollope J. Social behaviour of stump-tailed macaques in captivity. *Primates* 9, 365–394, 1968.

55. Reynolds, H. H. (Ed.) *Primates in Medicine, Vol. 4, Chimpanzee: Central Nervous System and Behavior; A review*. New York, Karger, 1969.

56. Rumbaugh, D. M. Learning skill of Anthropoids. In: Rosenblum, L. A. (Ed.) *Primate Behavior: Developments in Field and Laboratory Research*. New York, Academic Press, 1970.

57. Rumbaugh, D. M. Competence, cortex, and primate models. In Krasnegor, N. A., Lyon, G. R., and Goldman-Rakic, P. S. (Eds.) *Development of the Prefrontal Cortex: Evolution, Neurobiology, and Behavior*. Baltimore, P. H. Brookes Publishing, 1997.

58. Le Gros Clark, W. E. *The antecedents of Man*. Edinburgh, Edinburgh University Press, 1959.

59. Jackson, W. J., Reite, M. L., and Buxton, D. F. The chimpanzee central nervous system: A comparative review. In: Reynolds, H. H. (Ed.) *Primates in Medicine, Vol. 4, Chimpanzee: Central Nervous System and Behavior; A Review*. New York, Karger, 1969.

60. Harlow, H. F. The formation of learning sets. *Psychological Review* 56, 51–65, 1949.

61. Rumbaugh, D. M., Savage-Rumbaugh, E. S., and Washburn, D. A. Toward a new outlook on primate learning and behavior: complex learning and emergent processes in comparative perspective. *Japanese Psychological Research* 38, 113–125, 1996.

62. Nissen, H. W. Phylogenetic comparison. In: Stevens, S. S. (Ed.) *Handbook of Experimental Psychology*. New York, John Wiley & Sons, 1962.

63. Bartus, R. T. and Dean, R. L. Developing and utilizing animal models in the search for an effective treatment for age-related memory disturbances. In: Gottfries, C. G. (Ed.) *Normal Aging, Alzheimer's Disease and Senile Dementia: Aspects on Etiology, Pathogenesis, Diagnosis and Treatment*. Brussels, Editions de l'Universite de Bruxelles, 1985.

64. Long, J. A. and Evans, H. M. The oestrous cycle in the rat and its associated phenomena. *University of California, Number 6*. Berkely, University of California Press, 1922.

65. Overstreet, D. H., Halikas, K. A., Seredemom, S. B., Kampov-Polevoy, A. B., Viglin-skaya, I. V., Kashevskaya, O., Badishtov, B. A., Knapp, D. J., Mormede, P., Kalervo, K., Ting-Kai, L., and Rezvani, A. H. Behavioral similarities and differences among alcohol-preferring and nonpreferring rats: Confirmation by factor analysis and exten-sion to additional groups. *Alcoholism: Clinical and Experimental Research* 21, 840–848, 1997.

66. Rumbaugh, D. M., Washburn, D. A., and Hillix, W. A. Respondents, operants, and emergents: Toward an integrated perspective on behavior. In: Pribram, K. and King, J. (Eds.) *Learning as a self-organizing process.* Hillsdale, Lawrence Erlbaum Asociates, 1996.

67. Small, W. S. An experimental study of the mental processes of the rat. *American Journal of Psychology* 11, 133–165, 1900.

Chapter 2

The Behavioral Assessment of Sensorimotor Processes in the Mouse: Acoustic Startle, Locomotor Activity, RotaRod, and Beam Walking

Gerard B. Fox, Peter Curzon, and Michael W. Decker

Contents

I.   Introduction

Assessment of sensorimotor competence is an important part of the evaluation of animal behavior. Measurement of sensorimotor performance is of obvious importance in investigations of sensory or motor processes, but the effects of experimental manipulations on sensorimotor performance have broader implications for behavioral neuroscience. This is because behavioral experiments typically measure motor responses to sensory information. Thus, the results of behavioral experiments designed to assess other neurobiological processes often cannot be properly interpreted without considering concombinant effects on sensorimotor function. For example, if a lesion or genetic manipulation impairs performance on a spatial memory test, such as the radial arm maze (see Chapters 12 and 13), this impairment cannot be interpreted as evidence of cognitive dysfunction unless it is first established that it is not the result of sensorimotor deficits. Moreover, sensorimotor effects of manipulations can often be used in animal models as surrogates for effects that are more difficult to measure, and relatively simple variations of sensorimotor measures can be used as indices of performance in other behavioral domains, including cognition and emotion.

A number of behavioral tasks have been designed to assess sensorimotor performance in rodents, and this chapter will focus on four general procedures — acoustic startle, open field exploration, RotaRod, and beam walking.

II.   Acoustic Startle

The startle reflex is a stereotyped motor response to a sudden, intense stimulus that has been assessed experimentally in a variety of species, including rats, mice, cats, monkeys, and humans.[1,2] In rodents, the startle response is typically evoked using either acoustic or tactile stimuli and is characterized by contractions of the major muscles of the body, generally leading to extension of the forepaws and hind paws followed by muscle flexion into a hunched position. Mapping studies have demonstrated that the acoustic startle reflex is mediated by a specific neural pathway with acoustic information entering the CNS through auditory nerve input to the cochlear nucleus, which projects to the reticular pontine nucleus via the lateral lemniscus. Motor outputs are generated in the reticular pontine nucleus, which projects to the ventral spinal horn through the reticulospinal tract. Although the basic reflex pathway appears to be relatively simple, the reflex is subject to modulatory influences from higher brain centers.[3]

Measurement of acoustic startle responses can provide general information regarding sensorimotor processing, but measurement of the reflex under conditions that engage the influence of higher brain centers provides an even richer source of data. For example, presentation of lower intensity acoustic stimuli immediately prior to the acoustic startle stimulus attenuates the response to the startle stimulus.[4] This phenomenon, called prepulse inhibition (PPI), is influenced by forebrain centers. Deficits in PPI observed in schizophrenics are thought to reflect disease-related deficits in sensorimotor gating, and measurement of PPI in rodents has been proposed as a model of this feature of schizophrenia.[5-7] The use of this technique in mice is described in detail below.

An alternative use of the startle reflex that will not be described in detail in this chapter is the fear-potentiated startle technique.[8] This procedure takes advantage of the fact that startle responses are augmented in the presence of cues associated with aversive stimuli. Fear-potentiated startle has been used extensively with rats in studies of classical conditioning and anxiety. Interested readers should consult the work of Davis and colleagues for additional information on fear-potentiated startle.[8-11]

A. Acoustic Startle Methods

1. Equipment

The equipment used to measure the startle response has varied from simple lab-made devices[12,13] to more sophisticated units available from various commercial suppliers. We have used S-R Lab equipment sold by San Diego Instruments (www.sd-inst.com), but acoustic startle equipment can also be obtained from Coulbourn Instruments (www.coulbourn.com) and Hamilton Kinder (Julian, CA). The ability to deliver stimuli for a duration of 5 to 1000 ms with consistent intensity is important. Each device is typically enclosed in a larger soundproof cubicle which isolates the animal in the presence of background noise. This also serves to protect the animals in the immediate vicinity from being exposed to the acoustic startle stimulus. A simpler, cheaper S-R LAB screening system is available from San Diego Instruments, but this has no enclosure and the animals must obviously be isolated by location from other test animals. The magnitude of the response of an animal depends on the size of the animal, which means that the assessment of the acoustic startle reflex in the mouse requires more sensitive equipment than the assessment of the reflex in the rat.

The San Diego Instruments equipment we use includes a separate isolation chamber for each individual startle unit. The outer sound attenuating chamber is illuminated and ventilated with a small fan that also provides some level of background noise. In the upper part of this chamber is the acoustic sound source, a loudspeaker which produces a full spectrum white noise that is computer controlled for duration and decibel level. The startle unit is available in assorted sizes to accommodate mice or various sizes of rats. For mice, a clear plastic tube (5.0 cm in diameter by 12.8 cm long) is mounted on a rigid plastic base onto which is attached a piezoelectric device that detects both the acceleration and movement of the animal. The signal from the piezoelectric device is sent to a computer for digital transformation. The piezoelectric device supplied by San Diego Instruments has an adjustable potentiometer on the underside. For mice, the device should be set for maximum sensitivity. San Diego Instruments also provides a standardized force generator that can be attached to the chamber for calibrating the device. More recently, a S-R LAB calibrator-standardization unit has become available which allows for the dynamic measurement of the response in Newtons.

2.   Calibration and Setup

The initial setup of the equipment is fairly simple and the programming of the experiment easily accomplished by following the manual provided. The sound levels in each enclosure need to be calibrated. A Radio Shack (Allied Electronics) Sound Level Meter #33-2050 set on slow response ("A" weighting) is a simple and inexpensive device to use for measuring the intensity of the acoustic stimuli. The startle unit sound duration must be set at the maximum for accurate calibration and the sound meter should be placed in the position normally occupied by the animal holder. If isolation chambers are used, calibration should be done with all of the chamber doors closed to reduce the likelihood that sound from the other chambers will influence the reading made in the chamber being calibrated.

3.   Stimulus Parameters

Startle responses can be measured over a period of time up to 1 s after the presentation of the startle stimulus. However, since the startle response is typically over within 100 ms of stimulus presentation, we typically adjust the window to include only movements generated within the first 100 ms following the startle stimulus. We use the maximal force generated during this 100 ms period for data analysis, but it is also possible to use an area under the curve measure if desired. For mice, stimuli with intensities of 90 dB or higher will typically produce startle responses, although there is some strain-dependent variability startle responses.[14] The magnitude of the startle response varies as a function of the intensity of the startle stimulus, so more reliable responses are often obtained at higher intensities. Acoustic stimuli intensities should not be set higher than 120 dB to avoid producing damage to the ear and the loudspeakers.

4.   Testing Location

Animals are brought to a convenient holding area near the room containing the startle chambers for acclimation. Thus, the startle equipment is best located within an inner room with a sturdy door. This provides additional sound attenuation and keeps animals held in the vicinity of the testing room from being exposed to the startle stimulus.

5.   Subjects

We generally use male CD-1 mice in this procedure (Charles River, Portage, MI). We have also used other mouse strains, but there are clear strain-related differences in startle. As an example, Figure 2.1 shows the results obtained with three strains, DBA-2, C57/BL6, and CD-1. The mice can weigh between 28 to 40 g at the time of testing and are group housed 8 to 14/cage with water and food available *ad libitum*. Given the relationship between body size and the magnitude of the startle response, it is important to control for body weight in conducting experiments. If there are aggressive dominant males, it is best to remove them. In general, the best results are obtained from mice which are protected from stressors and habituated to the laboratory/animal quarters for at least 7 days.

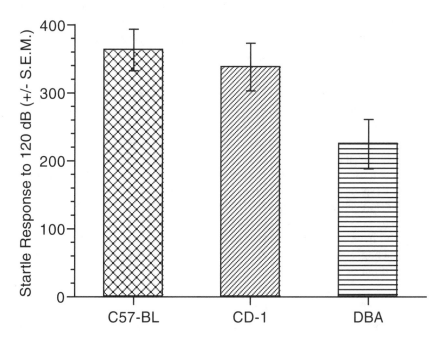

FIGURE 2.1
A comparison of the startle responses displayed by C57BL/6, CD-1, and DBA-2 mice to a 120 dB stimulus. Shown are mean (± s.e.m.) in arbitrary units.

B.   Specific Protocols

1.   Acoustic Startle

We allow the mice a 5 min adaptation period in the enclosure before the start of the session. Background white noise (65 dB) is present during this adaptation period and throughout the session. The session starts with three 120 dB, 40 ms sound bursts. These are not included in the analysis because the responses to the first few startle bursts are typically significantly different in magnitude from the rest of the trials. Thus, exposure to these initial bursts allows for the establishment of a stable baseline.

Following this, acoustic startle trials are initiated. For simple assessment of acoustic startle responses, we typically use 2 or 3 stimulus intensities (90 and 105 dB or 90, 105, and 120 dB). The stimuli are 40 ms in duration and are presented in a quasi-random order, such that an equal number of presentations of each stimulus intensity is included in each half of the session and no single intensity is presented more than two times in succession. The time between stimuli averages 15 s but this interval should be varied within a range of 5 to 30 s so that the animals do not anticipate the stimulus. At least 10 trials of each stimulus intensity should be conducted to obtain reliable results. An alternative to evaluating the magnitude of the startle response is to measure the startle threshold. Here, a wider range of stimulus intensities, e.g., from 70 to 120 dB, is used. Stimuli are presented in quasi-random order, and the lowest intensity producing a reliable response is determined. Since some habituation of the response can occur both within a session and between sessions (see below), trials at each intensity should always be evenly distributed within a session (but they should not, of course, be presented in a predictable sequence).

2.   Startle Habituation

Startle habituation is a simple measure of plasticity related to the gating of sensory information. This measure is of interest in the area of schizophrenia research, since schizophrenics show impaired startle habituation, an impairment that may be related to the hypervigilance characteristic of this condition.[5,6] For startle habituation, a single stimulus intensity is repeatedly presented using either a fixed or variable interval. Responses normally decline (habituate) over trials. An example of the data generated in such an experiment is shown in Figure 2.2. In this case, mice were exposed to a 40 ms, 120 dB stimulus presented every 15 s for 50 min (200 trials). Long-term habituation of the startle response can also be used as a memory index by conducting a second session at a later time (e.g., 24 h later) and assessing long-term retention of habituation.

3.   Prepulse Inhibition (PPI)

In this version of the test, the attenuation produced by a low intensity stimulus presented just before the startle stimulus is assessed. We sometimes assess PPI on the day following standard startle testing. This serves to habituate the animals to the basic handling procedures and, in the case of pharmacological studies of PPI, allows groups to be matched for baseline startle and elimination of animals that either startle excessively or do not respond. The PPI session starts with three 120 dB, 40 ms sound bursts that are not included in the analysis, followed by multiple presentations of three general trial types: a 40 ms 120 dB acoustic startle stimulus alone, the 120 dB stimulus preceded (by 100 ms) by acoustic prepulse stimuli (30 to 50 ms) at levels between 5 and 15 dB above background (i.e., 70 to 80 dB for the 65 dB background noise in our studies), and the prepulse stimuli alone (which should not produce a startle response). [Note: the interstimulus times are measured from stimulus offset to stimulus onset.]

Since the degree of PPI is related to the intensity of the prepulse (see Figure 2.3), it is often useful to include more than one level of prepulse intensity. It is also

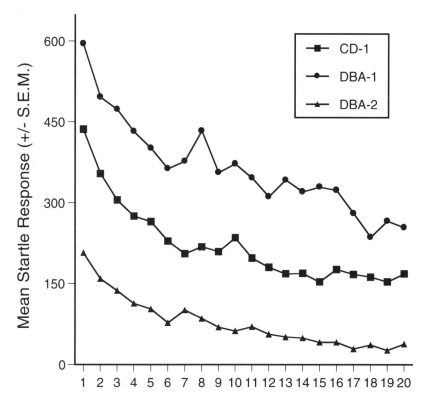

FIGURE 2.2

A comparison of the startle responses displayed by CD-1, DBA-1, and DBA-2 mice to repeated exposures to a 120 dB stimulus. Shown are mean responses across consecutive blocks of 10 trials each in arbitrary units. Note the habituation of the response.

sometimes useful to include a "no stimulus" trial to assess the influence of background movement on startle measures. To reduce variability, at least 12 trials of each type should be conducted within a single session. As with standard startle testing, trial types and intertrial intervals should be presented in a quasi-random, balanced manner with equal representations of trial types and intervals in each half of the session. This allows the session to be analyzed in two blocks to assess any changes over time.

4. Statistical Analysis

The results can be either printed out or, more practically, assembled in a computer text file. Microsoft Excel or similar spreadsheet software easily assembles the data into the various trial types. What is very evident from the mouse data is that there is a great deal of variability between animals that needs to be addressed. Using groups sizes of 12 to 14 mice, there are typically one or two animals with very low startle responses in the control group. This may reflect a " behavioral freezing" that is common to both rats and mice. As a matter of standard practice we therefore

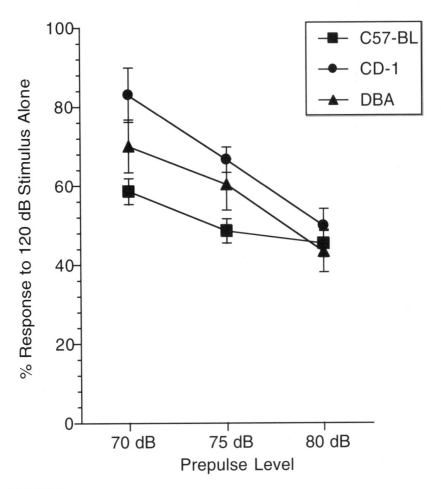

FIGURE 2.3
The effect of prepulses of 70, 75, and 80 dB on the startle responses displayed by C57BL/6, CD-1, and DBA-1 mice to a 120 dB stimulus. Shown are mean (± s.e.m.) percent of the response to the 120 dB stimulus alone. Note that increasing the intensity of the prepulse stimulus decreases the startle response (i.e., increasing the prepulse stimulus increases the magnitude of prepulse inhibition).

eliminate the lowest two responders in each experimental group. The same can be done with the upper end of the scale if that is a source of high variability in the groups.

The data from prepulse trials are expressed as a mean percent of baseline startle, calculated as [(startle response with prepulse/the startle response without PPI) × 100]. Thus, a higher percent represents a disruption of PPI. (Alternatively, data can be represented as the percent decrease in the response in the presence of the prepulse stimulus.) Data are analyzed with repeated-measures ANOVA using a program such as Statview (SAS Institute, Cary, NC). Interpretation of the PPI data must also include an evaluation of effects of treatments on baseline startle. Substantial effects of an experimental treatment on baseline startle responses make the interpretation of PPI more tenuous. For example, a treatment that greatly attenuates the acoustic startle response may preclude assessment of PPI. The use of a range of startle stimuli can be helpful in this regard by allowing one to match baseline startle responses across groups (e.g., when a treatment reduces baseline startle, PPI in the presence of a higher intensity startle stimulus in the treated group could be compared to PPI in the presence of a lower intensity startle stimulus in the control group).

5.   Example PPI Experiment

An experiment was conducted to evaluate the effects of scopolamine, a muscarinic cholinergic receptor antagonist, on prepulse inhibition. In pharmacological experiments, it is often important to get the maximum number of trials in the shortest time because the time course of some drugs is of fairly short duration. Session duration will obviously increase as more trial types/numbers are added, and this was taken into account when we set up our standard mouse testing paradigm, which is illustrated schematically in Figure 2.4. Since we did not observe responses in CD-1 mice to the intensities used for prepulse alone or when no stimuli were presented, these trial types were eliminated from our standard testing. The 75-trial session used ran for approximately 30 min and included three trial types — a 120 dB startle stimulus alone, the 120 dB stimulus preceded by a 75 dB prepulse stimulus, and the 120 dB stimulus preceded by an 80 dB prepulse stimulus (the trial sequence is shown in Table 2.1). In this experiment, naive male, CD-1 mice were used (i.e., they had not been previously exposed to the startle apparatus).

Scopolamine tended to increase responses to 120 dB stimuli in the absence of the prepulse, an effect that was only significant at the lowest dose tested (0.25 μmol/kg, i.p.; data not shown), and produced a dose-related impairment of prepulse inhibition (Figure 2.5). In vehicle-treated mice, the presentation of the prepulse reduced responding to the startle stimulus to 37% and 22% of baseline (i.e., the response was decreased by 63% and 78% by the 75 and 80 dB prepulse, respectively). Scopolamine produced a dose-related reduction in prepulse inhibition that was statistically significant at both prepulse levels at 0.5 and 1.0 μmol/kg, i.p. Pharmacological disruption of PPI in mice can also be obtained with compounds that influence dopaminergic (e.g., apomorphine, amphetamine) and glutaminergic (e.g., phencyclidine, MK-801) neurotransmission.[15]

III.   Motor Function and Spontaneous Exploration

Although many experimenters view locomotor activity as an overly simplistic measure that provides only limited information, alterations in this behavior can reveal important information on potential mechanisms of drug action. Moreover, locomotor activity may influence functional outcome in many models of CNS injury or disease. For example, many different psychoactive drugs can act at neuronal receptor sites and directly affect motor function. Similarly, brain injury models employed by many researchers can produce subtle or sometimes pronounced alterations in motor behavior. Furthermore, genetically altered animals have become popular in an attempt at unmasking the molecular and cellular correlates of such behaviors as learning and

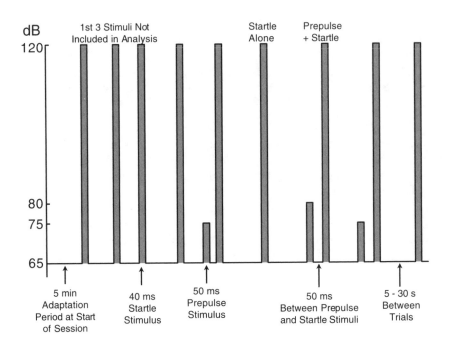

FIGURE 2.4
A schematic showing a sample protocol for stimulus presentation in a prepulse inhibition experiment. The parameters in this figure were used to obtain the experimental results illustrated in Figure 2.5. This figure is not to scale and illustrates only the first few trials of the 75-trial session.

memory in addition to numerous disease states. These animal models, however, are not without their drawbacks, including profound changes in motor function that can confound the interpretation of behavioral results. Therefore, it is important for the neuroscientist to be aware of, and to characterize, these changes carefully. The following will examine several methods for assessing motor and exploratory activity in the adult mouse.

Locomotor and exploratory behavior may also be influenced by several other factors such as time of day (rodents are nocturnal animals and are therefore significantly more active during periods of darkness), anxiety (animals may be more or less active depending on the situation to which they are exposed), state of wakefulness or arousal (stimulants will tend to increase activity whereas sedatives will tend to decrease activity, although the magnitude of these effects can often depend upon the strain of animal used), environmental novelty (mice tend to exhibit increased exploratory behavior when exposed to a novel environment and decreased activity upon re-exposure to the same environment, i.e., habituation), motivation (food-deprived mice may show increased activity), age (younger rodents are more active than aged animals), general health, and genetic strain (C57BL/6 mice are more active than 129-derived animals). All of these factors necessitate careful experimental design and it is therefore prudent to control and maintain consistency of these factors from the onset. However, natural variations in activity and stress levels often exist between mice of the same strain despite controlling for all of the factors described above; hence the need for adequate group sizes that will accommodate appropriate statistical analyses.

A. Methods: Spontaneous Activity

1. Open Field (Non-automated)

Perhaps the simplest and most economical method for assessing both exploratory and locomotor activity is the open field apparatus. As the name suggests, this generally consists of a square (or circular, depending on personal preference) arena, surrounded by walls to prevent the animal from escaping, and of adequate size (e.g., 100 cm by 100 cm or 100 cm diameter). In its simplest form, the floor is divided into equally spaced regions by marker pen which has been allowed sufficient time to dry so as not to produce unwanted olfactory effects. The box itself may be composed of either wood or plastic, although the latter is preferred to reduce olfactory issues and for ease in cleaning.

a. Typical Protocol

1. The open field should be located in a quiet room with controlled temperature and ventilation. A low-level illumination is preferred to reduce anxiety, unless this is a component of the task that you wish to study. The observer should be seated comfortably at a distance from the apparatus, ideally watching a monitor fed by a video camera positioned above the open field. (If a video camera is used, the grid lines can be drawn on the monitor screen rather than on the apparatus itself.) If visible to the behaving rodent, the investigator should be consistent with seating position, clothing, and potential olfactory cues.

2. If the stimulant activity of a drug is to be examined, the rodent should first be habituated to the apparatus for 3 or more 5 min sessions to reduce baseline activity. Do not habituate if anxiety or response to novelty is being studied. For brain lesion or injury studies, you may wish to examine performance at discreet times before and after surgery. Bear in mind, however, that activity and/or exploration following repeated exposure to the open field may be affected by differential habituation and cognition.

3. On the test day, administer the test drug, if required, at an appropriate time prior to placing the rodent into the center of the open field. The investigator records the following specific behaviors using prepared data sheets and appropriate counters over a specified period of time, usually 5 to 10 min.

    Parameters to record: locomotion (number of square crossings within the specified time), rearing, grooming, and stereotypical behaviors (such as licking, biting, and head weaving). These activities may be recorded separately for peripheral regions or in the center of the arena, the latter thought to reflect the degree of anxiety experienced by the rodent (i.e., animals with higher levels of activity in the center of the arena are less anxious). Defecation frequency may also be recorded as a measure of fear, but this tends to be more variable due to the relatively small numbers involved.

TABLE 2.1
Session Protocol for Prepulse Inhibition Experiment Shown in Figure 2.5

Trial #	ITI[a] (s)	Prepulse[b] dB	Trial #	ITI[a] (s)	Prepulse[b] dB
	5 min acclimation		38	20	–
1	20	–	39	25	75
2	25	–	40	20	–
3	30	–	41	10	75
4	20	–	42	20	–
5	10	75	43	10	80
6	20	–	44	20	75
7	10	80	45	10	–
8	20	75	46	20	75
9	10	–	47	20	80
10	20	75	48	15	–
11	20	80	49	10	80
12	15	–	50	25	–
13	10	80	51	30	75
14	25	–	52	25	–
15	30	75	53	25	80
16	25	–	54	20	75
17	25	80	55	10	–
18	20	75	56	30	80
19	10	–	57	10	75
20	30	80	58	15	80
21	10	75	59	15	75
22	15	80	60	30	–
23	15	75	61	25	80
24	30	–	62	20	80
25	20	80	63	30	75
26	20	80	64	15	–
27	30	75	65	15	80
28	15	–	66	30	75
29	15	80	67	30	–
30	30	75	68	20	80
31	30	–	69	25	75
32	20	80	70	30	–
33	25	75	71	30	80
34	30	–	72	15	75
35	30	80	73	30	80
36	15	75	74	20	–

TABLE 2.1 (CONTINUED)
Session Protocol for Prepulse Inhibition Experiment Shown
in Figure 2.5

Trial #	ITI[a] (s)	Prepulse[b] dB	Trial #	ITI[a] (s)	Prepulse[b] dB
37	30	80	75	25	75

[a] ITI, intertrial interval, the delay (in seconds) prior to the initiation of the trial

[b] Each prepulse trial consists of a 50 ms prepulse stimulus and a 40 ms, 120 dB startle stimulus (separated by 100 ms); each startle stimulus alone trial (designated by the dash in this column) consists of a 40 ms presentation of the 120 dB startle stimulus by itself.

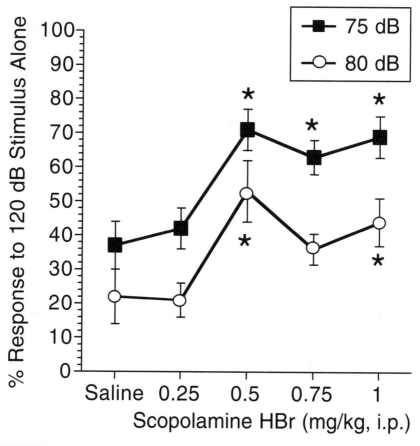

FIGURE 2.5
The effect of scopolamine (injected i.p. 15 min prior to testing) on the magnitude of prepulse inhibition produced by prepulse stimuli of 75 and 80 dB in CD-1 mice. Data are expressed as a percent of the response to the 120 dB stimulus alone (mean ± s.e.m). *Different from respective saline control, $p < 0.05$.

2.   Open Field (Automated)

For large studies it is often impractical to directly observe each animal individually. This can be resolved by making use of automated activity boxes consisting of arenas similar to those described above, but with regions demarcated by infrared beams instead of marker pens. Each box is connected individually to a low-end personal computer, generally running DOS, which collates all data from up to 30 or more boxes at a time. Although the cost can be somewhat prohibitive for smaller laboratories (upward of $50,000 for a set of 12 boxes), the major advantage of such systems is that they allow the collection of both vertical (rearing) and horizontal activity from the periphery and center of the apparatus over time periods which could not be accurately completed by a manual observer. For example, the computer can be programmed to accept data every 15 s and to calculate a mean for each 1 min time bin for up 2 hr or more. Clearly the amount of data collected rises considerably with increased time. In the authors' laboratory, a 12-box system is employed (San Diego Instruments, San Diego, CA) in a dedicated quiet room with dimmed lighting. Each arena is 40 by 40 cm in size with removable clear Plexiglass walls for ease of cleaning. Two sets of infrared photocells (one to detect rearing, the other for locomotion) are fixed to a rack which surrounds the Plexiglass and which can be adjusted in the vertical plane to allow measurements from rats or mice. The software can be set to exclude or include repetitive movements in front of a single beam to distinguish between different types of motor activity.

a.   Typical Protocol

1.   If a stimulant activity is expected, habituate the mice to the apparatus for several 1 hr sessions. Do not habituate for drugs expected to decrease activity.

2.   Program the computer to record activity as desired. An example would be to bin data every 5 min for up to 1 hr and to distinguish horizontal activity from vertical in both the peripheral and central regions of each arena.

3.   Administer test drug at the appropriate time. If decreased activity is expected, inject before placing mice into the center of the arena. Conversely, if increased activity is expected, place mice into the center of the arenas and allow to habituate for at least 30 min before drug administration. If animals are subjected to brain injury or other surgery, allow sufficient time for recovery (at least 24 h, if not more, depending on severity) before placement into the arenas.

4.   Record data for predetermined period of time, usually 30 to 120 min. Print out all raw data as a hard copy backup and convert the data file produced into a form suitable for analysis using software programs such as Microsoft Excel (Microsoft Corporation, Seattle, WA) and Statview (SAS Institute, Cary, NC).

Arenas employed by even more advanced systems (AccuScan Instruments, formerly Omnitech, Columbus, OH; www.accuscan-usa.com) can be subdivided into zones both physically by using a Plexiglass insert and virtually by specifying different software parameters. This allows many additional animals to be assessed simultaneously. In addition, this system can allow measurements of distance traveled, movement time, rearing duration, etc. If the additional cost of such systems is prohibitive, but something more sophisticated than the manual version of the open field apparatus is desired, other equipment can be modified to produce measures of locomotion. For example, many video tracking systems currently in use in labs investigating spatial memory (see Chapter 11) can track movement in several animals simultaneously. These can produce measures of distance and orientation over time, although rearing activity would have to be recorded separately. Therefore, if cost is an important factor, the investigator should determine whether existing equipment could be easily modified to measure motor activity. In addition, some photocell systems have been designed to measure activity in the home cage, where response to a novel environment is undesirable or for monitoring over the light:dark cycle. Such systems tend to come in specially designed cages and racks, and can be expensive.

3.   Variations

Locomotor activity can also provide indices of learning and memory and anxiety. Habituation of locomotor activity in a novel environment can be used to assess memory in mice.[16] For this procedure, the mouse is briefly exposed (e.g., 5 min) to a novel open field, and locomotor activity is assessed. Memory for the novel experience is then tested at a later time by re-exposing the mouse to the same open field. Activity during the second exposure is used as an index for assessing memory, with lower activity indicative of better memory for the open field. Of course, it is important that the treatments evaluated with this method do not have direct effects on locomotor behavior.

The pattern of exploration can also be an important index of anxiety. Informal assessment of anxiety can be derived by comparing time spent in the periphery of the arena relative to time spent in the center. Anxious animals tend to spend more time in the periphery. In addition, initial freezing in an open field is an index of anxiety, so the latency to move a given distance (or to move through a given number of squares) can also be used to assess fear and anxiety. A more formal assessment of anxiety can be made using a modified open field apparatus. The open field is separated into a well-lit area and a dark area and the relative time and activity in these two zones is compared. Anxiolytics increase the time spent in the well-lit zone in this light:dark test.[17]

4.   Example Experiment

Data from a typical automated experiment are presented in Figure 2.6. In this study, vertical (or rearing) (Figure 2.6A) and horizontal (Figure 2.6B) activity was assessed for 3 mouse strains, with data collected in 5 min bins for 30 min.[18] Note that BALB/c mice appear significantly less active than animals from the other strains. Note also the habituation response indicated by decreased activity over time, for most groups. It is important that appropriate statistical methods be used for analysis of behavioral data. Locomotor data are generally normally distributed so parametric analyses of variance (ANOVAs) are used routinely. A repeated measures ANOVA should be considered in most cases when a time course is employed. *Post hoc* tests that examine the mean square error relative to the overall analysis (e.g., Tukey's) can then be used for multiple comparisons between groups. Individual *t* tests should not be used unless

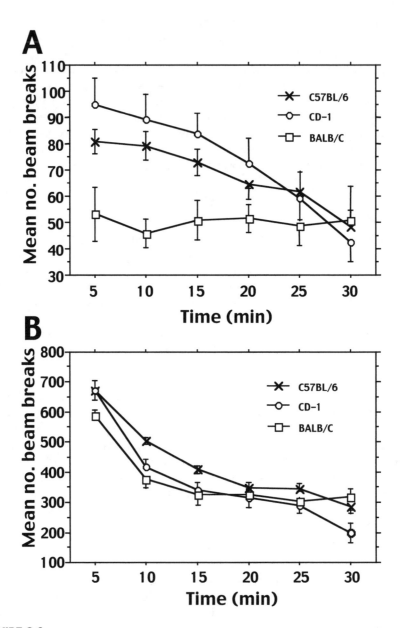

FIGURE 2.6
Automated measure of vertical (a) and horizontal (b) activity in the same animals from 3 different mouse strains. Data were collected over 5 min intervals for a total of 30 min. Most mice habituated to the test environment, as evidenced by the decline in activity over the duration of the experiment. Note that the BALB/c mice were less active than mice from the other 2 strains. (Statistical significance described in detail in main text.) (*Author's unpublished data.*)

they are corrected for multiple comparisons. For the data presented in Figure 2.6B, a repeated measures ANOVA yielded a significant Group effect [$F(2,26) = 4.382$, $p < 0.0229$], indicating overall differences between the strains in the study; Time effect [$F(5,130) = 126.103$, $p < 0.0001$], reflecting the decreased activity with time (habituation) overall; and Group X Time interaction [$F(10,130) = 3.049$, $p < 0.0017$], indicating significant differences between groups over time. *Post hoc* analysis with Tukey's pairwise comparisons detected significant differences for the 5, 10, and 15 min time points between C57BL/6 and BALB/C mice ($p < 0.01$).

B.   Methods: Motor Function

Motor function can be differentially affected depending on experimental parameters. For example, unilateral brain injury models often produce hemiparesis-like effects that may be reflected by deficits in grip strength, balance, and turning behavior, or may induce forepaw flexion. Many drugs can have either sedative or stimulant properties. Consequently, several models have been developed to examine specific motor deficits such as these. Two commonly used procedures are thus described.

1.   RotaRod

The ability of a rodent to maintain balance and keep pace with a rotating rod has been used with varying degrees of success over the years to assess motor function. Several versions of this test (commonly referred to as the RotaRod test) have been described over the years. Most require the mouse to walk on a rotating rod of fixed diameter (3.5 cm for the apparatus we use) which increases in speed over a predetermined period of time until the animal can no longer maintain its position. The RotaRod apparatus employed by this laboratory consists of a central drive rod connected to a stepper motor (AccuScan Instruments, formerly Omnitech, Columbus, OH; www.accuscan-usa.com) which is divided into four separate testing stations. The speed at which the rod rotates can be ramped up from 0 rpm to over 100 rpm.

a.   **Typical Protocol**

1.   Administer drug, as appropriate. For lesioned or injured animals, wait at least 24 hr following the surgical procedure.

2.   Set the apparatus to ramp up to approximately 40 rpm over 60 s. This is a good standard for young adult mice, although it should be noted that juvenile and older animals perform poorly at this task.

3.   Place four mice on the RotaRod, one per testing station, and simultaneously start the stepper motor and timer. Many models come equipped with a timer that begins automatically when the motor is switched on and stops when the animal falls down to the floor of the apparatus, as detected by interruption of an infrared beam.

4.   As the speed increases, the mouse is required to walk faster to remain in a stationary position. The latency to fall from the rotating rod is determined and taken as a measure of motor function. It is generally a good idea to take the mean of at least 2 to 3 measures from each animal.

b.   Variation   Some investigators[19,20] modify the rod itself by enclosing the core of the rod with a series of stainless steel bars of a specific diameter (Figure 2.7A). In this instance the time either to fall (Figure 2.7B) or to cling and make two full rotations is recorded as the outcome measure. This design may offer some advantages over the more traditional relatively smooth rod in that data, particularly in brain injury studies, may be more consistent within groups. With rodent strains that exhibit a poor baseline performance in this task, it is usually beneficial to pretrain these animals at least 2 to 3 times before commencing the study proper.

c.   Example Experiment   Data from a typical experiment are presented in Figure 2.7C, where the effect of sham surgery and controlled cortical impact (CCI) brain injury on time spent on the rotating rod is shown for three mouse strains. Sham operated controls exhibited a stable performance over the 4 weeks of testing, whereas a decrease in time spent on the RotaRod device was observed in injured mice from all strains for up to 7 days following injury. Once again a repeated measures ANOVA is appropriate for comparing groups over time as RotaRod data tend to be normally distributed and this test was conducted repeatedly over a 4-week period in this study. A significant Group effect [$F(5,57) = 16.601$, $p < 0.0001$], indicating overall differences between the strains in the study, Time effect [$F(7,399) = 47.183$, $p < 0.0001$], reflecting the attenuation of the deficits with time in the injured groups, and Group X Time interaction [$F(35,399) = 6.480$, $p < 0.0001$], indicating significant differences between groups over time, were observed. Using a *post hoc* test (Tukey's pairwise comparison), a significant impairment was detected among CCI injured mice from all 3 strains when compared with their respective surgery controls on days 1, 2, and 3 ($p < 0.05$) following surgery. A one-way ANOVA would be suitable for comparing these groups in instances where no time component is involved. A *t* test may also be appropriate in such instances. No significant difference was observed between strains for either treatment group.

2.   Beam Balance/Walking

While the RotaRod is useful for determining gross motor deficits in the rodent, the detection of more subtle motor effects requires a different approach. Fine motor coordination, for example, can be assessed by using a beam walking or balance task. This test essentially examines the ability of the animal to remain upright and to walk on an elevated and relatively narrow beam (Figure 2.8A) without falling to the cushioned pads below or slipping to one side of the beam. Again, unilateral brain injury models tend to induce a hemiparesis-like effect which can cause the rodent to slip to one side, usually that which is contralateral to the injury site (Figure 2.8B).

a.   Typical Protocol

1.   For mice, set up a beam approximately 6 mm wide and 120 mm in length, and suspended about 600 mm above some foam pads. (A larger beam, approximately 18 mm wide and 240 mm in length, in addition to a flat platform at one end to rest between trials is required for rats.)

2.   Place the animal on one end of the beam (for the rat this would be farthest from the platform). Animals from active strains such as the C57BL/6 mouse or the Long-Evans

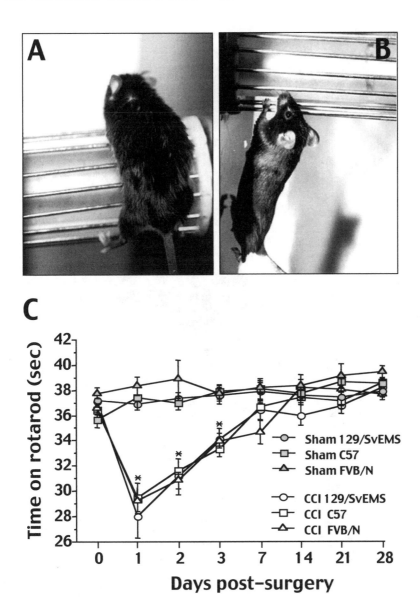

FIGURE 2.7

The effect of moderate controlled cortical impact (CCI) brain injury on rotarod performance. As the device gradually ramped up to speed (35 rpm), the mouse was required to walk faster to maintain a stationary position on the rod, which has been modified here to include a series of stainless steel bars (A). When the mouse can no longer keep up with the speed of rotation, it either falls from the bars (B) or clings tight and begins to rotate with the rod (not shown). Uninjured or sham-operated mice are generally adept at this task. However, a significant deficit can be seen for approximately 7 days following CCI brain injury, shown for 3 different mouse strains in (C). Photos depict an adult male C57BL/6 mouse. Data reproduced with permission from Fox G. B., LeVasseur R. A., Faden A. I., Behavioral responses of C57BL/6, FVB/N and 129/SvEMS mouse strains to traumatic brain injury: Implications for gene targeting approaches to neurotrauma, *Journal of Neurotrauma*, 16, 377, 1999.

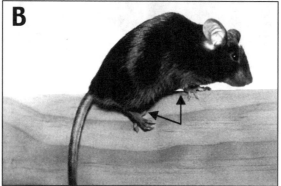

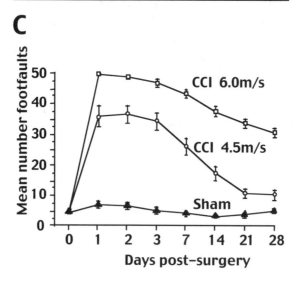

rat will instinctively walk along the beam to reach the opposite end. Once at this point they will generally turn 180° and continue to walk on to the opposite end. Establish a basal level of performance before surgery or treatment, and allow sufficient time for recovery (at least 24 hr) before retesting.

3. Count the number of footfaults, defined as the number of times the forepaws and/or hindpaws slip from the horizontal surface of the beam over a predetermined number of steps (50 is usually adequate). Allow the performing animal sufficient time (approximately 5 min) to complete this task. [It is useful to use a mirror on the side of the beam opposite the observer and to videotape the performance for scoring.]

4. Remove to home cage and retest as appropriate. It should be noted that mice, and especially rats, will become tired and reluctant to move if exposed to this test repeatedly over a short period on the same day.

b.   Variation   This task works well for active rodent strains and may not be suitable for less active animals. Another variation partly designed to address this issue in the rat involves training animals to walk across the beam to a "safe" dark box; the cognitive requirements for this version, however, may influence motor outcome to some degree so care should be taken here. A more simple approach measures the time taken to fall down onto the foam pads. In this instance, the investigator should vary the beam width until an acceptable latency is found for the particular strain to be used. Attention should also be paid to the body weight of the animal, as the suitable width of the beam may change according to the mouse's ability to grip the edge of the beam. For example, mice heavier than 35g will generally require a beam approximately 9 mm thick.

c.   Example Experiment   Data from a typical experiment are reproduced in Figure 2.8C. In this experiment, adult C57BL/6 mice were subjected to mild (4.5 m/s) or moderate (6.0 m/s) unilateral controlled cortical impact (CCI) brain injury and the number of contralateral hindlimb footfaults recorded over a 4-week period. An obvious deficit, dependent on injury severity, was observed when compared with sham-operated controls. For statistical analysis, beam walking data are generally normally distributed so parametric analyses of variance (ANOVAs) are advised.

FIGURE 2.8

The effect of moderate controlled cortical impact (CCI) brain injury on beam walking performance for the C57BL/6 mouse. Surgery-naive or sham-operated mice perform well on this task, traversing the beam several times, gripping its horizontal edge with the innermost digits (arrow in (A) illustrates this point). Footfaults, defined as hindlimb slipping from the horizontal surface of the beam, (arrow in (B) shows contralateral hindlimb slipping down the side of the beam) are generally counted over a total of 50 steps and a frequency of 15% or less is normal for this strain (C). However, mice subjected to mild (low velocity, 4.5 m/s) or moderate (higher velocity, 6.0 m/s) unilateral CCI brain injury exhibit a highly significant deficit (statistical significance described in detail in main text) in this task which is dependent on injury severity and persists for an extended period (C). Data reproduced with permission from Fox G. B., LeVasseur R. A., Faden A. I., Behavioral responses of C57BL/6, FVB/N and 129/SvEMS mouse strains to traumatic brain injury: Implications for gene targeting approaches to neurotrauma, *Journal of Neurotrauma*, 16, 377, 1999.

A repeated measures ANOVA should be considered in most cases when a time course such as that presented above is employed. *Post hoc* tests that examine the mean square error relative to the overall analysis (e.g., Tukey's) can then be used for multiple comparisons between groups. Individual *t* tests should not be used. For the data presented in Figure 2.8C, a repeated measures ANOVA yielded a significant Group effect [$F(2,33) = 94.265$, $p < 0.0001$], indicating overall differences between the different treatment groups in the study, Time effect [$F(7,231) = 89.383$, $p < 0.0001$], indicating significant overall changes in performance over the duration of the study, and Group X Day interaction [$F(14,231) = 20.995$, $p < 0.0001$], indicating significant performance differences between groups over time. *Post hoc* analysis with Tukey's pairwise comparisons detected significant differences for days 1 to 28 between sham controls and CCI-injured mice from both groups ($p < 0.001$). There were no significant differences between groups before injury (Day 0; $p > 0.05$).

References

1. Davis M., Neurochemical modulation of sensory-motor reactivity: acoustic and tactile startle reflexes, *Neuroscience & Biobehavioral Reviews*, 4, 241, 1980.
2. Davis M., The mamalian startle response, in *Neural Mechanisms of Startle Behavior*, Eaton R. C., Ed., Plenum Press, New York, 1984, 287.
3. Koch M., Schnitzler H. U., The acoustic startle response in rats — circuits mediating evocation, inhibition and potentiation, *Behavioural Brain Research*, 89, 35, 1997.
4. Ison J. R., Taylor M. K., Bowen G. P., Schwarzkopf S. B., Facilitation and inhibition of the acoustic startle reflex in the rat after a momentary increase in background-noise level, *Behavioral Beuroscience*, 111, 1335, 1997.
5. Geyer M. A., Braff D. L., Startle habituation and sensorimotor gating in schizophrenia and related animal models, *Schizophrenia Bulletin*, 13, 643, 1987.
6. Geyer M. A., Swerdlow N. R., Mansbach R. S., Braff D. L., Startle response models of sensorimotor gating and habituation deficits in schizophrenia, *Brain Research Bulletin*, 25, 485, 1990.
7. Swerdlow N. R., Geyer M. A., Using an animal model of deficit sensorimotor gating to study the pathophysiology and new treatments of schizophrenia, *Schizophrenia Bulletin*, 24, 285, 1998.
8. Davis M., Animal models of anxiety based on classical conditioning: the conditioned emotional response (CER) and the fear-potentiated startle effect, *Pharmacology & Therapeutics*, 47, 147, 1990.
9. Davis M., Neural systems involved in fear-potentiated startle, *Annals of the New York Academy of Sciences*, 563, 165, 1989.
10. Davis M., The role of the amygdala in fear -potentiated startle: implications for animal models of anxiety, *Trends in Pharmacological Sciences*, 13, 35, 1992.
11. Davis M., Falls W. A., Campeau S., Kim M., Fear-potentiated startle: a neural and pharmacological analysis, *Behavioural Brain Research*, 58, 175, 1993.
12. Hunter K. P., Willott J. F., Effects of bilateral lesions of auditory cortex in mice on the acoustic startle response, *Physiology & Behavior*, 54, 1133, 1993.

13. Weiss G. T., Davis M., Automated system for acquisition and reduction of startle response data, *Pharmacology, Biochemistry & Behavior*, 4, 713, 1976.
14. Crawley J. N., Belknap J. K., Collins A., Crabbe J. C., Frankel W., Henderson N., Hitzemann R. J., Maxson S. C., Miner L. L., Silva A. J., Wehner J. M., Wynshaw-Boris A., Paylor R., Behavioral phenotypes of inbred mouse strains: implications and recommendations for molecular studies, *Psychopharmacology*, 132, 107, 1997.
15. Curzon P., Decker M. W., Effects of phencyclidine (PCP) and (+)MK-801 on sensorimotor gating in CD-1 mice, *Progress in Neuro-Psychopharmacology & Biological Psychiatry*, 22, 129, 1998.
16. Platel A., Porsolt R. D., Habituation of exploratory activity in mice: a screening test for memory enhancing drugs, *Psychopharmacology*, 78, 346, 1982.
17. Costall B., Kelly M. E., Naylor R. J., Onaivi E. S., Actions of buspirone in a putative model of anxiety in the mouse, *Journal of Pharmacy & Pharmacology*, 40, 494, 1988.
18. Fox G. B., LeVasseur R. A., Faden A. I., Behavioral responses of C57BL/6, FVB/N and 129/SvEMS mouse strains to traumatic brain injury: Implications for gene targeting approaches to neurotrauma, *Journal of Neurotrauma*, 16, 377, 1999.
19. Fox G. B., LeVasseur R. A., Faden A. I., Sustained sensory/motor and cognitive deficits with neuronal apoptosis following controlled cortical impact brain injury in the mouse, *Journal of Neurotrauma*, 15, 599, 1998.
20. Hamm R. J., Pike B. R., O'Dell D. M., Lyeth B.G., Jenkins, L. W., The rotarod test: an evaluation of its effectiveness in assessing motor deficits following traumatic brain injury in the rat, *Journal of Neurotrauma*, 13, 187, 1994.

Chapter 3

Fundamentals, Methodologies, and Uses of Taste Aversion Learning

T. Edward Orr, Paul A. Walters, and Ralph L. Elkins

Contents

I.   Introduction

When many animals, including humans, consume a particular food or drink prior to experiencing gastrointestinal distress involving nausea, they typically avoid that substance upon subsequent exposure. This response, termed a conditioned taste aversion (CTA), occurs in humans when stimuli related to the consumed substance become associated with the experienced nausea. A similar process is presumed to account for the development of CTAs in animals.

CTAs represent a class of learning that is exceptionally efficient and is considered to be an adaptive specialization that has been demonstrated across a wide range of animal species. There are several distinguishing characteristics of CTA learning.

1. Many, but not all, pharmacologic agents can produce a CTA.[1]

2. CTAs can be developed in many humans and animals with as few as a single pairing of the flavor of an ingested substance (i.e., the conditioned stimulus [CS]) with nausea (the unconditioned stimulus [UCS]).[2,3] In most learning paradigms, animals require multiple sessions of CS-unconditioned stimulus (UCS) pairings before learning is demonstrated.

3. The learning can occur when presentation of the target flavor precedes the subsequent bout of nausea by many hours as opposed to the much shorter temporal constraints to successful learning that usually are associated with other conditioning methodologies.[4]

4. Neither intentional nor conscious mediation is necessary for CTA acquisition. Humans develop aversions to flavors that precede nausea even if they have a strong basis for believing that the flavor and illness were not causally linked,[3,5] and CTAs can be produced in deeply anesthetized rats.[6]

5. CTAs are readily produced with pairings of internal distress (i.e., nausea) with gustatory stimuli but not with audio-visual stimuli. Conversely, environmental avoidance responses are readily established by pairing audio-visual stimuli with cutaneous pain but not with nausea. These different associative efficiencies are putative specialized capabilities of separate gut-defense and skin-defense systems in vertebrates.[7]

6. Typically, emetic drugs are utilized as UCS agents within the CTA paradigms. This reliance on emetic agents is consistent with Garcia's gut-defense system and an extensive literature that advances nausea as a highly efficient UCS for CTA induction. However, nausea may not be necessary for CTA induction. There are compelling data indicating that some drugs that are abused by humans (e.g., amphetamines, cocaine) and self-administered by rats may produce rejection of a previously preferred CS flavor when administered as the UCS agent in a CTA conditioning paradigm.[8] However, taste reactivity tests indicate that learned taste rejection based on a drug of abuse may occur without the negative hedonic shift in palatability that is the major functional adaptation of the gut defense system.[9,10] The theoretical implications of these findings have led to considerable debate over the nature and underlying mechanisms of TA learning.[8,10]

II.   TA Conditioning and Assessment Methodology

The TA literature includes numerous variations of CTA induction methodologies. The preponderance of experimental taste aversions (TAs) have been induced by following the subject's ingestion of a distinctive CS flavor with administration of an emetic class UCS agent, usually via intraperitoneal injection. The basic CTA design utilized in our laboratory to produce TA learning in rats will be described for purposes of illustration. The methodologies can be extended to other species and modified as needed to suit the individual researcher's objectives. For present purposes, it is assumed that the CS is a distinctive flavor and the UCS is the post-injection gastrointestinal consequence of an emetic class agent injection.

Two of the most basic decisions prior to CTA induction are whether the study is to involve one or several conditioning trials and if the magnitudes of any resultant aversions will be assessed via single-bottle (forced drinking) or two-bottle (free-choice) drinking tests.

The basic CTA paradigm will be illustrated by describing an early comparison of the effects of single trial conditioning as assessed via one-bottle and two-bottle test methodologies. The objective was to develop parametric data concerning the magnitude and longevity of cyclophosphamide (Cytoxan®, Mead Johnson Laboratories) induced aversions to a saccharin-flavored CS solution.[11] Cyclophosphamide is a nitrogen-mustard derivative that produces strong nausea in humans when used during cancer chemotherapy, and presumably has some comparable effect in rats where it can function as a potent CTA inducer. Saccharin solutions are frequently used as CSs during TA experiments with rats. Saccharin is a nonnutritive and therefore does not introduce a caloric source of variability. Additionally, palatable saccharin solutions can be prepared that are usually accepted by fluid-deprived rats without any disruptive neophobia, which is operationally defined as a rejection of unfamiliar flavors.[12] The use of highly acceptable (i.e., palatable) CS solutions is important, because initially unacceptable solutions may require a period of preconditioning familiarization, and familiarity with the CS flavor can markedly impede CTA acquisition, as will be elaborated later in this discussion.

A.   Example Two-Bottle Experiment

1. Sprague-Dawley derived rats were randomly assigned to four groups of five subjects each and fluid deprived for 24 hours.

2. A novel 0.1% saccharin solution was then presented individually in standard drinking bottles attached to subject's home cages. Ingestion onset was noted for each subject, and the flavored solution was available for a minimum of 10 min and for at least 5 min after drinking began.

3. Five min after fluid removal, experimental subjects received a cyclophosphamide injection (25, 12.5, or 6.25 mg/kg, I.P.), and controls were injected with vehicle (i.e., saline).

4. Saccharin bottles were then replaced as the only available fluid source for the remainder of the 24-hour period following their initial presentation.

5. The next day marked the onset of a 50-day extinction test during which each subject had continuous access to separate bottles containing saccharin-flavored water and plain tap water.

6. Fluid ingestion was derived from daily bottle-weight data, and saccharin preference scores were obtained for each subject by computing the percentage of total daily fluid intake accounted for by saccharin flavor ingestion. Bottle positions were reversed in a nonsystematic manner to hinder the development of position habits.[11]

The results of the two-bottle experiment are summarized graphically in Figure 3.1. The different cyclophosphamide doses produced TA of differing magnitude and longevity. It may be important to note that the 12.5 mg/kg and the 25 mg/kg doses both produced initially strong aversions that were of similar magnitude over the first 10 days

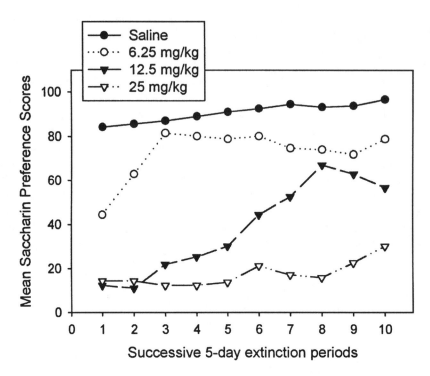

FIGURE 3.1
Group mean saccharin preference scores from 2-bottle data averaged over successive 5-day periods.

of extinction testing. However, the 12.5 mg/kg dose produced aversions that extinguished to a greater degree than those of the 25 mg/kg dose over the remaining 40 days of preference testing. These data indicate that resistance to extinction can be a useful indicator of aversion strength. The figure also reveals the remarkable longevity of TAs that were produced by the 25 mg/kg cyclophosphamide dose when aversion strength was measured by two-bottle tests, which afforded the subjects access to plain water and did not force them to endure fluid deprivation or to drink the saccharin solution.

From a practical standpoint, post-conditioning CS availability is unnecessary for the induction of strong CTAs as assessed via two-bottle preference testing. We now routinely provide subjects with ad lib access to water during the interval between the end of conditioning and the beginning of two-bottle testing. Additionally, we now wait at least two days following conditioning before reintroducing the CS flavor as part of the two-bottle testing procedure. The optimal waiting period is dependent on both the UCS dose and variety of UCS agent and must be determined by the use of a pseudo-conditioning (sensitization) control procedure as illustrated elsewhere.[13] Pseudoconditioning differs from conditioning only in that plain water is substituted for the CS solution prior to injection of the UCS agent. Pseudoconditioned control subjects encounter the CS solution as a novel substance at the onset of aversion assessment. The pseudoconditioning control procedure is needed to confirm an associative learning interpretation of TA acquisition.

B.   Example Single-Bottle Experiment

1.   Five rats were randomly assigned to one of 5 groups, and rats were trained to consume their total daily water intake within their home cages during a 10-min access interval.

2.   Following stabilization of drinking under the restricted schedule, a 0.1% solution of sodium saccharin in tap water was then introduced to all subjects in lieu of plain water during a standard access interval. Ingestion onset was noted for each subject, and the flavored solution was available for a minimum of 10 min and for at least 5 min after drinking began.

3.   Five min after removal of the saccharin solution, each rat received a designated drug or saline injection. Cyclophosphamide doses were 50, 25, 12.5, and 6.25 mg/kg.

4.   The subjects were subsequently maintained on the preinjection drinking schedule except that every third fluid-access period was an extinction test during which the saccharin solution was substituted for water.

5.   Daily pre- and post-access bottle weights were recorded. The absolute saccharin consumption computed from these values constituted the dependent variable of this single-bottle experiment.[11]

Dose-dependent differences in initial aversion magnitude as measured by single-bottle post-conditioning consumption of the saccharin CS solution were demonstrated (see Figure 3.2). The most noteworthy feature of the single-bottle results is the rapid extinction that occurred at all dose levels. Even the CTA induced by a 50 mg/km dose extinguished completely following two test sessions. This rapid extinction is not necessarily a disadvantage, depending on the objective of the researcher. However, it is apparent that the obtainable strength and longevity of CTAs induced by injection of an emetic class agent would not have been observed if all CTA methodology had featured fluid-deprivation driven single-bottle tests of aversive strength to the exclusion of two-bottle free-choice testing methodology.

Our single-bottle and the two-bottle findings extend prior reports that two-bottle tests are more sensitive than single-bottle procedures.[14,15] However, the single- vs. two-bottle study from our laboratory does not provide a complete comparison of the two assessment procedures. The two-bottle subjects had CS flavor access for 24 hours following injection and prior to assessment, but single-bottle subjects were deprived over this period.

C.   Preconditioning CS and UCS Familiarity Effects

Preconditioning familiarity with the CS flavor can profoundly attenuate CTA induction during one trial conditioning (Figure 3.3).[16] Rats were familiarized with a saccharine solution for 1, 3, 10, or 20 days prior to conditioning. During their respective familiarization periods, the subjects had *ad libitum* access to the saccharine solution and to plain tap water. The study also had a no-exposure group and a saline-injected control group. After the familiarization period, rats were injected with 25 mg/kg cyclophosphamide following the ingestion of the saccharin CS. As revealed by excluded two-bottle extinction tests, the magnitude of the saccharine

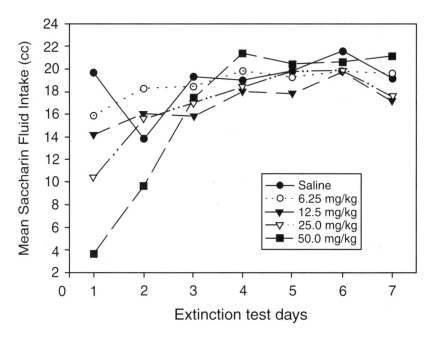

FIGURE 3.2
Mean single-bottle dose response relationships showing saccharin fluid consumption during the extinction of cyclophosphamide-induced taste aversions.

aversion was inversely related to duration of preconditioning saccharine familiarity. One day of CS pre-exposure failed to influence initial CTA magnitude, but resulted in an increased rate of extinction. In contrast, 20 days of flavor pre-exposure completely blocked CTA formation, whereas intermediate pre-exposure resulted in graded degrees of CTA attenuation.

As with pre-exposure to the CS stimulus, a history of familiarity with UCS effects can also weaken CTA acquisition as described elsewhere.[17] Therefore, in many studies, the effects of subject preconditioning exposure to the CS flavor and the UCS for CTA induction must be controlled. Additionally, experimental manipulation of CS and UCS familiarity can be used to influence CTA magnitude for specific purposes.

D.   Forgetting

Both of the previously summarized experiments involved extinction testing. Extinction refers to loss of aversion magnitude across a period of time during which contact with the CS flavor is provided. In addition to extinction, aversion strength may diminish over time through a process of forgetting. TA forgetting is operationally defined as aversion diminution during post-conditioning periods in which the subject has no contact with the CS flavor. In a demonstration of forgetting within a CTA paradigm, TAs of varying strengths were induced in groups of rats that received low

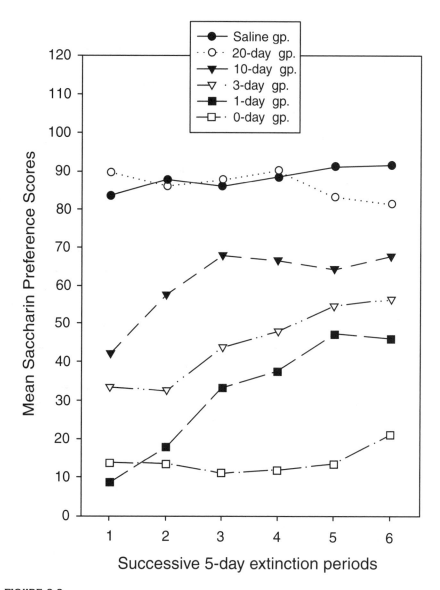

FIGURE 3.3

Taste aversion magnitude and resistance to extinction as functions of differing amounts of CS exposure prior to conditioning.

(6.25 mg/km), medium (12.5 mg/km), or high (25 mg/km) doses of cyclophosphamide following saccharine solution ingestion.[18] CTA retention was assessed following saccharine-free intervals of 2, 10, 20, or 40 days. The retention interval of these subjects was followed by 30 days of two-bottle preference testing, thereby additionally permitting an assessment of TA extinction following different degrees of TA forgetting. As depicted in Figure 3.4, the low-dose subjects displayed moderate strength CTAs that were forgotten within 20 days and had little resistance to

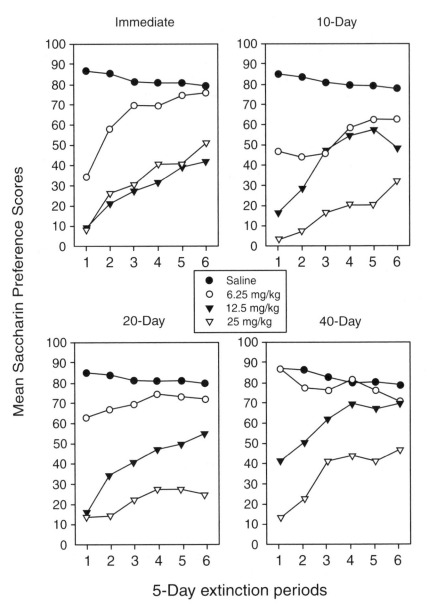

FIGURE 3.4

Mean saccharin preference score of saline-control and drug-injected groups of each of the four delay conditions. Preference scores were averaged over 6 successive 5-day extinction periods. Controls--●, 6.25 mg/kg cyclophosphamide--○, 12.5 mg/kg cyclophosphamide--▲, 25.0 mg/kg cyclophosphamide--▽.

extinction when testing began 2 days after conditioning (the immediate condition on the graph). The medium-dose subjects displayed stronger CTAs having greater resistance to both forgetting and extinction. Unlike the low- and medium-dose subjects, the high-dose subjects displayed no aversion diminution as a result of forgetting over the 40-day time limit of the study. There are survey data that indicate that some humans have maintained TAs over intervals that span many years.[3] The limits and variability of CTA retention in the absence of contact with the CS flavor are not known for any species.

E.   Additional Considerations

There are a number of additional variables (e.g., CS quality or salience, CS quantity, route of CS and UCS administration, and contextual effects) and control procedures (e.g., sensitization or pseudoconditioning) that merit attention during the design and interpretation of CTA experiments. Important information can be found through two bibliographies of the TA literature[19,20] and in edited proceedings of three conferences.[21-23]

III.   Use of CTA in Drug Discrimination Learning

Drug discrimination learning (DDL) is a well-established behavioral pharmacology paradigm that is used to characterize and classify the subjective effects (stimulus properties) of pharmacological agents. Although there are many possible variations of the standard procedure, animals typically are trained to respond in a dichotomous manner depending on their presumed subjective experience of an injected agent(s). In most procedures, food-deprived rats are trained to bar press for food on one (the drug lever) of two levers in sessions preceded by injections of the training drug, and to bar press on the other lever (the vehicle lever) when injected with the drug vehicle prior to the session. The differential responding that occurs as a consequence of training establishes some salient property of the drug as a discriminative stimulus (DS) that signals to the rat which lever will yield food pellets. Once training has been accomplished, manipulations of the procedure can be used to characterize the training drug or other drugs. Two such manipulations allow for the study of drug generalization and drug antagonism.

In drug generalization studies, the training drug is replaced by other drugs or by different doses of the training drug. If administration of the test drug results in bar pressing on the drug lever and not the vehicle lever, it is assumed that the rat experiences the test drug to be similar to the training condition. Similarly, by varying the dose of the training drug, the procedure offers a behavioral indication of the dose required to produce discriminatively different effects within the animal.[24]

In studies of drug antagonism, injection of the DS training drug is accompanied by administration of a putative antagonist of the drug. If the antagonist sufficiently attenuates the training drug effects, the rat should respond to the drug-antagonist treatment as a vehicle injection and show preference for the vehicle lever.

The CTA paradigm has been successfully modified for use in DDL studies to investigate the properties of a wide range of pharmacologic agents, including serotonin agonists,[25] opiate agonists,[26] opiate antagonists,[27] cholecystokinin,[28] phencyclidine,[29] benzodiazepines,[30] alprazolam,[31] and pentobarbital.[32] Available evidence indicates that pharmacologic agents that function as DS in operant procedures can function as DS in the CTA design. Furthermore, the CTA design has been successfully extended to studies of drug-drug antagonism.[33]

Both traditional and CTA-based DDL studies employ a training drug as the DS. CTA-based DDL studies are different from traditional DDL studies in that (1) the DS signals that CS consumption is associated with illness and (2) the behavioral measure in the CTA paradigm is a drink/don't drink choice. These differences are summarized in Table 3.1.

TABLE 3.1
Differences Between Traditional and CTA-Based DDL Studies

Method	DS	Learned contingency	Behavioral measure
Traditional	Training drug	Correct bar $\Rightarrow$ food delivery	Left vs. right bar pressing
CTA	Training drug	CS consumption $\Rightarrow$ illness	Amount of drinking

The basic paradigm for CTA-based DS studies is as follows:

1. Groups of 6 or more animals are assigned to either the experimental or unconditioned control group, and body weights are determined and recorded.

2. All rats are allowed only limited access to water bottles (e.g., 20 to 30 min/day). This technique is commonly used to assure that rats will drink heavily and reliably during the time of limited fluid access that will later be used in testing.

3. Once the rats have adjusted their daily fluid intake to accommodate the limited access to fluids (usually after about 7 days), the rats are allowed 2 to 3 sessions of access to the CS. In most studies, the CS is 0.1 to 0.25% (w/v) saccharin in water. On the third day of CS adaptation, the rats are given a drug vehicle injection prior to CS presentation to adapt them to the stress of the injection.

4. On the first training session, rats are injected with the training drug in a manner such that the drug is expected to be active during the presentation of the CS (e.g., 15 min prior to CS presentation). The choice of dose or ranges of doses of the training drug should be based on the findings from other DS studies obtained through the current literature or via preliminary parametric studies.

5. The rats are then allowed 20 to 30 min exposure to the CS. The CS solution may be contained in drinking bottles or in calibrated tubes.

6. At the end of the 20 to 30 min exposure period, the bottles or tubes containing the CS are removed, and the rats in the conditioned group (i.e., the experimental group) are injected with an emetic agent. The most common agent used in these studies is LiCl (75 to 90 mg/kg [or 1.8 to 2.1 mEq/kg] in a volume of approximately 12 ml/kg). A useful parametric evaluation of CTA induction via different LiCl doses is available elsewhere (Referemce 34).

7. Rats in the unconditioned control group receive the same drug/vehicle pre-injections but receive saline (i.e., LiCl vehicle) injections instead of LiCl injections after the CS bottles are removed, regardless of the type of pre-injection received.

8. Fluid (i.e., CS) consumption per body weight is calculated by either determining the difference in weight of the drinking bottles pre- to post-session or by determining the pre- to post-fluid volume difference in the calibrated tubes.

9. On the subsequent 3 training days, before the session, the rats are injected with drug vehicle in volume equal to the drug injection. [Note: A post-session LiCl vehicle injection is not necessary.]

10. This cycle of one training day followed by 3 recovery days is repeated until some pre-determined training criterion is met (e.g., 50% suppression relative to controls, less than 5 ml per session during drug sessions; statistically significant separation of saline and drug sessions).

11. The amount of CS consumption is determined for each session. If the drug acts as a DS, rats will gradually consume less of the CS during drug sessions than during vehicle sessions.

12. Testing begins by substituting the training drug with (1) different doses of the training drug, (2) another drug in varying doses, or (3) a combination of different pharmacologic agents.

13. Rats are allowed 20 to 30 min access to the CS solution. LiCl injections are suspended during testing.

14. The volume of CS consumption is determined, and data are summarized as total CS consumed/body weight or as preference scores (amount of CS consumed as a percent of total fluid consumption).

A typical acquisition training schedule with 3 cycles of one training day and 3 recovery days is shown in Table 3.2.

TABLE 3.2
Typical Acquisition Training

	Conditioned group	Unconditioned control group
Days 1–7	20–30 min limited access to water	20–30 min limited access to water
Days 8–9	20–30 min access to CS; no water	20–30 min access to CS; no water
Day 10	Training drug — CS – LiCl	Training drug — CS – saline
Days 11–13	Vehicle — CS	Vehicle — CS
Day 14	Training drug — CS – LiCl	Training drug — CS – saline
Days 15–17	Vehicle — CS	Vehicle — CS
Day 18	Training drug — CS – LiCl	Training drug — CS – saline
Days 19–21	Vehicle — CS	Vehicle — CS

This acquisition training utilizes two within group contingencies as shown below in Table 3.3.

Acquisition training should yield the following:

1. A significant reduction in responding when training drug vs. vehicle injection precedes CS exposure in the conditioned group (1 vs. 2 in Table 3.3). Specifically, discriminative

TABLE 3.3
Acquisition Training Contingencies

Conditioned group		Unconditioned control group	
1.	Training drug-CS exposure-LiCl-discomfort	3.	Training drug-CS-exposure-saline-no discomfort
2.	Vehicle-CS exposure-saline-no discomfort	4.	Vehicle-CS-exposure-saline-no discomfort

responding is presumably indicated within the conditioned group if the animals consume less of the CS when the training drug precedes CS exposure.

2. No within group difference in responding when training drug vs. vehicle injection precedes CS exposure in the unconditioned control group (3 vs. 4 in Table 3.3). A difference resulting from these two procedures would indicate a direct hypodipsic or hyperdipsic effect (as opposed to a conditioned effect) of the training drug on CS consumption.

3. A significant difference in responding between the conditioned group and the unconditioned control group when injections of the training drug precede CS exposure (1 vs. 3 in Table 3.3). If this difference is not found, then the reduction in CS consumption in the conditioned group may be attributable to a hypodipsic effect of the drug rather than to a discriminative conditioning effect.

4. No between group difference in responding between the conditioned group and the unconditioned control group when injections of the vehicle precede CS exposure (2 vs. 4 in Table 3.3). A difference in these conditions would indicate a residual effect of LiCl on CS consumption.

The CTA-based DS study offers the following advantages and disadvantages over traditional bar-press DS studies.

Advantages

1. The CTA discrimination training is more rapid than bar-press DDL training and can usually be accomplished in a few weeks.

2. Expensive and sophisticated equipment is not needed for CTA DDL training. For the behavioral measure (i.e., drinking), the investigator needs only a scale to weigh drinking bottles for pre-post weight quantification of drinking or graduated cylinders used to measure consumption.

3. The CTA paradigm may be more sensitive to some types of drugs than more traditional paradigms. For example, the CTA paradigm appears to be particularly sensitive to discrimination using opioid antagonists.[35]

Disadvantages

1. The CTA paradigm puts considerable stress on the animal in that it involves multiple injections, fluid deprivation, and presumed LiCl-induced illness. However, other DDL designs, which utilize drug injections and food deprivation, may be as aversive as the CTA design.

2. Reports are not presently available assessing the use of the CTA paradigm in complex discrimination designs (e.g., two- or three-drug discriminations).

3. Many agents can produce hypodipsia or hyperdipsia.[36] However, this limitation can be easily circumvented by procedural modifications as discussed below.

There are several control factors that should be considered in the experimental design:

1. **Novelty of the UCS** — Prior exposure to the CTA-inducing UCS agent (e.g., LiCl) can attenuate or eliminate its ability to produce a CTA when subsequently paired with a CS.[1] Therefore, investigators utilizing the CTA paradigm should ensure that subjects are naïve to the conditioning agent.

2. **CS pre-exposure** — The strength of a CS within the CTA paradigm to control responding is dependent upon the salience of the CS, and the salience of the CS is in turn influenced by the animal's past experience with the CS.[37] Thus, for rapid acquisition of a CTA, a CS with which the animal has no previous experience is preferred. However, the strength of the CS to control responding may overshadow the discriminative properties of a DS. For example, if novel exposure to saccharin is paired with illness, rats may learn to avoid the saccharin altogether despite training with a DS. In order to shift control of responding from the CS to the DS, rats are exposed to the CS for several sessions prior to the introduction of DS training, thereby reducing the salience and behavioral control of the CS.

3. **Hypodipsic/hyperdipsic drug effects** — Without careful control, the direct, unconditioned effects of the training or test agents on fluid consumption could contaminate data. There are several steps that can be used to control for unconditioned effects. First, the investigator could incorporate the learned discomfort and learned safety procedures in the same experiment. The simultaneous use of the drug signaling discomfort in one group of rats and safety in another group can be used to control for drug hypodipsic/hyperdipsic effects. However, performance in both the learned discomfort and learned safety groups can still be altered by the unconditioned effects of the drugs, and data interpretation can be difficult.[36] Second, rats could be presented with two bottles during testing: one containing the CS and one containing plain tap water. Data are collected as the percentage of CS consumed vs. total fluid consumption (CS consumption/total fluid consumption). This two-bottle choice procedure is robust to hypodipsic/hyperdipsic effects because consumption of the CS and water are reduced similarly, leaving the preference to the CS unchanged.

4. **Interference of the training drug with LiCl action** — Some agents, such as antiemetics or analgesics, may inhibit the effects of LiCl or other UCS agents. A potential method of reducing the effects of the training drug on LiCl action is to increase the training drug-LiCl interval. It is well established that animals acquire CTAs despite long intervals between the CS and UCS. Thus, LiCl administration could be delayed for up to several hours without significantly diminishing the acquisition of the CTA.

    Alternatively, the learned safety approach may be more appropriate for use with these types of agents. For example, if the training drug suppresses the effects of LiCl, then rats may more quickly learn the discrimination when the training drug signals safety and is not used in conjunction with LiCl.

5. **Housing conditions** — Healthy rats can develop food aversions when in the presence of sick or nauseated rats. This robust phenomenon is known as the poisoned partner effect.[38] Thus, control subjects injected with the vehicle (for LiCl) should be housed separately from rats injected with the LiCl.

6. **Learned safety design** — In the basic CTA-based DS design described above, drug injections signal that consumption of the CS is to be followed by discomfort. However, the drug can also be used to signal the absence of discomfort, or safety. Accordingly, LiCl injections would be paired with the CS when saline injections preceded CS presentation; during drug sessions, saline injections would follow CS presentation. In this case, the discrimination is measured by the extent to which test drugs increase CS consumption above levels seen during saline sessions.[39]

TABLE 3.4*

1.	Training drug-CS exposure-saline-no discomfort	3.	Drug-CS-exposure-saline-no discomfort
2.	Saline-CS exposure-LiCl-discomfort	4.	Saline-CS-exposure-saline-no discomfort

From Elkins, R. L., Individual differences in bait shyness: Effects of drug dose and measurement technique, *The Psychological Record*, 23, 249, 1975. Used with permission.

IV. Drug Toxicity

Standard taste aversion conditioning has been proposed as one measure of drug toxicity (e.g., Reference 40). The assumption is that toxic substances will produce TAs in animals whereas, nontoxic substances will not produce aversions. Riley and Tuck[20] argued that TA conditioning might serve as one element of a panel of toxicity tests but noted several problems with the use of this methodology. The problems center on two factors. The first problem is the demonstration that several toxic substances fail to produce TAs or produce only mild aversions. Screening with this methodology could result in a given drug falsely classified as nontoxic (a false negative). Drugs known to produce false negatives include gallamine, sodium cyanide, warfarin, and aluminum chloride. A second problem is the production of strong TAs by substances with little or no known toxicity at the doses administered. Screening with this methodology could result in drugs falsely classified as toxic (a false positive). Drugs producing false positives include amphetamine, scopolamine, and ethanol. (Possible interpretations of these finding are found in Riley and Tuck.[20]) In summary, classical TA conditioning can be used as one of a panel of screens for toxicity, but should not be relied upon as a single measure of toxicity.

Riley[41] proposed that a procedure similar to DDL be used to test for substance toxicity. As discussed in the drug discrimination procedure, animals are trained to avoid a saccharin solution when one drug is present systemically and to drink the saccharin solution when they are drug-free or when a different drug is present. The same control procedures are used as in the TA DDL methodologies explained previously in this chapter. In the proposed behavioral toxicology procedure, specific toxins are substituted for the drugs used to train discrimination. Specifically, animals would be trained to avoid a saccharin solution in the presence of a particular toxin and to drink the saccharin solution in its absence or when a second type of toxin is present. In theory, by training animals to accept the saccharin solution in the presence of a toxin specific to one organ system or a set of systems, it should be possible to determine if novel substances are toxic to the same systems. It is assumed that toxins with similar mechanisms of action will produce similar internal cues; thus, animals will demonstrate this similarity by responding appropriately to the saccharin solution. However, this methodology has not been rigorously tested.

V. Selection Breeding for Efficient and Inefficient CTA Conditionability

Researchers who envision an extended program of CTA studies may wish to consider devoting part of their resources to the selective breeding of subjects to increase the statistical power of their experiments and to enhance the likelihood of detecting a significant effect against a background of diminished variability. The selective breeding of strains of TA prone (TAP) and TA resistant (TAR) rats will be described to demonstrate the utility of the approach and to emphasize issues that may be of interest to neuroscientists who plan to use TA methodologies in their research.

The findings of wide individual differences in TA conditionability in studies of outbred rats[11] and humans,[42,43] provided the impetus for a selective breeding based on TA conditionability. The objectives were to determine if TA conditionability could be manipulated via selective breeding and, if so, to produce rat lines suitable for studies of biological bases of individual differences in TA conditionability. Selection was based on results of our typical saccharine CS and cyclophosphamide UCS conditioning procedures. Aversion acquisition was evaluated with two-bottle *ad libitum* preference tests that typically lasted for 15 days. By selecting from within the two extremes of TA conditionability and by mating without permitting sibling pairings, we have developed essentially non-overlapping lines of TAP and TAR rats.[44,15] The two lines display minimal within-lines variability in TA conditionability coincident with a marked between-lines CTA difference that approached the maximum attainable with two-bottle preference measurements. The results of selection across 25 generations are depicted graphically below (Figure 3.5). The selection was undertaken in the hope of developing, to the degree that is possible in a limited breeding population, two lines of rats that were essentially randomly bred except for respect to the biological basis of TA conditionability. Studies to date indicate this effort has met with considerable success. The selectively bred differences in CTA learning ability have not generalized to other conventional learning tasks including shock-motivated place avoidance,[44] food reinforced operant bar-press responding under different schedules of reinforcement,[45] or the efficient harvesting of food on an elevated spiral arm maze.[46] It appears that the selective process has exerted an effect that is highly specific to TA conditionability. Moreover, the line differences are not restricted to the CS and UCS stimulus of line selection. The TAP and TAR differences in TA conditionability have been maintained with other CSs,

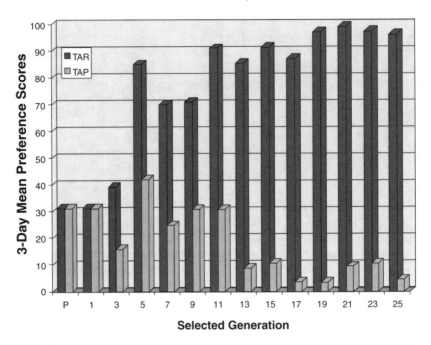

FIGURE 3.5

The selective breeding of strains of taste aversion prone (TAP) and TA resistant (TAR) rats expressed as a 3-way mean preference score over 25 generations. From Elkins, R. L., Individual differences in bait shyness: Effects of drug dose and measurement technique, *The Psychological Record*, 23, 249, 1975. Used with permission.

including several flavors of fluids and live crickets.[47] (Rats are voracious predators of crickets, and the use of crickets as a highly palatable CS source for CTA studies is recommended as a viable alternative to flavored solution CSs.) Likewise, the line differences in CTA conditionability have been maintained with other emetic UCS agents including lithium chloride and emetine hydrochloride,[15] as well as with the use of rotational stimulation (motion sickness) as a UCS agent.[48] Of considerable additional importance, when ethanol was injected as a UCS agent, TAP but not TAR rats developed dose-dependent CTAs to a saccharin solution CS. Furthermore, TAP rats drank little ethanol, but TAR rats drank large amounts of ethanol when given free access to separate sources of ethanol solutions and plain water.[49] This selective breeding outcome has advanced a genetically mediated propensity for TA conditionability as a possible deterrent to pathological levels of ethanol intake. The TAP and TAR rats can provide useful subjects for studies of the biological bases of this hypothesized CTA protective mechanism and for more general analyses of TA conditionability as an intriguing learning capability. A limited number of TAP and TAR subjects may be provided to other interested researchers in the future.

Acknowledgments

This work was supported in part by the Medical Research Service of the Department of Veterans Affairs. The authors wish to express appreciation to Anthony L. Riley, Ph.D. for his contributions and input on drug discrimination learning.

References

1. Gamzu, E., Vincent, G., and Boff, E., A pharmacological perspective of drugs used in establishing conditioned food aversions, *Ann NY Acad Sci*, 443, 231, 1985.
2. Garcia, J. and Ervin, F. R., Gustatory-visceral and teleceptor-cutaneous conditioning. Adaptation in the internal and external milieus, *Comm Behav Biol*, 1, 389, 1968.
3. Logue, A. W., Conditioned food aversion learning in humans, *Ann NY Acad Sci*, 443, 316, 1985.
4. Domjan, M., Cue-consequence specificity and long-delay learning revisited, *Ann NY Acad Sci*, 443, 54, 1985.
5. Garcia, J., Palmerino, C. C., Rusiniak, K. W., and Kiefer, S. W., Taste aversions and the nature of instinct, in McGaugh J. and Thompson, R., Eds., *The Neurobiology of Learning and Memory*, Plenum Press, New York, 1985.
6. Roll, D. and Smith, J. C., Conditioned taste aversion in anesthetized rats, in M. Seligman and J. Hager, Eds., *Biological Boundaries of Learning*, New York, Appleton, 1972.
7. Garcia, J., Lasiter, P. S., Bermudez-Rattoni, F., and Deems, D. A., A general theory of aversion learning, *Ann NY Acad Sci*, 443, 8, 1985.
8. Hunt, T. and Amit, Z., Conditioned taste aversion induced by self-administered drugs: Paradox revisited, *Neurosci Biobehav Rev*, 11, 107, 1987.
9. Grill, H. J., Introduction: Physiological mechanisms in conditioned taste aversions, *Ann NY Acad Sci*, 443, 67, 1985.
10. Parker, L. A., Taste reactivity responses elicited by cocaine-, phencyclidine-, and methamphetamine-paired sucrose solutions, *Behav Neurosci*, 107, 118, 1993.
11. Elkins, R. L., Individual differences in bait shyness: Effects of drug dose and measurement technique, *The Psychological Record*, 23, 349, 1973.
12. Carroll, M. E., Dinc, H. I., Levy, C. J., and Smith, J. C., Demonstrations of neophobia and enhanced neophobia in the albino rat, *J Com Physiol Psychol*, 89, 457, 1975.
13. Grote, F. W., Jr. and Brown, R. T., Conditioned taste aversions: Two-stimulus tests are more sensitive than one-stimulus tests, *Behav Res Meth Instr*, 3, 311, 1971.
14. Dragoin, W., McCleary, G. E., and McCleary, P., A comparison of two methods of measuring conditioned taste aversions, *Behav Res Meth Instr*, 3, 309, 1971.
15. Elkins, R. L., Walters, P. A., and Orr, T. E., Continued development and unconditioned stimulus characterization of selectively bred lines of taste aversion prone and resistant rats, *Alcohol Clin Exper Res*, 16(5), 928, 1992.
16. Elkins, R. L., Attenuation of drug-induced bait shyness to a palatable solution as an increasing function of its availability prior to conditioning, *J Behav Biol*, 9(2), 221, 1973.

17. Elkins, R. L., Bait shyness acquisition and resistance to extinction as functions of UCS exposure prior to conditioning, *Physiological Psychology*, 2(3A) 341, 1974.

18. Elkins, R. L., Taste aversion retention: An animal experiment with implications for consummatory-aversion alcoholism treatments, *Behav Res Ther*, 22, 179, 1984.

19. Riley, A. L. and Clarke, C. M. Conditioned taste aversions: A bibliography, in *Learning Mechanisms in Food Selection*, Barker, L. M., Best, M. R., and Domjan, M., Eds., Baylor University Press, 1977, chap 13.

20. Riley, A. L. and Tuck, D. L., Conditioned food aversions: A bibliography, *Ann NY Acad Sci*, 443, 381, 1985.

21. Barker, L. M., Best, M. R., and Domjan, M., Eds., *Learning Mechanisms in Food Selection*, Baylor University Press, 1977.

22. Braveman, N. and Bronstein, P., Eds., *Experimental Assessments and Clinical Applications of Conditioned Taste Aversions*, Vol. 443, New York Academy of Sciences, New York, 1985.

23. Burish, T. G., Levy, S. M., and Meyerowitz, B. E., *Cancer, Nutrition and Eating Behavior: A Biobehavioral Perspective*, Erlbaum, Hillsdale, NJ, 1985.

24. Colpaert, F. C., Drug discrimination: Behavioral, pharmacological, and molecular mechanisms of discriminative drug effects, in *Behavioral Analysis of Drug Dependence*, Goldberg, S. R. and Stolerman, I. P., Eds., Academic Press, New York, 1986, chap. 5.

25. Lucki, I. and Marcoccia, J. M., Discriminated taste aversion with a 5-HT$_{1A}$ agonist measured using saccharin preference, *Behav Pharmacol*, 2, 335, 1991.

26. Smurthwaite, S. T. and Riley, A. L., Nalorphine as a stimulus in drug discrimination learning: Assessment of the role of μ- and κ-receptor subtypes, *Pharmacol Biochem Behav*, 48, 635, 1994.

27. Smurthwaite, S. T., Kautz, M. A., Geter, B., and Riley, A. L., Naloxone as a stimulus in drug discrimination learning: Generalization to other opiate antagonists, *Pharmacol Biochem Behav*, 41, 43, 1991.

28. Melton, P. M., Kopman, J. A., and Riley, A. L., Cholecystokinin as a stimulus in drug discrimination learning, *Pharmacol Biochem Behav*, 44, 249, 1993.

29. Mastropaolo, J. P., Moskowitz, K. H., Dacanay, R. J., and Riley, A. L., Conditioned taste aversion as a behavioral baseline for drug discrimination learning: An assessment with phencyclidine, *Pharmacol Biochem Behav*, 32, 1, 1989.

30. Rowan, G. A. and Lucki, I., Discriminative stimulus properties of the benzodiazepine receptor antagonist flumazenil, *Psychopharmacology*, 107, 103, 1992.

31. Glowa, J. R., Jeffreys, R. D., and Riley, A. L., Drug discrimination using a conditioned taste-aversion paradigm in rhesus monkeys, *J Exper Analysis Behav*, 56, 303, 1991.

32. Riley, A. L., Jeffreys, R. D., Pournaghash, S., Titley, T. L., and Kufera, A. M., Conditioned taste aversions as a behavioral baseline for drug discrimination learning: Assessment with the dipsogenic compound pentobarbital, *Drug Devel Res*, 16, 229, 1989.

33. Stevenson, G. W., Pournaghash, S., and Riley, A. L., Antagonism of drug discrimination learning with the conditioned taste aversion procedure, *Pharmacol Biochem Behav*, 41, 245, 1991.

34. Nachman, M. and Ashe, J. H., Learned taste aversions in rats as a function of dosage, concentration, and route of administration of LiCl, *Physiol Behav*, 10, 73, 1973.

35. Kautz, M. A., Geter, B., McBride, S. A., Mastropaolo, J. P., and Riley, A. L., Naloxone as a stimulus for drug discrimination learning, *Drug Devel Res*, 16, 317, 1989.

36. Lucki, I. and Marcoccia, J. M., Discriminated taste aversion with a 5-HT$_{1A}$ agonist measured using saccharin preference, *Behav Pharmacol*, 2, 335, 1991.

37. Schwartz, B., *Psychology of Learning and Behavior*, 2nd ed. W.W. Norton and Company, Inc. New York, 1984.

38. Bond, N. W., The poisoned partner effect in rats: Some parametric considerations, *Anim Learning & Behav*, 12, 89, 1984.

39. Jaeger, T. V. and Mucha, R. F., A taste aversion model of drug discrimination learning: Training drug and condition influence rate of learning, sensitivity and drug specificity, *Psychopharmacol*, 100, 145, 1990.

40. MacPhail, R. C., Studies on the flavor aversions induced by trialkyltin compounds, *Neurobehavioral Toxicology & Teratology*, 4, 225, 1982.

41. Riley, A. L., Use of drug discrimination learning in behavioral toxicology: classification and characterization of toxins, in *Neurotoxicology: Approaches and Methods*, Chang, L. W. and Slikker, W., Eds., Academic Press, San Diego, 1995, chap. 14.

42. Elkins, R. L., Covert sensitization treatment of alcoholism: Contributions of successful conditioning to subsequent abstinence maintenance, *Addictive Behavior*, 5, 67, 1980.

43. Elkins, R. L., An appraisal of chemical aversion (emetic therapy) approaches to alcoholism treatment, *Behav Res Ther*, 29, 387, 1991.

44. Elkins, R. L., Separation of taste-aversion-prone and taste-aversion-resistant rats through selective breeding: Implications for individual differences in conditionability and aversion-therapy alcoholism treatment, *Behav Neurosci*, 100, 121, 1986.

45. Hobbs, S. H. and Elkins, R. L., Operant performance of rats selectively bred for strong and weak acquisition of conditioned taste aversions, *B Psychon S*, 21, 303, 1983.

46. Hobbs, S. H., Walters, P. A., Kolbe, E. F., and Elkins, R. L., Radial maze performance in taste aversion prone and taste aversion resistant strains of rats, *Psychonomic Science*, 31, 171, 1993.

47. Elkins, R. L., Gerardot, R. J., and Hobbs, S. H., Differences in cyclophosphamide induced suppression of cricket predation in selectively bred strains of taste aversion prone and resistent rats, *Behav Neurosci*, 103, 112, 1989.

48. Elkins, R. L., Walters, P. A., Harrison, W. R., and Albrecht, W., Congruity of rotational and pharmacological taste aversion (TA) conditioning with strains of selectivity bred TA prone and TA resistant rats, *Learning Mot*, 21, 190, 1990.

49. Orr, T. E., Walters, P. A., and Elkins, R. L., Differences in free-choice ethanol acceptance between taste aversion (TA)-prone & TA–resistant rats, *Alcohol Clin Exp Res*, 21, 1491, 1997.

Chapter **4**

Drug Discrimination

Richard Young, John R. James, and
John A. Rosecrans

Contents

I.   Introduction

Discrimination is a process that involves making distinctions between situations in order to respond appropriately to each. In a drug discrimination procedure the effect(s) produced by injection of drug and vehicle are different so that an organism can distinguish between the two; the organism can be taught to respond appropriately to each effect. The fact that drugs can serve as discriminative stimuli has been demonstrated by many investigators. The paradigm has been shown to be a highly sensitive and very specific drug detection method that provides both quantitative and qualitative results concerning drugs. While the technique used to demonstrate this phenomena has varied, it is typical that an organism is required to distinguish between drug and non-drug conditions in order to receive a reinforcer, such as food. This procedure permits the experimenter to determine if an organism has perceived a drug effect by the response the subject makes. Thus, this procedure is analogous to asking a human to report whether he or she has perceived a drug effect. Using S(+)-amphetamine (at 1.0 mg/kg) as a training model we demonstrate our approach to the study of psychoactive agents using drug discrimination methodologies. S(+)-amphetamine was chosen because it has an extensive history of being a reliable discriminative stimulus in animals and humans (see[1] for review). Other drugs, however, can also be employed (see Table 4.1).

TABLE 4.1
Examples of Drugs Which Have Been
Shown to Serve as Discriminative
Stimuli

Drugs	Reference
Amphetamine	Harris and Balster[3]
Apomorphine	Colpaert et al.[4]
Buspirone	Hendry et al.[5]
Chlordiazepoxide	Colpaert et al.[6]
Clozapine	Browne and Koe[7]
Cocaine	Jarbe[8]
Cyclazocine	Teal and Holtzman[9]
Desipramine	Shearman[10]
DOM	Young et al.[11]
LSD	Hirschhorn and Winter[12]
MDA	Glennon and Young[13]
MDMA	Glennon and Misenheimer[14]
Morphine	Hirschhorn and Rosecrans[15]
Nicotine	Meltzer et al.[16]
Pentazocine	Kuhn et al.[17]
Pentobarbital	Herling and Shannon [18]
Phencyclidine (PCP)	Shannon[19]
Tetrahydrocannabinol ($\Delta°$)	Jarbe et al.[20]

II   Method

A.   Subjects

Twelve experimentally naive, male albino Sprague-Dawley rats (Charles River Labs) weighing 325 to 350g at the beginning of the experiment are used. Rats are housed individually and have free access to water, but are gradually food restricted to approximately 80% of their free-feeding weights before training begins. The colony in which the animals are housed is kept at a constant temperature (~21 to 23°C) and humidity (40 to 50%); lights are on from 0600 to 1800 hrs.

B.   Apparatus

Behavioral testing is conducted in four standard two-lever operant chambers (Coulbourn Instruments, Lehigh Valley, PA, model E10-10) housed within light-and sound-attenuating outer chambers. One wall of each operant chamber is fitted with two levers and a food delivery device centered equidistant between the levers for delivery of reinforcers (0.01ml of sweetened milk). Illumination of each chamber is provided by an overhead 28-V house light. Solid-state and electromechanical programming and recording equipment (or computers) can be used.

C.   Initial Training

Initial training usually begins with "magazine training," which is training an animal to eat from the food tray and, consequently, to recognize the noise made by the food mechanism as indicative of the imminent presentation of food. If an animal depresses a lever, the food mechanism releases a pellet of food into the tray or produces a small well of liquid (e.g., sweetened milk). In such a situation the effect may be an increase in the probability of lever-pressing behavior. In this case, the food is a positive reinforcer.

Lever pressing is usually brought about through shaping. An experimenter using the technique of shaping, or successive approximations, reinforces every step that takes an organism closer to the final response. For example, an experimenter may begin by reinforcing (i.e., presenting food) a rat every time it turns toward a lever. Later, only movements that bring it closer to a lever are reinforced. After the rat has learned to approach a lever, it is not reinforced until it touches the lever. Eventually, the rat's behavior will have been so shaped that it will readily press a lever when put in the chamber. When every response is followed by reinforcement, the organism is said to be on a continuous reinforcement schedule. As a consequence of such a schedule, the number of reinforcers could become quite high and might lead to decreased responding over time because of satiation. However, it is not necessary to reinforce every response in order to maintain responding. That is, one can reinforce the animal only part of the time, or intermittently. The intermittent schedule can be based on a proportion of responses or on a time interval. The two most common schedules of reinforcement are ratio and interval schedules, each of which can be either fixed (unvarying) or variable (random). In drug discrimination studies the fixed-ratio (FR) and variable (random) interval (VI or RI) schedules of reinforcement are extensively used and will be discussed here. In a fixed ratio schedule the subject must complete a fixed number of responses in order to obtain each reinforcement. In a FR10 schedule, for example, every tenth response is reinforced. In variable-interval (VI) schedules, the length of time elapsing before reinforcement is delivered varies around the mean value specified by the schedule. On a VI 15 sec schedule, for example, reinforcement is available on the average after 15 sec have elapsed since the last reinforcement but may be available as shortly as 2 sec later, or not until 60 sec have elapsed. The first response after a time interval has elapsed produces a reinforcer for the organism.

D. Discrimination Procedure

The illumination of the house light signals the beginning of an experimental session. After lever responding is established, each daily session is preceded by an intra-peritoneal (IP) injection of 1.0 mg/kg of S(+)-amphetamine sulfate (dose is based on weight of salt) or saline (0.9% NaCl) with only the stimulus-appropriate lever present (i.e., left or right lever). A pretreatment interval of 15 min is used; during this interval the animals are in their home cages. Training sessions are of 15 min duration 5 days per week. For six of the rats, responses on the right-side lever are reinforced after drug administration while responses on the left-side lever are rein-forced after saline administration; these conditions are reversed for the remaining six rats. S(+)-amphetamine or saline are administered on a random schedule with the constraint that no more than two consecutive sessions with the drug or vehicle can occur. Initially, the animals must develop behavior tolerance to the disruptive effects of S(+) amphetamine in order to perform the task. However, the animals do not develop tolerance to the stimulus effect of S(+)-amphetamine or else the drug could not serve a discriminative function.

E. Discrimination Data

The degree to which stimulus control has been established is determined by evaluating response selection data independently of the reinforcer. When using an FR10 schedule of reinforcement, for example, discrimination learning is assessed for each subject by dividing the number of responses occurring on the drug-designated lever by the total number of responses occurring on both levers prior to obtaining the first reinforcer. This value is then multiplied by 100 to obtain percent amphetamine-appropriate lever responding. For instance, let us assume that a rat has the right-side lever designated as the amphetamine-appropriate lever. On a Monday, the animal is injected with 1.0 mg/kg of S(+)-amphetamine and 15 minutes later is placed in an operant chamber and proceeds to press the left-side lever nine times and the right-side lever ten times; food reward is presented after the tenth right-side lever press. For this day, discriminative control would be assessed at 53% S(+)-amphetamine-appropriate responding (i.e., 10 divided by 19 times 100). On Tuesday this same rat receives a saline injection and is put into the chamber and presses the right-side lever four times and the left-side lever ten times; food is presented after the tenth left-side lever press. On this day, discriminative control would be assessed at 29% S(+)-amphetamine-appropriate responding (i.e., 4 divided by 14 times 100). When using the variable interval schedule of reinforcement, discrimination performance is evaluated during a short period (e.g., 2.5 min) of nonre-inforced responding (referred to as extinction) at the beginning of a session (usually once or twice per week). Data collected during the extinction period is evaluated the same way as it is under the FR schedule of reinforcement. In addition, under both FR and VI schedules of reinforcement the animal's response rate is also calculated. For example, under the FR schedule, for each 15 min behavioral session, the overall response rate for both levers (e.g., responses/second) can be calculated for each rat. Under the VI schedule, the total number of responses made during the 2.5 min extinction session can be recorded as responses per minute. As might be expected, the administration of drug or vehicle during initial training sessions under either FR or VI schedules of reinforcement usually results in the animals dividing their responses equally (i.e., 50% S(+)-amphetamine-appropriate responding after injection of drug or saline) between the two levers. However, as training sessions with drug and vehicle progress, the animals gradually learn to respond on the drug-designated lever when given drug, and on the saline-designated lever when given saline. A generally accepted guideline is that after 6 to 9 weeks of training, animals consistently make ≥80% of their responses on the drug-appropriate lever after administration of 1.0 mg/kg of S(+)-amphetamine and ≤20% of their responses on the same lever after administration of saline.

III.   Applications

A.   Stimulus Generalization

Maintenance of the S(+)-amphetamine/saline discrimination is ensured by continuation of training sessions throughout generalization studies. Discrimination training sessions are conducted with S(+)-amphetamine or saline on the 4 days prior to a stimulus generalization test session. On at least one of those days, six of the animals receive 1.0 mg/kg of S(+)-amphetamine and the other six rats receive saline; percent S(+)-appropriate responding is then determined under the FR and VI schedules of reinforcement as described above. Animals not meeting the above criteria (i.e., ≥80% drug-appropriate responding after drug administration and ≤20% drug-appropriate responding after saline injection) are not used in that week's stimulus generalization test. During generalization investigations, test sessions are interposed between discrimination training sessions. With these test sessions, the rats are given a test treatment and then allowed to select one of the two levers in a 15 min session (FR procedure). The lever on which the animal first totals ten responses is regarded as the selected lever. Subsequent reinforcement is delivered for responses on this lever according to the FR10 schedule of reinforcement. Under the VI schedule, the animals are injected with test treatment, given a 2.5 min extinction session, and then removed from the operant chambers. The phenomenon of generalization involves engaging in previously learned behaviors in response to new situations that resemble those in which the behaviors were first learned. In the drug discrimination procedure, subjects respond to other drug stimuli that are more or less similar to those present during discrimination training. Stimulus generalization studies are used to determine whether a discriminative stimulus will generalize (i.e., substitute) to other drugs. The rationale of this approach is that an animal trained to discriminate a dose of training drug will display stimulus generalization only to agents having a similar effect (though not necessarily an identical mechanism of action). An example of the results obtained in generalization tests performed in rats trained to discriminate 1.0 mg/kg of S (+)-amphetamine from saline is shown in Table 4.2. Stimulus generalization is said to have occurred when the animals, after being administered a given dose of a challenge drug, make ≥80% of their responses on the S(+)-amphetamine-appropriate lever. Where stimulus generalization occurs, an effective dose 50 (ED50) value can be calculated by the method of Finney[2] and reflects the dose at which the animals would be expected to make 50% of their responses on the S(+)-amphetamine-appropriate lever. Besides complete stimulus generalization, two other results might be obtained: partial generalization and saline-appropriate responding. Partial generalization is said to have occurred when the animals, after being administered a thorough dose-effect test, make approximately 40% to 70% of their responses on the S(+)-amphetamine-appropriate lever. In this case, percent S(+)-amphetamine lever responding is not fully appropriate for either training condition. Data of this type are very difficult to interpret. However, it is our position that partial generalization may occur with a test compound because there are pharmacological effects that are common to both the training drug and the challenge drug. Full stimulus generalization does occur because the overlap is incomplete. For example, one explanation for partial generalization results may be that low doses of a test compound are similar to low doses of the training drug. However, as the dose of challenge drug is increased, another kind of pharmacological effect emerges. Lastly, the administration of various doses of a challenge drug may result in less than 20% S(+)-amphetamine-appropriate responding. Such a result does not necessarily mean that a challenge drug is inert, but does suggest that the effect(s) of the challenge drug are different from that produce by the training drug. For example, in rats trained to discriminate S(+)-amphetamine at 1.0 mg/kg from saline (Table 4.2), we have found that the administration of 0.5 to 2.5 mg/kg of fenfluramine, an amphetamine analog that is a nonstimulant anorectic agent, produces saline-appropriate responding, while the administration of 3.0 mg/kg produces disruption of behavior (i.e., no responding). Similarly, the administration of 1.0 to 5.0 mg/kg of buspirone, a serotonin 5-HT1A receptor anxiolytie that is structurally unrelated to amphetamine, produces saline-appropriate responding, while the administration of 5.5 mg/kg produces disruption of behavior. Since percent S(+)-amphetamine-appropriate lever responding is fairly low, it can be stated that the stimulus effect produced by 1.0 mg/kg of S(+)-amphetamine is quite different from those produced by fenfluramine or buspirone. However, the fact that the latter two compounds can serve as training drugs themselves indicates that they are not actually saline-like agents (see[5] for example).

B.   Drugs as Stimuli

Table 4.1 lists drugs from several different pharmacological classes that have been shown to serve as stimuli. In most of these studies either the FR of the VI schedule of reinforcement has been used to establish the discrimination. While the techniques used to demonstrate this phenomena have varied, it is typical that an organism is taught

TABLE 4.2
Results of Stimulus Generalization
Studies with S(+)-Amphetamine
(1.0 mg/kg) as the Training Drug

Drug	Dose	% S(+)-amphetamine-appropriate responding
(-)Ephedrine	1.0	17%
	2.5	21%
ED50 = 2.7	3.0	63%
	4.0	95%
Fenfluramine	0.5	4%
	1.5	22%
	2.5	20%
	3.0	Disruption (i.e., no responding)
Buspirone	1.0	4%
	3.0	0%
	5.0	16%
	5.5	Disruption (i.e., no responding)

to respond on one lever (e.g., right-side lever) when a dose of a training drug is administered before a training session and on another lever (e.g., left-side lever) when vehicle (saline) is given. Correct responses are intermittently reinforced by delivery of food (e.g., pellet, sweetened milk). The organism's learning to respond appropriately to the two dissimilar situations involved in the administration of a dose of training drug on certain days and the injection of vehicle on other days establishes the discrimination. Organisms from a number of species have served as subjects in discrimination studies (see,[1,9] and [20] for example): rat, mouse, gerbil, pigeon, cat, monkey, and humans (money as reinforcer).

C.   Time Course

Once a dose of drug has been established as a training drug, its time-course of action can be measured. This test would investigate the effects of changing the pretreatment interval of the training dose of S(+)-amphetamine (i.e., 1.0 mg/kg) and the beginning of a test session. For example, in addition to the standard 15-min delay, we could examine the effects of 1, 5, 10, 30, 60, and 90-min pretreatment intervals. The stimulus can be characterized by its latency of onset, its latency to peak effect, and its duration of action. The latency of onset refers to the time between the administration of the training drug and the first observable indications that it is beginning to have some effect on drug-appropriate responding. The latency to peak effect refers to the time from the administration of the drug until percent drug-appropriate responding is maximal (i.e., 80 to 100%). Lastly, the duration of action refers to the time from the onset of action until the drug no longer produces significant percent drug-appropriate responding. Clearly, familiarity with the time-course of action of the training drug and challenge compounds in tests of stimulus generalization and/or stimulus antagonism is of great importance, so that drug responses are not measured too soon or too long after drug administration.

D.   Stimulus Antagonism

An effective strategy for determining drug mechanisms involved in the stimulus effects of psychoactive agents is to study drugs which antagonize their effects. The rationale of this procedure is that the training agent will only be blocked by antagonists that interfere with the training drug's mechanism of action. Three approaches could be utilized. With the first approach it can be determined whether the stimulus effect of 1.0 mg/kg of S(+)-amphetamine can be attenuated when various doses of a suspected antagonist, such as the dopamine D_2 receptor antagonist haloperidol, are combined with the training stimulus. A dose-response antagonism of percent S(+)-amphetamine appropriate responding will occur in the case of haloperidol blocking the amphetamine stimulus. With the second approach the dose-response of S(+)-amphetamine is determined both in the presence and absence of a constant dose of haloperidol; a shift of the amphetamine dose-response curve to the right could occur. A parallel shift to the right suggests a competitive interaction between S(+)-amphetamine and haloperidol. Finally, with the third procedure, various doses of S(+)-amphetamine can be combined with various doses of haloperidol. This approach will generate a series of S(+)-amphetamine-haloperidol dose-response curves and may provide a comprehensive illustration of the drug interactions that are taking place. Regardless of the approach used, the results of antagonism tests, as with generalization tests, typically fall into one of three categories: 1) complete antagonism (i.e., saline-like responding); 2) partial antagonism (i.e., ~40 to 70% S(+)-amphetamine appropriate responding); and 3) drug-like responding (i.e., ≥80% S(+)-amphetamine appropriate responding).

IV.   Summary

The drug discrimination assay can be viewed as a drug detection procedure whereby an animal must be able to recognize a drug state in order to choose the correct response for a reinforcer. Many agents from different psychoactive drug classes have been shown to serve as discriminative stimuli. Once trained, the animals can be essentially "asked" whether they recognize a novel agent as producing a stimulus effect similar to that produced by the training drug. Drugs that produce similar effects in humans often generalize (substitute, transfer) to one another in tests of stimulus generalization in animals. Finally, the drug antagonism approach has been used by numerous investigators to determine whether a drug-induced stimulus can be attenuated when various doses of an antagonist are combined with the training stimulus. This strategy has been shown to be particularly effective in elucidating neurochemical mechanisms involved in the discriminative stimulus properties of drugs.

References

1. Young, R. and Glennon, R.A., Discriminative stimulus properties of Amphetamine and structurally related phenylalkylamines. *Med. Res. Rev.*, 6, 99, 1986.
2. Finney, D., *Probit Analysis*. London: Cambridge University Press, 1952.
3. Harris, R.T. and Balster, R.L., An analysis of the function of drugs in the stimulus control of operant behavior. In: *Stimulus properties of drugs*, III, T. Thompson and R. Pickens, Eds. New York: Appleton-Century-Croft, 1971.
4. Colpaert, F.C., Niemegeers, C.J.E., Kuyps, J.J. M.D., and Janssen P.A.J., Apomorphine as a discriminative stimulus and its antagonism by haloperidol. *Eur. J. Pharmacol.*, 32, 383, 1975.
5. Hendry, J.S., Balster, R.L., and Rosecrans, J.A., Discriminative stimulus properties of buspirone compared to central nervous system depressants in rats. *Pharmacol. Biochem. Behav.*, 19, 97, 1983.
6. Colpaert, F.C., Desmedt, L.K.C., and Janssen, P.A.J., Discriminative stimulus properties of benzodiazepines, barbiturates, and pharmacologically related drugs: Relation to some intrinsic and anticonvulsant effects. *Eur. J. Pharmacol.*, 37, 113, 1976.
7. Browne, R.G. and Koe, B.K., Clozapine and agents with similar behavioral and biochemical properties. In: *Drug Discrimination: Applications in CNS Pharmacology*, 241, F.C. Colpaert and J.L. Slangen, Eds. Amsterdam: Elsevier Biomedical Press, 1982.
8. Jarbe, T.U.C., Cocaine as a discriminative cue in rats: interactions with neuroleptic and other drugs. *Psychopharmacology*, 59, 183, 1978.
9. Teal, J.J. and Holtzman, S.G., Stereoselectivity of the stimulus effects of morphine and cyclazocine in the squirrel monkey. *J. Pharmacol, Exp. Ther.*, 215, 369, 1980.
10. Shearman, G., Miksic, S., and Lal, H., Discriminative stimulus properties of desipramine. *Neuropharmacology*, 17, 1045, 1978.
11. Young, R., Glennon, R.A., and Rosecrans, J.A., Discriminative stimulus properties of the hallucinogenic agent DOM. *Commun. Psychopharmacol.*, 4, 501, 1980.
12. Hirschhorn, I.D. and Winter, J.C., Mescaline and lysergic acid diethylamide (LSD) as discriminative stimuli. *Psychopharmacologia*, 22, 64, 1971.
13. Glennon, R.A. and Young, R., MDA: A psychoactive agent with dual stimulus effects. *Life Sci.*, 34, 379, 1984.
14. Glennon, R.A. and Misenheimer, B.R., Stimulus effects of N-monoethyl-l-(3,4-methylenedioxyphenyl)-2-aminopropane (MDE) and N-hydroxy-1-(3,4-methylenedioxyphenyl)-2-aminopropane (N-OH MDA) in rats trained to discriminate MDMA from saline: *Pharmacol. Biochem. Behav.*, 33, 909, 1989.
15. Hirschhorn, I.D. and Rosencrans, J.A., Generalization of morphine and lysergic acid diethylamide (LSD) stimulus properties to narcotic analgesics. *Psychopharmacology*, 47, 65, 1974.
16. Meltzer, L.T., Rosencrans, J.A., Aceto, M.D., and Harris, L.S., Discriminative stimulus properties of (-)-and (+)-nicotine. *Fedn. Proc. Am. Socs. Exp. Biol.*, 38, 863 (Abstract).
17. Kuhn, D.M., Greenberg, I., and Appel, J.B., Stimulus properties of the narcotic antagonist pentazocine: Similarity to morphine and antagonism by naloxone. *J. Pharmacol. Exp. Ther.*, 196, 121, 1976.

18. Herling, S. and Shannon, H.E., Ro 15-1788 Antagonize the discriminative stimulus effects of diazepam in rats but not similar effects of pentobarbital. *Life Sci.*, 31, 2105, 1982.

19. Shannon, H.E., Evaluation of phencyclidine analogs on the basis of their discriminative stimulus properties in the rat. *J. Pharmacol. Exp. Ther.*, 216, 543, 1981.

20. Jarbe, T.U.C., Henriksson, B.G., and Ohlin, G.C., Delta-9-THC as a discriminative cue in pigeons: Effects of delta-8-THC, CBD, and CBN. *Arch. Int. Pharmacodyn. Ther.*, 228, 68, 1977.

Chapter **5**

Conditioned Place Preference: An Approach to Evaluating Positive and Negative Drug-Induced Stimuli

John R. James, Richard Young, and
John A. Rosecrans

Contents

0-8493-0704-X/01/$0.00+$.50

I.   Introduction

Central to the development of conditioned place preference (CPP) techniques is the observation that animal subjects such as the rat will return to an environment associated with the repeated administration of a powerful Central Nervous System (CNS) stimulant such as cocaine.[1] The inference is that the animal prefers to enter the cocaine-associated chamber because of cocaine's positive reinforcement stimulus properties. In contrast, a drug that has few CNS effects would not be expected to show any preference as to where the drug in question was administered. Drugs that have aversive behavioral effects such as lithium, on the other hand, would be expected to induce the opposite effect; animal subjects would be expected to enter an environment not associated with drug administration. Thus, we appear to have a relatively simple means of evaluating whether a drug has potential reward and/or aversive effects.

An example of one of the approaches currently being used involves evaluating the time an animal spends in each of two behavioral chambers separated by a passable partition (shuttle arrangement) following the administration of a specific drug after being placed in one of these chambers. Animals are then administered no drug, drug-vehicle, or saline, and placed in the other chamber. If the animal prefers to spend most of its time in the chamber associated with drug administration the stimulus is considered to be rewarding. If the animal spends less time in the chamber associated with the drug after drug administration (test session) the drug stimulus is considered to be aversive in nature. This result is very similar to an animal pressing a specific lever to receive a drug injection as in self-administration studies. There is no specific apparatus that must be used for CPP and CPA. Researchers have used two-choice, three-choice, four-choice, and open field apparatus (corral).[2,3] Several factors will determine the choice of apparatus: (1) the number of subjects per group needed to achieve statistical significance; (2) the need for a novel environment;[4] and (3) the desire to eliminate a forced choice paradigm. The researcher has the option of including a central choice area between the experimental chambers, or in the case of a two-choice or open field no central choice area may be desired.

This approach has been especially important in the substance abuse area. However, while a seemingly simple behavioral technique that does not involve a complicated methodology (one needs a minimum of an environment presenting a choice for preference), the results from these experiments are not always easy to interpret. This is especially evident for drugs that have weak positive reinforcing qualities, or drugs that induce mixed reward and aversive stimulus effects. Many dependent variables are important in using a CPP design. Species, strain, dose, route of administration, time-interval of drug administration, and the CPP apparatus used can influence the results and interpretations made when such experiments are conducted.

II. Experimental Design

A. Subjects

Rats have been and continue to be the most widely used subject in CPP and CPA. The Sprague-Dawley rat (outbred), once the gold standard in behavioral research, is not always the optimum subject. The Long-Evans rat (outbred) is being utilized specifically in drug abuse research. The Fischer-344 and Lewis rats (inbred) are examples of strains that allow evaluation of differences between genetic and environmental aspects of behavior.[5] The wide range of mice with specific genetic traits or the even more specific knockouts makes the mouse a desirable subject when genotype and not phenotype is the focus of research.

B. Drugs

The paradigm is a valuable tool for evaluation of drugs of abuse. The most commonly used drugs are alcohol, amphetamine, cocaine, and morphine.[6] Any drug with the potential to produce a reinforcing/rewarding or aversion/punishment discriminative stimulus is a candidate for CPP or CPA. Nicotine presents an interesting enigma, as low doses appear to induce positive rewarding effects while high doses appear to be aversive[7-10] in a two or three-chamber CPP apparatus. Nicotine's effects also vary in relation to the strain of animal studied, suggesting that these approaches are not as easy to interpret unless dealing with a powerful reinforcing agent such as cocaine.

CPP is a comparative technique that involves a comparison between the drug and the no-drug condition, which is usually drug vehicle or saline. It would also be interesting to compare injection vs. no injection.

C. Route of Administration

The route of administration depends on the properties of the drug and the route of administration used in humans. Most investigators use the intraperitoneal (IP) route of administration when evaluating drugs of abuse. Subcutaneous routes are preferred with some drugs such as nicotine, but is used infrequently with cocaine because of its irritant effects.

D.   Drug Pharmacokinetics: Dose-Response and Time-Duration

Conditioning paradigms such as CPP are dependent on matching time of maximal pharmacological effect with establishing the conditional association with an environment. As with humans, the rewarding qualities are indirectly related to time of onset of the pharmacological effects. For example, a drug with a slow onset and a long duration of action (phenobarbital) is not a good reinforcer. The onset of the drug is to make an association between the drug effect and a specific behavioral state. Cocaine, on the other hand, has a rapid onset, especially when inhaled or injected intravenously, and thus is very readily used because of its rewarding stimulus effects.

III.   Apparatus Design: Space, Size, Shape, and Number of Individual Chambers, Other Environmental Considerations

A.   Two or More Chambered Apparatus

Apparatus may be constructed of as many chambers as your research design requires (forced choice, novel chamber). The size of the chambers will also depend on the subject's species. The chambers need to allow motility, but not be so large that the subject requires several minutes to explore the entire chamber.

B.   Open Field Apparatus

These are generally 8 sq. ft. or larger. Several designs can be used but one has to ensure that the subject will have a large enough area to be able to detect an animal's place preference. Generally, animals prefer to avoid the central area of an open field apparatus.

C.   Center-Starting Chamber

Eliminates bias created when an animal is placed directly into an experimental chamber at the start of a testing session.

D.   Radial Arm Design

A maze design can be used when up to eight chamber options are required.

E.   Sensory Modalities

Visual Cues — Vision cues are a poor choice for albino rodents, but pigeons are an excellent subject for visual cues. Rats prefer a dark environment vs. a light environment. The use of different shapes and colors are good variables for pigeons or primates, while simple light and dark choices are appropriate for rodents. The preferred chamber design for rodents is a patterned design, not a solid black or white configuration which increases the strong preference for the dark environment.

   Olfaction — Olfaction is one of the best exteroceptive cues (no odor, female or male scent, and food) for rodents.

   It is worth noting that failure to clean the chambers after each test session, during habituation and during training, has the potential to confound results.

   Auditory Cues — Developmental rodent studies utilizing powerful exteroceptive cues often use recordings of ultrasonic chatter between mother and pup. Tones and clicking sounds provide means of creating distinct auditory environments.

   Tactile Cues — One should allow for a wide range and degree of specificity between chambers by using different flooring materials (solid floor [smooth, textured], spaced bars, wire mesh, wood chips).

IV. Experimental Designs

A.   Number and Duration of Conditioning Trials

Drugs with powerful reinforcing properties (amphetamine, cocaine, and morphine) require fewer and shorter conditioning trials then a weaker reinforcing drug (nicotine). Drugs with a very short half-life may allow a researcher to run a drug session in the morning and a vehicle session in the afternoon.

   The number of drug and vehicle training sessions should be equal. Number of testing sessions in CPP or CPA is one. The test session is in a non-drugged state.

   Comment — A criticism of CPP or CPA is testing in a non-drug state when conditioning in a drugged state creates the potential of confounding by state dependent learning. State dependent learning takes place when an animal is conditioned to the interaction between exteroceptive cues and interoceptive cues, but then tested with only exteroceptive cues. This dilemma potentially arises for weak reinforcing drugs. Adding a test session in the drugged state will determine if state dependent learning was a factor in the conditioning. Additional test sessions for evaluation of agonist and antagonist are helpful in developing a complete behavioral and pharmacological profile of the test drugs.

B.   Biased and Unbiased Designs

In addition to the above considerations, two conditioning techniques can be used in CPP and CPA design, biased and unbiased training. In biased conditioning the

Single Alternation Drug Schedule

	Drug days	Vehicle days
WEEK 1	Monday	Tuesday
	Wednesday	Thursday
	Friday	
WEEK 2	Tuesday	Monday
	Thursday	Wednesday
	Friday	

Double Alternation Drug Schedule

	Drug days	Vehicle days
WEEK 1	Monday	Wednesday
	Tuesday	Thursday
	Friday	
WEEK 2	Monday	Tuesday
	Thursday	Wednesday
	Friday	

subject's baseline preference is first determined for the chosen apparatus before experimental manipulation. In unbiased conditioning the subject is trained to associate the experimental manipulation with a particular environment. Both the biased and unbiased conditioning techniques in a two-chamber apparatus require a forced choice decision by the animal. Incorporating a three-chamber, four-chamber, or open field apparatus eliminates the potential confound of forced choice. The subject of choice for years was the rat, but labs frequently now use the mouse. The Syrian hamster and primates have also been used in CPP and CPA.[11,12] Thus, there are a variety of approaches that can be used to investigate CPP and the reader is referred to reviews by Schechter[13,14] for additional views of the field.

V.   Training and Testing

A.   Training

Several days of free access to all environments allow the animal to habituate to the apparatus, eliminating novelty as a confounding variable. Baseline data should be determined as an average of time spent in each chamber over 3 to 5 days. The length of time necessary to determine baselines depends on the environmental differences between apparatus chambers. Distinct environments require fewer baseline sessions while less distinct or ambiguous environments require more baseline sessions. Durazzo et al.[1] found 20-min habituation sessions just as statistically significant as

40-min habituation sessions. Test sessions are generally 30 to 40 min in length. In CPP a drug or environmental manipulation thought to have reinforcing properties is paired with the less preferred environment and vehicle, or no environmental manipulation is paired with the preferred environment.

B.   Testing

The testing of subjects is performed in a non-manipulated state. The percentage of time spent in each chamber is tabulated during the test session. If the animal spends more time in the less preferred (baseline data) chamber associated with the experimental manipulation, the researcher can infer the experimental manipulation had a rewarding effect on the affective state of the animal. In CPA a drug or environmental manipulation thought to have aversive properties is paired with the preferred environment and vehicle or no environmental manipulation is paired with the less preferred environment. If the animal spends more time in the preferred (baseline data) chamber associated with vehicle, or no environmental manipulation, the researcher can infer the experimental manipulation had a punishing effect on the affective state of the animal. Designs that include a novel chamber do not allow free access to the novel chamber during the habituation period.

C.   Concurrent Measurements

A variety of other measures can be used to obtain information about the drug being studied. Spontaneous activity and location of subject within a chamber can be determined via infrared beams and computer technology.[15] In addition, video taping can also be utilized to determine other behavioral characteristics of the subjects under drug and non-drug states.

VI.   Research Objectives: Pharmacological and Neuroscience Objectives

A.   Mechanisms of Drug Action

The CPP design is a simple procedure that has the potential for learning more about the positive reinforcing properties of a specific drug. This is especially important in research where the goal is to determine mechanisms of drug action using neuroscience approaches. It is extremely difficult using brain site techniques (lesions, microdialysis, and other neurophysiolgical procedures) in animals self-administering a drug or other operant procedures. The CPP procedure in many cases is a useful and manageable tool in evaluating drug action at different brain sites.

B.   Drug Design and Discovery

The CPP design has the potential to evaluate new compounds suspected of having the same effects as a given agent. Thus, animal subjects trained with morphine should generalize to meperdine or methadone.

C.   Antagonist Studies

A major issue in substance abuse research is the ability to detect new drugs that may be useful as antagonists to drugs such as cocaine that may help cocaine addicts in their rehabilitation. Thus, rats trained to cocaine can be challenged with other agents to determine the relative ability to attenuate cocaine's reward stimulus. In addition, antagonism studies can provide additional information as to how a specific drug is acting (receptor classification).

VII. Conclusion

CPP and CPA are behavioral paradigms that are best used with experimental manipulations that produce a distinct discriminative stimulus. As mentioned previously, a drug may have both rewarding and punishing potential. Are you evaluating the rewarding or punishing effect of the drug? Preliminary studies are necessary when a literature review fails to provide pharmacokinetic information about the test article. When deciding upon the "biased" or "unbiased" conditioning method, a researcher might consider combining the two methods, first using the "biased" method where the preferred environment is determined and then using the "unbiased" method where the subjects are split between both environments.

References

1. Durazzo, T.C., Gauvin, D.V., Goulden, K.L., Briscoe, R.J., and Holloway, F.A., Cocaine- induced conditioned place approach in rats: The role of dose and route of administration, *Pharmacol. Biochem. Behav.* 49, 1001, 1994.
2. Hasenohrl, R.U., Oitzl, M.S., and Huston, J.P., Conditioned place preference in the corral: A procedure for measuring reinforcing properties of drugs, *J. Neurosci. Methods*, 30, 141, 1989.
3. Parker, L.A., Place conditioning in a three- or four-choice apparatus: Role of stimulus novelty in drug-induced place conditioning, *Behav. Neurosci.*, 106, 294, 1992.
4. Klebaur, J.E. and Bardo, M.T., The effects of anxiolytic drugs on novelty-induced place preference, *Behav. Brain Res.*, 101, 51, 1999.

5. Kosten, T.A., Miserendino, M.J., Chi, S., and Nestler, E.J., Fischer and Lewis rat strains show differential cocaine effects in conditioned place preference and behavioral sensitization but not in locomotor activity or conditioned taste aversion, *J. Pharmacol. Exp. Ther.*, 269, 137, 1994.

6. Tzschentke, T.M., Measuring reward with the conditioned place preference paradigm: A comprehensive review of drug effects, recent progress and new issues, *Prog. Neurobiol.*, 56, 613, 1998.

7. Fudala, P.J., Teoh, K.W., and Iwamoto, E.T., Pharmacologic characterization of nicotine-induced conditioned place preference, *Pharmacol. Biochem. Behav.*, 22, 237, 1985.

8. Fudala, P.J. and Iwamoto, E.T., Further studies on nicotine-induced conditioned place preference in the rat, *Pharmacol. Biochem. Behav.*, 25, 1041, 1986.

9. Clarke, P.B. and Fibiger, H.C., Apparent absence of nicotine-induced conditioned place preference in rats, *Psychopharmacol.*, 92, 84, 1987.

10. Jorenby, D.E., Steinpreis, R.E., Sherman, J.E., and Baker, T.B., Aversion instead of preference learning indicated by nicotine place conditioning in rats, *Psychopharmacol.*, 101, 533, 1990.

11. Meisel, R.L., Joppa, M.A., and Rowe, R.K., Dopamine receptor antagonists attenuate conditioned place preference following sexual behavior in female Syrian hamsters, *Eur. J. Pharmacol.*, 309, 21, 1996.

12. Pomerantz, S.M., Wertz, I., Hepner, B., Wasio, L., and Piazza, I., Cocaine-induced conditioned place preference in rhesus monkeys, *Soc. Neurosci.*, Abstr., 18, 1572, 1992.

13. Schechter, M.D. and Calcagnetti, D.J., Trends in place preference conditioning with a cross-indexed bibliography; 1957-1991, *Neurosci. Biobehav. Rev.*, 17, 21, 1993.

14. Schechter, M.D. and Calcagnetti, D.J., Continued trends in the conditioned place preference literature from 1992 to 1996, inclusive, with a cross-indexed bibliography, *Neurosci. Biobehav. Rev.*, 22, 827, 1998.

15. Brockwell, N.T., Ferguson, D.S., and Beninger, R.J., A computerized system for the simultaneous monitoring of place conditioning and locomotor activity in rats, *J. Neurosci. Meth.*, 64, 227, 1996.

Chapter 6

Intravenous Drug Self-Administration in Nonhuman Primates

Leonard L. Howell and Kristin M. Wilcox

Contents

I.   Introduction

The abuse of psychoactive drugs such as cocaine and heroin has spanned several decades and continues to be widespread in the U.S. Currently, research efforts have focused on the development of therapeutics to treat drug abuse. Drug self-administration studies have done much to help us understand the behavioral and pharmacological mechanisms underlying drug abuse. An understanding of these mechanisms will in turn aid in the development of effective therapeutic agents.

Important to the study of drug effects on behavior is the understanding that drugs function as stimuli to control behavior.[1] Based on the principles of operant conditioning, presentation of a stimulus as a consequence of behavior may either increase or decrease the probability that a behavior will occur again.[2] If a stimulus increases the probability that a behavior will recur, then that stimulus is defined as a positive reinforcer.[2] Stimuli such as food and water function as positive reinforcers, and data from self-administration studies indicate that drugs also function as positive reinforcers. Drug self-administration procedures in animals have been used extensively to evaluate the reinforcing effects of drugs. The first studies examining the reinforcing effects of drugs in the 1950s and 1960s focused on morphine self-administration in morphine-dependent animals.[3,4] Later studies demonstrated that dependence was not necessary to initiate self-administration.[5,6] Since these early studies, self-administration procedures have been used as a model of drug taking that can be studied under controlled laboratory conditions and applied to human drug use.

Drug self-administration studies in animals have contributed substantially to our knowledge of the neuropharmacological mechanisms controlling drug abuse. For example, studies with opioids have shown that drugs with high affinity for μ opioid receptors function as positive reinforcers,[7,8] whereas opioids with high affinity for κ opioid receptors generally do not.[8,9] Additionally, the self-administration of psychomotor stimulants and opioids has been found to be affected by the administration of antagonists either systemically or centrally (cf,[10,11]). If administration of an antagonist shifts the dose-response function for a self-administered drug to the right, then it can be assumed that the site of action of the antagonist is important for the reinforcing effects of the drug.[12-15] In addition to the administration of antagonists, self-administration of drugs has been affected by lesions of certain brain neurotransmitter systems (cf,[10,11]). Studies have also found that animals will self-administer drugs directly into certain brain areas, suggesting a neuropharmacological mechanism for their reinforcing effects (cf,[10]).

Over the years, drug self-administration procedures in animals have been found to be valid and reliable for determining the abuse liability of drugs in humans. It is well established that animals will self-administer most drugs that are abused by humans.[16,17] In particular, studies in nonhuman primates have made a significant contribution to the field of drug abuse research. Nonhuman primates are ideal subjects since they are phylogenetically more closely related to humans than are other species. Thus, we can apply information obtained from nonhuman primates to problems of human drug abuse with greater accuracy. This chapter will discuss methods of self-administration in nonhuman primates, including different preparations and schedules of drug reinforcement. While the focus will be primarily on self-administration of psychoactive stimulants such as cocaine, the methodology and general principles apply to other pharmacological classes including opiates, benzodiazepines, and alcohol.

II. Surgical Procedures

Protocols for intravenous drug self-administration require the surgical implantation of a chronic intravenous catheter to permit infusion of the drug solution. Typically, a superficial vessel, such as the external jugular or femoral vein, is accessed via a surgical cut-down procedure.[18] Using appropriate anesthesia — either inhaled anesthetics (e.g., isoflurane) or injected ketamine in combination with a benzodiazepine — and under aseptic conditions, one end of a catheter is implanted into the vessel while the other end is routed subcutaneously to a point of access. If the distal end of the catheter is externalized, an appropriate jacket is used to prevent the animal from damaging the preparation,[19] and the catheter is sealed with a stainless-steel obturator when not in use. An alternative means of access involves the attachment of the distal end of the catheter subcutaneously to a vascular access port.[20] A Huber needle designed to minimize insult to the skin or port membrane is inserted perpendicular to the port to allow for injection of drug solution. Lastly, a tethering system can be used to protect the catheter while providing convenient access.[5,21] The preparation requires continuous housing in an experimental chamber, and restraint by a harness and a spring arm attached to the top or back of the chamber. However, movement of the animal within the chamber is not restricted by the tethering system. The distal end of the catheter is routed subcutaneously to exit between the monkey's scapulae, and is threaded through the spring arm. For each of the preparations, the catheter is connected via plastic tubing to a motor-driven syringe located outside the test chamber during experimental sessions. At least twice weekly, catheters are flushed with sterile saline or water, and filled with heparinized saline (20 units of heparin per ml of saline). All solutions that come in contact with the catheter are prepared with sterile components and stored in sterile glassware.

III. Schedules of Drug Delivery

A. Initial Training

Drug self-administration involves operant behavior that is reinforced and maintained by drug delivery. Animals acquire drug injections by emitting a discrete response, such as pressing a lever or key. The number and pattern of responses required for each injection are defined by the schedule of reinforcement. Availability of drug under a given schedule typically is signaled by an environmental stimulus, such as the illumination of a stimulus light located proximal to the response lever. Schedule parameters and stimulus conditions are controlled by computers while responses emitted by the animal are recorded simultaneously. The primary dependent measures are number of drug injections and rate of responding during each session. As with behavior maintained by nondrug reinforcers such as food, responding maintained by drug injections is determined by the schedule parameters and the behavioral history of the animal.

Daily experimental sessions are conducted in the home cage or in a standard primate chair either custom designed[22] or commercially available.[23,24] If a primate chair is utilized, location of the chair within a ventilated, sound-attenuating chamber will minimize distractions and interference from daily laboratory activities. Typically, responding is initiated using a 1-response, fixed-ratio schedule so that each response in the presence of a stimulus light will result in the intravenous injection of a drug solution. The drug dose is determined by the concentration and volume of solution, and should be sufficient to maintain reliable drug self-administration in a well-trained animal. The saliency of the drug injection is enhanced by a change in the stimulus lights during the injection period. It is critical to avoid excessive drug intake and toxicity during training sessions. Drug intake can be limited by scheduled timeout periods following each injection, during which stimulus lights are extinguished and responding has no scheduled consequence. Defined limits on the number of injections per session are also recommended. Once the animal has acquired the lever press response and reliably receives all scheduled injections, the response requirement can be gradually increased under a variety of intermittent schedules of reinforcement.

B.   Fixed-Ratio Schedules

The most basic schedule of reinforcement is the fixed-ratio (FR) schedule which defines the number of responses required per drug injection. Once responding on the lever is engendered, the response requirement is gradually increased to the terminal value. Nonhuman primates rapidly acquire drug self-administration behavior under FR schedules, and stable daily performances can be obtained in several weeks. FR schedules typically generate high response rates and a "break and run" pattern of responding characterized by a brief pause in responding after each drug injection, followed by an abrupt change to a steady high rate of responding until the next FR is completed. It is important to note that total session intake of drug is a direct function of response rate under FR schedules. Drug intake can be limited by scheduling timeouts following each injection and by restricting the total number of drug injections per session.

C.   Fixed-Interval Schedules

In contrast to FR schedules, fixed-interval (FI) schedules are time based and specify a minimal inter-injection interval. They represent a suitable alternative to FR schedules because they engender high levels of behavioral output. Typically, a stimulus light is illuminated in the test chamber during the FI to serve as a discriminative stimulus. Once the FI has elapsed, a single response is required for drug delivery. A limited hold can be imposed following the FI to restrict the time period during which a response is reinforced, resulting in higher rates of responding. Temporal control over behavior is enhanced if a timeout is scheduled following each injection. Once responding on the lever is engendered at a very short FI (1 to 5 seconds), the interval is gradually increased to the terminal value. In contrast to the break-and-run pattern engendered by FR schedules, FI schedules engender a scalloped pattern of responding characterized by little or no responding early in the interval and increased rates of responding as the interval elapses. Nonhuman primates often require several months of training before a stable pattern of responding develops. Also, response rate can vary markedly with little or no change in the total number of injections per session.

D.   Second-Order Schedules

Environmental stimuli that are paired with reinforcement can substitute for the reinforcer itself through Pavlovian conditioning. Moreover, these conditioned stimuli can reinforce behavior that results in their presentation through the process of second-order conditioning.[25] When a stimulus light that has been paired with drug injection is presented following an operant response, the frequency of that response increases.[26,27] Accordingly, second-order schedules of drug self-administration can generate very high behavioral outputs for a single injection of drug. If the stimuli are presented early in the drug-taking history of the animal, they can also enhance the acquisition of drug self-administration. Nonhuman primates are well-suited for studies using second-order schedules due to their complex behavioral repertoire. Typically, responding is initiated using a 1-response FR schedule so that each response in the presence of a stimulus light (e.g., red) will produce an intravenous drug injection and the brief illumination of a different stimulus light (e.g., white), followed by a timeout. The ratio value is increased gradually as responding increases. When the schedule value reaches a terminal value, drug injection no longer follows completion of each FR and, instead, is arranged to follow an increasing number of FR components during a predetermined interval of time. As the interval duration is extended during training, a greater number of FR components will be completed per drug injection. Ultimately, the terminal schedule will arrange for drug injection following the first FR component completed after the FI has elapsed. Drug administration is accompanied by a change in the stimulus light (e.g., from red to white), followed by a timeout. The drug-paired stimulus light also is presented briefly upon completion of each FR component. Daily sessions can consist of several consecutive FIs depending on the interval and session duration. By using this second-order procedure and limiting the daily session to approximately one hour, any direct effects the self-administered drug might have on rate and pattern of responding will be absent during the first component and minimized during the experimental session. Hence, performance measures can be related directly to the reinforcing effects of the drug. A cumulative record typical of performance engendered by a second-order FI schedule with FR components is shown in Figure 6.1.

E.   Progressive-Ratio Schedules

Progressive-ratio self-administration procedures are designed to quantify the reinforcing effects of drugs and to determine their reinforcing efficacy. Reinforcing efficacy is referred to as the maximum reinforcing effect of a drug or other

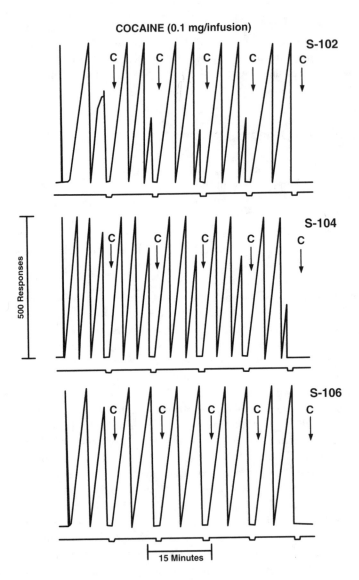

FIGURE 6.1

A cumulative record of lever pressing maintained in three squirrel monkeys under a second-order FI 600-second schedule of cocaine (0.1 mg/injection) intravenous self-administration with FR 20 components. A red light illuminated the test chamber during the 600-second FI, and every 20th response during the FI changed the red light to white for 2 seconds. The first FR completed after the 600-second FI elapsed produced a drug injection and changed the red light to white for 15 seconds. Abscissa: time. Ordinate: cumulative number of responses. Pen deflection on the time axis indicates the end of the interval and the beginning of a timeout. The response pen reset vertically upon completion of each FI or when the pen reached the top of the paper.

reinforcer.[28] Generally, it has been inferred from the strength of the behavior maintained by the drug. In a progressive-ratio procedure, the number of responses required to obtain a reinforcer progressively increases over the duration of the session. Eventually, responding for the reinforcer will cease when the response requirement becomes too great. This point, termed a breakpoint, is a measure of a drug's reinforcing efficacy. Under a progressive-ratio schedule, the response requirement can either increase following the delivery of the drug[29] or at the beginning of each daily session.[30,31] If the response requirement increases following the delivery of the drug, it will be incremented within a daily session, and completion of the response requirement one time will result in the delivery of drug. If the response requirement increases at the beginning of a daily session, it will be fixed over the duration of a daily session. The same response requirement will be in effect for several days (i.e., until stability criteria are met) before progressing to the next response requirement, allowing multiple determinations of self-administration at a particular response requirement. More recent progressive-ratio procedures combine both of these approaches and require that the animal respond a given number of times at a particular response requirement before proceeding to the next response requirement within a daily session.[32,33] Thus, these procedures have the advantage of collecting multiple determinations of self-administration at a particular response requirement within a daily session while still allowing the response requirement to be increased within the session. A cumulative record typical of performance engendered by a progressive-ratio schedule is shown in Figure 6.2.

IV.   Research Application and Data Interpretation

The primary focus of drug self-administration research in nonhuman primates has been to establish the reinforcing properties of drugs of abuse and to identify neurochemical mechanisms underlying drug use. A better understanding of the neurochemical basis of drug self-administration is essential for the development of treatment medications for human drug abusers. The key feature of drug-reinforced behavior is control of behavior by response-contingent drug delivery.[34] Hence, drug-reinforced behavior should be distinguished from drug self-administration. A number of control procedures have been described to demonstrate that increases in behavior that result in drug delivery are due specifically to the reinforcing effects of the drug.[6] The most commonly used procedure is to substitute saline for the drug solution and determine whether the behavior undergoes extinction. The rate and pattern of responding maintained by drug delivery depends on a number of variables including the schedule of reinforcement, drug dose, the volume and duration of injection, and the duration of drug self-administration sessions. Drug self-administration studies have consistently obtained an inverted U-shaped dose-effect curve relating the unit dose of drug delivered per injection and response rate or number of injections delivered. The dose-effect function reflects a combination of reinforcing effects and unconditioned stimulus effects such as sedation or marked hyperactivity. Typically, the ascending limb of the dose-effect curve reflects the reinforcing effects, and

90-31

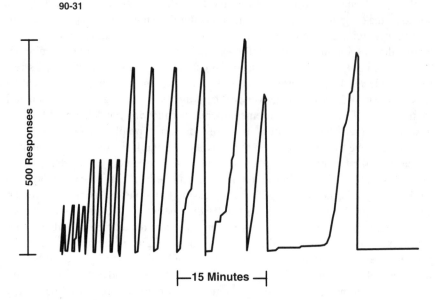

FIGURE 6.2

A cumulative record of lever pressing maintained in a rhesus monkey under a progressive-ratio schedule of cocaine (0.1 mg/kg/injection) self-administration over a daily session. The daily session consisted of five components, each made up of four trials at a particular response requirement. The response requirement began at an FR of 120 and doubled in subsequent components (i.e., 120, 240, 480, 960, 1920). A trial ended with a drug injection or the expiration of a limited hold. The session ended if the limited hold expired two consecutive times. Abscissa: time. Ordinate: cumulative number of responses. The response pen reset vertically upon completion of the FR or when the pen reached the top of the paper.

response rate increases with drug dose. In contrast, the descending limb of the curve reflects a nonspecific disruption of operant behavior as excessive drug accumulates over the session, and response rate decreases with drug dose. In addition, an inverse relationship has been obtained between infusion duration and reinforcing effects.[35,36] The longer the infusion time required to deliver a constant volume of drug solution, the less effective the drug functions as a reinforcer.

A study by Glowa et al.[37] illustrates the use of an FR schedule of drug self-administration to characterize the effectiveness of a dopamine reuptake inhibitor to alter the reinforcing effects of cocaine in rhesus monkeys. A standard tethered-catheter, home-cage system was used for intravenous drug delivery.[5] Animals were trained to self-administer cocaine under an FR 10 schedule of drug delivery during 90-minute daily sessions. Pretreatment with the high-affinity dopamine reuptake inhibitor, GBR 12909, dose-dependently decreased rates of cocaine-maintained responding, and the effect was larger when lower doses of cocaine were used to maintain responding. Moreover, the rate-decreasing effects of GBR 12909 were greater on cocaine-maintained responding than on food-maintained responding under a multiple schedule of drug and food delivery. The results obtained were consistent with previous reports demonstrating that drugs with dopamine agonist effects can decrease cocaine-maintained responding.[38,39] This type of drug interaction has been attributed to a satiation of cocaine-maintained responding by pretreatment with a drug having dopaminergic effects. The latter approach to cocaine medication development has been referred to as substitute agonist pharmacotherapy.[40] Hence, response-independent delivery of a dopamine reuptake inhibitor may have decreased cocaine self-administration by substituting for the reinforcing effects of response-produced cocaine. This interpretation is supported by studies showing that GBR 12909 will substitute for cocaine as a reinforcer in squirrel monkeys.[41-43]

Note that Glowa et al.[37] incorporated several important design features in their self-administration study. First, multiple unit doses of cocaine were self-administered on separate occasions in order to establish a complete dose-effect curve for cocaine. Hence, pretreatment effects of GBR 12909 could be assessed over a broad range of cocaine doses. Second, multiple pretreatment doses of GBR 12909 were administered in order to establish dose-dependency of pretreatment effects and to identify the optimal pretreatment dose that lacked overt behavioral toxicity. Lastly, the specificity of pretreatment effects on cocaine-maintained behavior was assessed by comparing drug effects on food-maintained behavior. The multiple schedule which alternated cocaine and food as maintaining events during separate components was well-suited for this application. Moreover, cocaine dose was manipulated to match response rate to that obtained during the food component of the multiple schedule. The finding that GBR 12909 suppressed cocaine-maintained responding at doses that had little or no effect on food-maintained responding under identical schedules and comparable response rates provides convincing evidence that GBR 12909 selectively attenuated the reinforcing effects of cocaine.

A study by Woolverton[44] provides another example of cocaine self-administration under an FR schedule in rhesus monkeys. The objective was to characterize the effectiveness of dopamine antagonists to alter the reinforcing effects of cocaine. A standard tethered-catheter, home-cage system was used for intravenous drug delivery. Animals were trained to self-administer cocaine under an FR 10 schedule of drug delivery during 2-hour daily sessions. When responding was stable, the animals were pretreated with the D_1 antagonist, SCH 23390, or the D_2 antagonist, pimozide. Intermediate doses of pimozide generally increased cocaine self-administration, whereas SCH 23390 either had no effect or decreased cocaine self-administration. High doses of both antagonists decreased the rate of cocaine self-administration, but also produced pronounced catalepsy. Hence, the latter effects could not be attributed to a selective interaction with the reinforcing effects of cocaine. The author concluded that the selective increase in responding maintained by cocaine following pimozide pretreatment suggested a role for the D_2-receptor in cocaine self-administration.

Strengths of the Woolverton[44] study design included multiple unit doses of cocaine and multiple pretreatment doses of both dopamine antagonists. Extinction of cocaine self-administration when saline was substituted for cocaine was also characterized. Note that response rate for cocaine increased following pretreatment with the D_2-selective antagonist. The latter effect is interpreted as a behavioral compensation to overcome the attenuation of the reinforcing effects of cocaine by pimozide. Since drug intake is a direct function of response rate under FR schedules, an increase in rate will result in greater session intake of cocaine which may effectively surmount the dopamine antagonist effects of pimozide. The finding that the pattern of responding following pimozide was virtually identical to that seen in the first session of extinction supports the view that pimozide was attenuating the reinforcing effects of cocaine. However, alternative interpretations were acknowledged, largely because specificity of pretreatment effects on cocaine-maintained behavior was not assessed by comparing drug effects on behavior maintained by nondrug reinforcers.

Nader et al.[45] used an FI schedule of drug self-administration to characterize the effectiveness of a novel cocaine analog to alter the reinforcing effects of cocaine in rhesus monkeys. A standard tethered-catheter, home-cage system was used for intravenous drug delivery. Animals were trained to self-administer cocaine under an FI 5-minute schedule during 4-hour daily sessions. Under this schedule, the first response after 5 minutes produced a 10-second cocaine injection. Pretreatment with the high-affinity dopamine reuptake inhibitor, 2β-propanoyl-3β-(4-tolyl)-tropane (PTT), dose-dependently decreased rates of cocaine-maintained responding and total session intake of cocaine. The reinforcing effects of PTT were also evaluated in a separate group of animals. When substituted for cocaine, PTT maintained response rates that were similar to those maintained by saline and significantly lower than rates maintained by cocaine. The results demonstrated that a long-acting dopamine reuptake inhibitor could effectively decrease cocaine self-administration in nonhuman primates. Also, failure of PTT to maintain rates of self-administration greater than those obtained during extinction conditions suggested that PTT may have limited abuse liability.

The study of Nader et al.[45] was consistent with previous findings using dopamine reuptake inhibitors to decrease cocaine self-administration under fixed-ratio schedules of drug self-administration.[37] Hence, a generality of pretreatment effects has been demonstrated across experimental conditions, further supporting a substitute agonist approach to cocaine medication development. Both studies included multiple unit doses of cocaine and multiple pretreatment doses of the dopamine reuptake inhibitors. In addition, assessment of the reinforcing properties of PTT provided critical information concerning the abuse liability of the candidate medication. The fact that PTT did not reliably maintain self-administration behavior, whereas GBR 12909 has been shown to function effectively as a reinforcer in monkeys,[41-43] illustrates the importance of pharmacokinetic factors in drug self-administration studies. It is possible that low rates of PTT self-administration were due to its relatively long duration of action at inhibiting dopamine uptake. Hence, its long duration of action and increased potency compared to cocaine may have required lower session intake to produce cocaine-like reinforcing effects.

Human drug use often involves a ritualized sequence of behaviors that occurs in a specific environment. The environmental stimuli associated with drug use are believed to play a major role in the maintenance of drug-seeking behavior.[27] Second-order schedules of drug self-administration have been used in nonhuman primates to maintain extended sequences of responding between drug injections[19,42,43,46] analogous to patterns of drug use in humans. The second-order schedule is well suited for drug-interaction and drug-substitution experiments because response rate is a direct function of the unit dose administered (Figure 6.3). Note that high doses of

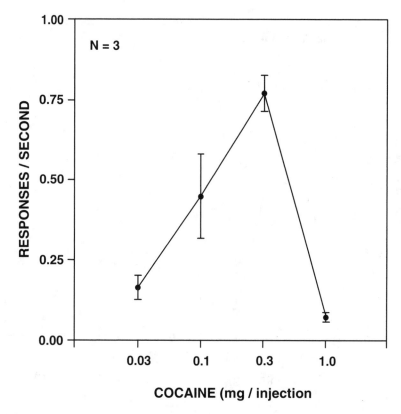

FIGURE 6.3

Mean (± S.E.M.) rate of responding maintained in a group of three squirrel monkeys under a second-order FI 900-second schedule of cocaine intravenous self-administration with FR 20 components. Data for each dose of cocaine were derived from at least five consecutive sessions on two separate occasions. Abscissa: dose, log scale. Ordinate: mean response rate expressed as responses per second.

drugs can disrupt performance during the latter components of a session as multiple doses accumulate. In drug-interaction experiments, changes in the positioning of the cocaine dose-effect curve leftward or rightward will indicate altered potency of cocaine to function as a reinforcer. A downward shift in the cocaine dose-effect curve will indicate an insurmountable attenuation of cocaine self-administration. In drug-substitution experiments, maximum rates of responding maintained over a range of drug doses can be used to compare drug efficacy.

A study by Howell et al.[47] provides an example of cocaine self-administration under a second-order FI schedule in squirrel monkeys. The objective was to characterize the effectiveness of a phenyltropane analog of cocaine to alter the reinforcing effects of cocaine. The distal end of the catheter was externalized and exited between the scapulae, and a nylon-mesh jacket protected the catheter when not in use. Animals were trained to self-administer cocaine under a second-order FI 15-minute schedule with FR 20 components during 1-hour daily sessions. Pretreatment with the dopamine reuptake inhibitor, RTI-113, significantly decreased rates of cocaine

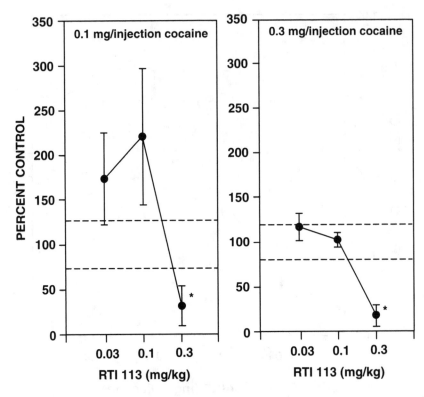

FIGURE 6.4
Mean (± S.E.M.) rate of responding in a group of three squirrel monkeys maintained under a second-order FI 900-second schedule of cocaine (0.1 and 0.3 mg/injection) intravenous self-administration with FR 20 components. Cocaine was self-administered alone or following pretreatment with RTI-113. Subjects were pretreated with each dose of RTI-113 for three consecutive sessions, and each subject received all drug combinations on two separate occasions. Abscissae: dose, log scale. Ordinates: mean response rate expressed as a percentage of control rate obtained when subjects were pretreated with saline. Asterisks indicate a significant ($p < 0.05$) effect of RTI-113 pretreatment.

self-administration, and the effect was not surmounted by increasing the unit dose of cocaine (Figure 6.4). The latter findings are consistent with previous studies using dopamine reuptake inhibitors to decrease cocaine self-administration under FR[37] and FI[45] schedules of drug self-administration. Hence, the generality of pretreatment effects has been demonstrated over a range of experimental conditions and in two different primate species. Note that low doses of RTI-113 actually increased rates of cocaine self-administration at the low unit dose of cocaine, providing evidence of additivity of effects. The latter finding is an important consideration when conducting drug interaction studies with two drugs having a similar mechanism of action.

Similar to other self-administration schedules, the reinforcing potencies of drugs can be determined under progressive-ratio schedules.[30,33] For example, cocaine is ten-fold more potent than the local anesthetic, procaine, under a progressive-ratio procedure.[33] However, progressive-ratio procedures are most useful for determining

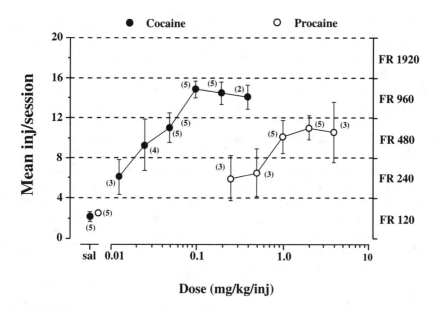

FIGURE 6.5

Breakpoint values and injections/session maintained by cocaine (closed symbols) and procaine (open symbols) under a progressive-ratio schedule in rhesus monkeys. Data are the mean (± S.E.M.) for the number of monkeys indicated in parentheses above each dose. The daily session consisted of five components, each made up of four trials at a particular response requirement. The response requirement began at an FR of 120 and doubled in subsequent components (i.e., 120, 240, 480, 960, 1920). Abscissae: dose, log scale. Left ordinate: mean injections/session. Right ordinate: breakpoint values. Dashed lines represent the mean number of injections/session taken at a particular breakpoint value.

the reinforcing efficacies of self-administered drugs. Drugs can be rank-ordered based on their relative reinforcing efficacy as determined by breakpoint in the progressive ratio.[30,33] For example, cocaine has been found to maintain higher breakpoint values than diethylpropion, chlorphentermine, and fenfluramine in baboons.[30] More recently, cocaine has been found to maintain higher breakpoint values than procaine in rhesus monkeys (Figure 6.5).[33]

As mentioned above, breakpoint typically is used as the dependent measure to assess reinforcing efficacy under a progressive-ratio procedure. However, breakpoint data violate the assumption of homogeneity of variance necessary for reliable statistical analysis. Variability is greater at high breakpoint values than at low breakpoint values, making it difficult to determine effects based on drug dose at high breakpoints.[48,49] Therefore, some researchers have applied a data transformation to breakpoint data before analysis (see Rowlett et al.[49]). In addition, the number of injections per session has been found to be a reliable measure of reinforcing efficacy, and does not violate the assumption of homogeneity of variance.[33,49] With intravenous self-administration paradigms, it is important to consider that effects other than reinforcing effects may influence responding for drug injections when high doses of a drug are available. Downward turns in dose-response curves have been explained by drug accumulation. To address this issue in the progressive ratio, researchers have used a timeout after each injection. The idea is that the timeout will allow the effects of the drug to dissipate before another injection can be obtained. A timeout length of 30 minutes is effective for studying cocaine in the progressive-ratio procedure.[33,49]

In addition to progressive-ratio procedures, choice procedures are used to study the reinforcing efficacy of drugs. The choice paradigm allows animals access to two reinforcers and evaluates the preference of one reinforcer over the other. Typically, animals choose between a food and a drug reinforcer or two drug reinforcers.[50,51] Reinforcing efficacy can be determined based on the preference of one reinforcer over the other. For example, rhesus monkeys given a choice between a high and a low dose of cocaine will prefer the higher cocaine dose.[51] A study by Johanson and Aigner[52] suggested a difference in the maximum reinforcing effects of cocaine and procaine using a choice procedure. They evaluated the preference for an intravenous injection of cocaine vs. an intravenous injection of procaine in rhesus monkeys. At equipotent doses for reinforcing effects, monkeys chose intravenous injections of cocaine greater than 80% of the time.[52] These results are consistent with those of the Woolverton[33] progressive-ratio study mentioned above. Thus, choice paradigms are reliable for studying reinforcing efficacy.

V.   Discussion

Nonhuman primate models of drug self-administration provide a rigorous, systematic approach to characterize the reinforcing effects of psychoactive drugs. The longevity of nonhuman primates is an important consideration, allowing for long-term studies to be conducted and repeated-measures designs to be employed. A single venous catheter can be maintained readily for over a year, and multiple implants permit the conduct of self-administration experiments for several years in individual subjects. Long-term studies with repeated measures are well suited for comprehensive drug-interaction experiments. While rodent models of drug self-administration have contributed substantially toward an understanding of neuropharmacology, the nonhuman primate represents an animal model with unique relevance to understanding the neurochemical basis of substance abuse in humans. For example, the complexity of the topographical organization of the striatum and its connections with surrounding areas in primates[53-55] complicates extrapolations from rodents to primates. Moreover, a large number of brain regions respond differently to acute drug administration in monkeys[56] compared to rodents.[57,58] Both the topography of altered brain metabolism and the direction of metabolic responses differ markedly.[56,58] The metabolic effects reported in monkeys are more consistent with data on functional activity in humans.[59,60] These findings, in conjunction with potential species differences in drug metabolism, illustrate the importance of nonhuman primate models in substance abuse research.

Research efforts that have utilized nonhuman primate models of drug self-administration have focused primarily on the identification of neurochemical mechanisms that underlie drug reinforcement, and the development of pharmacotherapies to treat drug addiction. In clinical evaluations of new medications, a decrease in drug self-administration is the goal of treatment.[61-63] Preclinical evaluations of pharmacotherapies require the establishment of stable baseline patterns of drug self-administration prior to drug-interaction studies. Subsequently, the treatment medication is administered as a pretreatment before the conduct of self-administration sessions. It is critical to study several doses of the treatment medication to determine an effective dose range and a maximally effective dose that lacks overt behavioral toxicity. The effects of the treatment medication typically are evaluated first in combination with a dose of the self-administered drug on the ascending limb of the dose-effect curve that maintains high rates of self-administration. However, a complete dose-effect curve should be characterized for the self-administered drug because pretreatment effects can differ depending on the unit dose of the drug self-administered. A rightward shift in the dose-effect curve suggests that drug pretreatment is antagonizing the reinforcing effects of the self-administered drug. A downward displacement of the dose-effect curve indicates an insurmountable attenuation of the reinforcing effects. Alternatively, a leftward shift is consistent with an enhancement of the reinforcing effects. Medications that shift the dose-effect curve downward and decrease self-administration over a broad range of unit doses are most likely to have therapeutic utility. Medications that shift the dose-effect curve to the right and simply alter the potency of the self-administered drug may prove to be ineffective at higher unit doses. Clinically, most medications are administered on a chronic basis and may require long-term exposure before therapeutic effects are noted.[64,65] Accordingly, preclinical studies should include repeated daily exposure to the medication to characterize peak effectiveness and to document continued effectiveness over multiple sessions. It is also critical to re-establish baseline levels of drug self-administration between successive exposures to the medication to ensure that the catheter preparation is functional and that persistent effects of the pretreatment drug do not interfere with the interpretation of drug interactions obtained.

The primary treatment outcome measures in drug self-administration studies are rate of responding and the number of drug injections delivered per session. Both measures are influenced by the schedule of reinforcement, drug dose, the volume and duration of injection, and the duration of the self-administration session. Moreover, most drugs that are self-administered have direct effects on rate of responding that may be distinct from their reinforcing effects. For example, cocaine injections may increase rate of responding early in the session, but suppress behavior later in the session as total drug intake accumulates. Another important consideration in evaluating medication effectiveness is the selectivity of effects on drug self-administration. If the drug pretreatment decreases drug self-administration at lower doses or to a greater extent than behavior maintained by a nondrug reinforcer such as food, the outcome is indicative of selective interactions with the reinforcing properties of the self-administered drug. In contrast, a general sedative effect will likely suppress drug- and food-maintained responding to a comparable extent. Lastly, the reinforcing properties and abuse potential of the medication should be evaluated by substituting a range of doses of the medication for the self-administered drug. Since reinforcing effects in preclinical studies are correlated with abuse liability in humans, reliable self-administration of the medication is usually considered undesirable.

While the present chapter has focused on the intravenous route of drug self-administration, the reinforcing properties of drugs have been studied effectively in nonhuman primates via the oral and inhalation routes. For example, orally delivered cocaine can function as a reinforcer in rhesus monkeys, and persistent and orderly responding is obtained when dose and FR size are varied.[66] Orally delivered phencyclidine and ethanol also maintain self-administration behavior in rhesus monkeys under progressive-ratio schedules.[67] However, establishing drugs as reinforcers via oral administration can be difficult due to metabolic effects associated with the gastrointestinal system, and delayed onset of CNS activity associated with slow absorption and distribution. In addition, drug solutions often have a bitter taste that may be aversive to nonhuman primates. Accordingly, complex induction procedures are used frequently to establish oral self-administration of drug solutions. Studies that have demonstrated cocaine's ability to function as a reinforcer have used a fading procedure from an initial baseline of ethanol-maintained responding,[66] although concurrent access to cocaine and vehicle solution is sufficient to establish oral self-administration.[68] Lastly, cocaine and heroin are self-administered by rhesus monkeys via smoke inhalation under FR and progressive-ratio schedules.[69-71] Although initial training is difficult due to the aversive characteristics of smoke, and drug dose is difficult to quantify, rhesus monkeys can rapidly learn to self-administer the drugs via the inhalation route. Given the above considerations, the advantages of intravenous self-administration procedures are clearly evident. Drug dose is easily manipulated and quantified, metabolic effects in the gastrointestinal system and slow absorption are avoided, and onset of CNS activity is rapid. Important, orderly, and reliable dose-effect curves are obtained that are sensitive to pharmacological manipulation.

References

1. Thompson, T. and Pickens, R., Eds., *Stimulus Properties of Drugs*, Appleton-Century-Crofts, New York, 1971.
2. Skinner, B. F., *The Behavior of Organisms*, Appleton-Century-Crofts, New York, 1938.
3. Headlee, C. P., Coppock, H. W., and Nichols, J. R., Apparatus and technique involved in laboratory method of detecting the addictiveness of drugs, *J. Am. Pharm. Assoc., Sci. Ed.*, 44, 229, 1955.
4. Thompson, T. and Schuster, C. R., Morphine self-administration, food-reinforced, and avoidance behaviors in rhesus monkeys, *Psychopharmacologia*, 5, 87, 1964.
5. Deneau, G., Yanagita, T., and Seevers, M. H., Self-administration of psychoactive substances by the monkey: A measure of psychological dependence, *Psychopharmacologia*, 16, 30, 1969.
6. Pickens, R. and Thompson T., Cocaine-reinforced behavior in rats: Effects of reinforcement magnitude and fixed-ratio size, *J. Pharmacol. Exp. Ther.*, 161, 122, 1968.
7. France, C. P., Winger, G. D., Medzihradsky, F., Seggel, M. R., Rice, K. C., and Woods, J. H., Mirfentanil: Pharmacological profile of a novel fentanyl derivative with opioid and nonopioid effects, *J. Pharmcol. Exp. Ther.*, 258, 502, 1991.
8. Young, A. M., Stephens, K. R., Hein, D. W., and Woods, J. H., Reinforcing and discriminative stimulus properties of mixed agonist-antagonist opioids, *J. Pharmacol. Exp. Ther.*, 229, 118, 1984.

9. Tang, A. H. and Collins, R. J., Behavioral effects of a novel kappa-opioid analgesic, U-50488, in rats and rhesus monkeys, *Psychopharmacology*, 85, 309, 1985.
10. Koob, G. F. and Bloom, F. E., Cellular mechanisms of drug dependence, *Science*, 242, 715, 1988.
11. Koob, G. F. and Weiss, F., Pharmacology of drug self-administration, *Alcohol*, 7, 193, 1990.
12. Bergman, J., Kamien, J. B., and Spealman, R. D., Antagonism of cocaine self-administration by selective dopamine D_1 and D_2 antagonists, *Behav. Pharmacol.*, 1, 355, 1990.
13. Bertalmio, A. J. and Woods, J. H., Reinforcing effects of alfentanil is mediated by *mu* opioid receptors: Apparent pA_2 analysis, *J. Pharmacol. Exp. Ther.*, 251, 455, 1989.
14. Dewit, H. and Wise, R. A., Blockade of cocaine reinforcement in rats with the dopamine receptor blocker pimozide, but not with the noradrenergic blockers phentolamine or phenoxybenzamine, *Can. J. Psychol.*, 31, 195, 1977.
15. Wilson, M. C. and Schuster, C. R., The effects of chlorpromazine on psychomotor stimulant self-administration in the rhesus monkey, *Psychopharmacologia*, 26, 115, 1972.
16. Griffiths, R. R., Bigelow, G. E., and Henningfield, J. E., Similarities in animal and human drug-taking behavior, in *Advances in Substance Abuse*, Vol 1., Mello, N. K., Ed., JAI Press, Greenwich, CT, 1980, 1.
17. Mello, N. K., Behavioral pharmacology of narcotic antagonists, in *The International Challenge of Drug Abuse, NIDA Research Monograph Series 19*, Peterson, R. C., Ed., U.S. Government Printing Office, Washington, D.C., 1979, 126.
18. Herd, J. A., Morse, W. H., Kelleher, R. T., and Jones, L. G., Arterial hypertension in the squirrel monkey during behavioral experiments, *Am. J. Physiol.*, 217, 24, 1969.
19. Howell, L. L. and Byrd, L. D., Serotonergic modulation of the behavioral effects of cocaine in the squirrel monkey, *J. Pharmacol. Exp. Ther.*, 275, 1551, 1995.
20. Wojnicki, F. H. E., Rothman, R. B., Rice, K. C., and Glowa, J. R., Effects of phentermine on responding maintained under multiple fixed-ratio schedules of food and cocaine presentation in the rhesus monkey, *J. Pharmacol. Exp. Ther.*, 288, 550, 1999.
21. Byrd, L. D., A tethering system for direct measurement of cardiovascular function in the caged baboon, *Am. J. Physiol.: Heart Circ. Physiol.*, 5, H775, 1979.
22. Byrd, L. D., The behavioral effects of cocaine: Rate dependency or rate constancy, *Eur. J. Pharmacol.*, 56, 355, 1979.
23. Howell, L. L. and Landrum, A. M., Behavioral and pharmacological modulation of respiration in rhesus monkeys, *J. Exp. Anal. Behav.*, 62, 57, 1994.
24. Howell, L. L. and Landrum, A. M., Effects of chronic caffeine administration on respiration and schedule-controlled behavior in rhesus monkeys, *J. Pharmacol. Exp. Ther.*, 283, 190, 1997.
25. Rescorla, R. A., *Pavlovian Second-Order Conditioning: Studies in Associative Learning*, Lawrence Erlbaum, Hillsdale, NJ, 1980.
26. Schindler, C. W., Katz, J. L., and Goldberg, S. R., The use of second-order schedules to study the influence of environmental stimuli on drug-seeking behavior, in *Learning Factors in Substance Abuse, NIDA Research Monograph Series 84*, Ray, B. A., Ed., U.S. Government. Printing Office, Washington, D.C., 1988, 180.

27. Katz, J. L. and Goldberg, S. R., Second-order schedules of drug injection: Implications for understanding reinforcing effects of abused drugs, *Adv. Subst. Abuse*, 4, 205, 1991.

28. Hodos, W., Progressive ratio as a measure of reward strength, *Science*, 134, 943, 1961.

29. Bedford, J. A., Baily, L. P., and Wilson, M. C., Cocaine reinforced progressive ratio performance in the rhesus monkey, *Pharmacol. Biochem. Behav.*, 9, 631, 1978.

30. Griffiths, R. R., Brady, J. V., and Snell, J. D., Progressive-ratio performance maintained by drug infusions: Comparison of cocaine, diethylproprion, chlorphentermine, and fenfluramine, *Psychopharmacology*, 56, 5, 1978.

31. Griffiths, R. R., Bradford, L. D., and Brady, J. V., Progressive-ratio and fixed-ratio schedules of cocaine-maintained responding in baboons, *Psychopharmacology*, 65, 125, 1979.

32. Rowlett, J. K. and Woolverton, W. L., Self-administration of cocaine and heroin combinations by rhesus monkeys responding under a progressive-ratio schedule, *Psychopharmacology*, 133, 363, 1997.

33. Woolverton W. L., Comparison of the reinforcing efficacy of cocaine and procaine in rhesus monkeys responding under a progressive-ratio schedule, *Psychopharmacology*, 120, 296, 1995.

34. Young, A. M. and Herling, S., Drugs as reinforcers: Studies in laboratory animals, in *Behavioral Analysis of Drug Dependence*, Goldberg, S. R. and Stolerman, I. P., Eds., Academic Press, San Diego, 1986, 9.

35. Balster, R. L. and Schuster, C. R., Fixed-interval schedule of cocaine-reinforcement: Effect of dose and infusion duration, *J. Exp. Anal. Behav.*, 20, 119, 1973.

36. Kato, S., Wakasa, Y., and Yanagita, T., Relationship between minimum reinforcing doses and injection speed in cocaine and pentobarbital self-administration in crab-eating monkeys, *Pharmacol. Biochem. Behav.*, 28, 407, 1987.

37. Glowa, J. R., Wojnicki, F. H. E., Matecka, D., Bacher, J., Mansbach, R. S., Balster, R. L., and Rice, K. C., Effects of dopamine reuptake inhibitors on food- and cocaine-maintained responding. I: Dependence on unit dose of cocaine, *Exp. Clin. Psychopharmacol.*, 3, 219, 1995.

38. Caine, S. B. and Koob, G. F., Modulation of cocaine self-administration in the rat through D_3 dopamine receptors, *Science*, 260, 1814, 1993.

39. Skjoldager, P., Winger, G., and Woods, J. H., Effects of GBR 12909 and cocaine on cocaine-maintained behavior in rhesus monkeys, *Drug Alcohol Depend.*, 33, 31, 1993.

40. Carroll, F. I., Howell, L. L., and Kuhar, M. J., Pharmacotherapies for treatment of cocaine abuse: Preclinical aspects, *J. Med. Chem.*, 42, 2721, 1999.

41. Bergman, J. Madras, B. K., Johnson, S. E., and Spealman, R. D., Effects of cocaine and related drugs in nonhuman primates. III. Self-administration by squirrel monkeys, *J. Pharmacol. Exp. Ther.*, 251, 150, 1989.

42. Howell, L. L. and Byrd, L. D., Characterization of the effects of cocaine and GBR 12909, a dopamine uptake inhibitor, on behavior in the squirrel monkey, *J. Pharmacol. Exp. Ther.*, 258, 178, 1991.

43. Howell, L. L., Czoty, P. W., and Byrd, L. D., Pharmacological interactions between serotonin and dopamine on behavior in the squirrel monkey, *Psychopharmacology*, 131, 40, 1997.

44. Woolverton, W. L., Effects of a D_1 and D_2 dopamine antagonist in the self-administration of cocaine and piribedil by rhesus monkeys, *Pharmacol. Biochem. Behav.*, 24, 351, 1986.
45. Nader, M. A., Grant, K. A., Davies, H. M. L., Mach R. H., and Childers, S. R., The reinforcing and discriminative stimulus effects of the novel cocaine analog 2β-propanoyl-3β-(4-tolyl)-tropane in rhesus monkeys, *J. Pharmacol. Exp. Ther.*, 280, 541, 1997.
46. Kelleher, R. T. and Goldberg, S. R., Fixed-interval responding under second-order schedules of food presentation or cocaine injection, *J. Exp. Anal. Behav.*, 28, 221, 1977.
47. Howell, L. L., Czoty, P. W., Kuhar, M. J., and Carroll, F. I., Comparative behavioral pharmacology of cocaine and the selective dopamine uptake inhibitor, RTI-113, in the squirrel monkey, *J. Pharmacol. Exp. Ther.*, 292, 521, 2000.
48. Depoortere, R. Y., Li, D. H., Lane, J. D., and Emmett-Oglesby, M. W., Parameters of self-administration of cocaine in rats under a progressive-ratio schedule, *Pharmacol. Biochem. Behav.*, 45, 539, 1993.
49. Rowlett, J. K., Massey, B. W., Kleven, M. S., and Woolverton, W. L., Parametric analysis of cocaine self-administration under a progressive-ratio schedule in rhesus monkeys, *Psychopharmacology*, 125, 361, 1996.
50. Iglauer, C. and Woods, J. H., Concurrent performances: Reinforcement by different doses of intravenous cocaine in rhesus monkeys, *J. Exp. Anal. Behav.*, 22, 179, 1974.
51. Johanson, C. E., Pharmacological and environmental variables affecting drug preference in rhesus monkeys, *Pharmacol. Rev.*, 27, 343, 1976.
52. Johanson, C.-E. and Aigner, T., Comparison of the reinforcing properties of cocaine and procaine in rhesus monkeys, *Pharmacol. Biochem. Behav.*, 15, 49, 1981.
53. Haber, S. N., Kunishio, K., Mizobuchi, M., and Lynd-Balta, E., The orbital and medial prefrontal circuit through the primate basal ganglia, *J. Neurosci.*, 15, 4851, 1995.
54. Lynd-Balta, E. and Haber, S. N., The organization of midbrain projections to the ventral striatum in the primate, *Neuroscience*, 59, 609, 1994.
55. Lynd-Balta, E. and Haber, S. N., Primate striatonigral projections: A comparison of the sensorimotor-related striatum and the ventral striatum, *J. Comp. Neurol.*, 3345, 562, 1994.
56. Lyons, D., Friedman, D. P., Nader, M. A., and Porrino, L. J., Cocaine alters cerebral metabolism within the ventral striatum and limbic cortex of monkeys, *J. Neurosci*, 16, 1230, 1996.
57. Porrino, L. J., Functional effects of cocaine depend on route of administration, *Psychopharmacology*, 112, 343, 1993.
58. Porrino, L. J., Domer, F. R., Crane, A. M., and Sokoloff, L., Selective alterations in cerebral metabolism within the mesocorticolimbic dopaminergic system produced by acute cocaine administration in rats, *Neuropsychopharmacology*, 1, 109, 1988.
59. London, E. D., Cascella, N. G., Wong, D. F., Phillips, R. L., Dannals, R. F., Links, J. M., Herning, R., Grayson, R., Jaffe, J. H., and Wagner, H. N., Cocaine-induced reduction of glucose utilization in human brain, *Arch. Gen. Psychiat.*, 47, 567, 1990.
60. Pearlson, G. D., Jeffery, P. J., Harris, G. J., Ross, C. A., Fischman, M. W., and Camargo, E. E., Correlation of acute cocaine-induced changes in local cerebral blood flow with subjective effects, *Am. J. Psychiat.*, 150, 495, 1993.

61. Mello, N. K. and Mendelson, J. H., Buprenorphine suppresses heroin use by heroin addicts, *Science*, 27, 657, 1980.
62. Mello, N. K., Mendelson, J. H., and Bree, M. P., Naltrexone effects on morphine and food self-administration in morphine-dependent rhesus monkeys, *J. Pharmacol. Exp. Ther.*, 218, 550, 1981.
63. Mello, N. K., Mendelson, J. H., and Kuehnle, J. C., Buprenorphine effects on human heroin self-administration, *J. Pharmacol. Exp. Ther.*, 230, 30, 1982.
64. Gawin, F. H., Cocaine addiction: Psychology and neurophysiology, *Science*, 251, 1580, 1991.
65. Gawin, F. H. and Ellinwood, E. H., Cocaine and other stimulants: Actions, abuse and treatment, *N. Engl. J. Med.*, 318, 1173, 1988.
66. Meisch, R. A., Bell, S. M., and Lemaire, G. A., Orally self-administered cocaine in rhesus monkeys: Transition from negative or neutral behavioral effects to positive reinforcing effects, *Drug Alcohol Depend.*, 32, 143, 1993.
67. Rodefer, J. S. and Carroll, M. E., Progressive ratio and behavioral economic evaluation of the reinforcing efficacy of orally delivered phencyclidine and ethanol in monkeys: Effects of feeding conditions, *Psychopharmacology*, 128, 265, 1996.
68. Macenski, M. J. and Meisch, R. A., Oral cocaine self-administration in rhesus monkeys: Strategies for engendering reinforcing effects, *Exp. Clin. Psychopharmacol.*, 3, 129, 1995.
69. Mattox, A. J., Thompson, S. S., and Carroll, M. E., Smoked heroin and cocaine base (speedball) combinations in rhesus monkeys, *Exp. Clin. Psychopharmacol.*, 5, 113, 1997.
70. Carroll, M. E., Krattiger, K. L., Gieske, D., and Sadoff, D. A., Cocaine-base smoking in rhesus monkeys: Reinforcing and physiological effects, *Psychopharmacology*, 102, 443, 1990.
71. Siegel, R. K., Johnson, C. A., Brewster, J. A., and Jarvik, M. E., Cocaine self-administration in monkeys by chewing and smoking, *Pharmacol. Biochem. Behav.*, 4, 461, 1976.

Chapter 7

Assessing Attention in Rats*

Philip J. Bushnell

Contents

I.   Introduction

Attention refers to a variety of hypothetical constructs by which the nervous system apprehends and organizes sensory input and generates coordinated behavior.

---

* This manuscript has been reviewed by the National Health and Environmental Effects Research Laboratory, U.S. Environmental Protection Agency and approved for publication. Approval does not signify that the contents necessarily reflect the views and policies of the Agency nor does mention of trade names or commercial products constitute endorsement or recommendation for use.

Whereas it has been a subject of psychological investigation since William James introduced it to the field in the late 19th century, systematic assessment of attention in animals has begun only recently. Assessment of attention (or other unobservable cognitive processes) requires quantification of some observable phenomenon, such as the behavior of the animal or the electrical activity of its nervous system. To the extent that these events can be measured objectively in both humans and in experimental animals, attention can be inferred as readily in any species. This chapter will discuss two behavioral methods by which different varieties of attention can be assessed in rodents; similar procedures can be implemented for the same purpose in humans.

Because important events may occur distributed both in time and space, attending to these events can be differentiated along both spatial and temporal dimensions. Attention may thus be either focused or distributed among the locations or physical characteristics of discrete stimuli (spatial selection), and also focused or distributed in time to a specified (and usually short) time interval or sustained for a long time interval. People may be instructed, and animals trained, to perform operant tasks that can be tuned to assess such attentional functions. Two such procedures are described here: a signal detection task for sustained attention and stressing temporal uncertainty, and the 5-choice serial reaction time (5-CSRT) task for spatio-temporal attention, which typically challenges spatial uncertainty, but also involves temporal uncertainty.

II.   Sustained Attention

A.   Introduction

The ability of human subjects to report the occurrence of rare and unpredictable signal events over prolonged periods of time has been extensively characterized (reviews by Parasuraman et al., 1987; Craig and Davies, 1991; Nachreiner and Hänecke, 1992). Accurate detection of such signals is assumed to depend upon maintaining attention to the task over time, and many of the factors that affect performance of humans on these tasks have been systematized (e.g., Parasuraman, 1984). Sustained attention tasks comprise an important and sensitive component of neurobehavioral test batteries used for assessing the effects of drugs in humans, e.g., benzodiazepines (Koelega, 1989), stimulants (Koelega, 1993), and ethanol (Koelega, 1995).

A major problem for tests of sustained attention involves quantifying and minimizing the false alarm rate. That is, a subject can successfully report many signals simply by responding frequently — though doing so will generate a large number of erroneous reports that a signal had occurred (false alarms). Human subjects can be instructed not to respond in this manner and will normally withhold most false alarms; animals can be trained to do so as well. However, manipulations that increase or decrease response rates are difficult to interpret if no independent measure of the false alarm rate is obtained.

Better estimates of the false alarm rate can be obtained by counting responses to specified non-signal events (blank trials). This approach has been used with both fixed and retractable response levers. The task described below employs a discrete-trial, two-lever approach which requires rats to report the occurrence or non-occurrence of a single, centrally-located signal. Thus, two retractable levers are inserted into the test chamber after a variable period of time to "ask" the rat to report whether a brief signal has been presented during that period. If a signal has been presented (signal trial), a press on one lever produces food and a press on the other lever produces a short timeout period without food. If no signal has occurred (blank trial), the converse contingencies apply. Because the levers are retracted between trials, no presses can occur during the inter-trial interval (ITI). Because an explicit response is required on each trial, the proportions of hits and false alarms [P(hit) and P(fa)] can be calculated in relation to the total number of completed signal and blank trials, respectively.

Validation of this method includes both studies of the effects of parameters known to affect human sustained attention and pharmacological and neurobiological manipulations. Parametric studies include observations that signal intensity, signal rate, and the type of task all affect response accuracy (Bushnell et al., 1994; McGaughy and Sarter, 1995; Bushnell, 1999), as predicted from studies of vigilance in humans (Parasuraman, 1984). Thus the parameters that affect the behavior of rats in this task closely parallel those which affect sustained attention in humans.

Pharmacological studies have shown dose-related impairment of signal detection in this task after a variety of nicotinic drugs (McGaughy and Sarter, 1995; Turchi et al., 1995; Bushnell et al., 1997; McGaughy et al., 1999), d-amphetamine (McGaughy and Sarter, 1995), the muscarinic drugs pilocarpine and scopolamine, and the α_2-adrenergic compounds clonidine and idazoxan (Bushnell et al., 1997). Further, a role for the cholinergic projections from the basal forebrain to the cortical mantle in sustained attention has been suggested by McGaughy et al. (1996), and this method has proved to be sensitive to inhalation of organic solvents as well (Bushnell et al., 1994; Bushnell, 1997).

B.   Materials

Subjects — Rats, male or female, of many strains and varieties, can learn to perform the task. The animals must be mildly hungry at the time of testing, which can be arranged by a number of standard methods (Ator, 1991). Methods are available to maintain adult rats at a constant body weight (Ali et al., 1992). See Preparation of the Subjects, below.

Apparatus — Assemble one or more standard operant conditioning chambers, equipped at minimum with a signal light, a food cup and food pellet dispenser, and two retractable response levers. A loudspeaker for presentation of masking noise may also be used. The two retractable levers should be mounted on either side of the food cup. Mount the signal light immediately above one of the levers at the start of training, and later move it to the top center of the wall above the food cup when the rat has learned the response rule required for the task. This equipment can be purchased from one of the vendors of behavioral test systems listed in Section V.

Assign the lever below the signal lamp as the signal lever and the other lever as the blank lever. Set up half of the chambers with the signal lever on the left and the other half with it on the right. Counterbalance all treatments for signal lever position.

A computer and interface for programming the stimulus events and recording the animals' responses are also necessary. Programming the procedures can be accomplished via a number of commercial packages, including state notation software such as Sked-11 by State Systems, Inc., which runs on old but versatile Digital PDP-11 hardware, and Med-PC, a similar Windows-based system produced by Med Associates.

Calibration devices should include a photometer for measuring the intensity of the light under various stimulus conditions and a sound level meter for measuring the intensity of the white noise.

C.   Preparation of the Subjects

1. Ensure that the rats are motivated (hungry). One method is to determine their free-feeding or *ad libitum* body weight, and then reduce that weight by 15%. Do not deprive the rats completely of food, but reduce their allotment such that target body weights are achieved within 5 to 10 days. If the rat is fully grown, this target weight can be maintained for the remainder of the experiment. On the other hand, if the rat is still growing, allow it to grow in parallel with free-fed rats to a maximum level (e.g., 350 g for an adult male rat and 250 g for an adult female).

2. Adapt the rats to the handling procedures (Ator, 1991) and to the food pellets that will be used to reinforce responding in the operant chambers. Typically, precision 45 mg pellets are used, or 25 mg pellets for mice or small rats, available commercially. This latter step can be accomplished by offering the rats 5 to 10 pellets each day in their home cages or in a holding cage for several days prior to beginning training. This adaptation will obviate possible bait-shyness that may accompany introduction of a novel food.

D.   Training Steps

1. Shape the rats to press the signal lever for food, by autoshaping (Davenport, 1974; Bushnell, 1988), long (e.g., overnight) sessions with a continuous reinforcement schedule in effect, or hand shaping. If an autoshaping procedure is used, turn on the signal lamp whenever the lever is extended into the chamber, and turn it off when the lever retracts (either when the rat presses it, or after 15 sec without a press). If overnight sessions are used, be sure to provide adequate water. Criterion: 1 session of 50 reinforced responses on this lever.

2. Shape the rats to press the other (blank) lever by the same means. However, do not turn on the signal lamp when shaping responses on this lever. Criterion: 1 session of 50 reinforced responses on this lever.

3. Begin training, using trials in which both levers are extended into the chamber simultaneously on each trial. Light the signal lamp in half the trials (signal trials), and not

in the other half (blank trials). Retract both levers as soon as one is pressed. Deliver a food pellet after a press on the signal lever in a signal trial and after a press on the blank lever in a blank trial. Turn off all lights for 3 sec after a press on the signal lever in a blank trial and after a press on the blank lever in a signal trial. In signal trials, turn on the signal light 2 sec before extending the levers, and leave it on until the rats press a lever. Use correction trials to reduce the likelihood of position habits: repeat the conditions presented in each trial that terminates in an incorrect response, up to a maximum of three such correction trials. If the rat makes three consecutive errors, extend only the other (correct) lever in the fourth trial, to force a correct response. Criterion: two 100-trial sessions with overall accuracy of 80% or better.

4. Remove the correction trials and increase the total number of trials to 120. Criterion: one 120-trial session with overall accuracy of 80% or better.

5. Turn off the signal when the levers extend (rather than when the rat makes a response) and increase the total number of trials to 150. Criterion: one 150-trial session with overall accuracy of 80% or better.

6. Reduce the duration of the signal from 2 sec to 0.3 sec in gradual steps (e.g., 1.5 sec, 1.0 sec, 0.7 sec, 0.5 sec, and 0.3 sec). The onset of the signal should occur 2 sec before insertion of the levers in all cases, leaving an empty period between offset of the signal and insertion of the levers. Increase the total number of trials in stages to 240. Criterion: one session with overall accuracy of 80% or better at each signal duration.

7. Move the signal lamp from its position above the signal lever to a position at the top of the panel, centered between the levers (above the food cup). Retrain to criterion accuracy.

8. Make the interval after the signal offset variable (select values of 2, 3, or 4 sec randomly on each trial). Maintain accuracy at the 80% criterion.

9. Make the interval before the signal onset variable. (Begin with a list of relatively short and homogeneous values, and work up to a list of values ranging from less than 1 sec to about 25 sec, selected randomly on each trial. A constant-probability list, such as that provided by Fleschler and Hoffman, 1963, is recommended for the final stage.) Maintain accuracy at the 80% criterion.

10. Vary the strength of the signal. Either the intensity or duration may be varied. Varying the intensity is preferred, but requires digital control of the voltage provided to the signal lamp (through a digital-to-analog converter). At least three signal strengths should be used, preferably more (up to seven). Administer at least 20 trials at each signal strength (10 signal and 10 blank). Maintain accuracy at the 80% criterion.

E.   Data Analysis and Notes

The proportion of correct detections of the signal [P(hit)] should increase with increasing signal strength. The signal strength should be adjusted such that the weakest signal produces a P(hit) about equal to the guessing rate, and the strongest signal produces a P(hit) of about 1.0. The guessing rate is given by the proportion of errors on blank trials, or false alarms [P(fa)]. P(fa) should be independent of signal strength and range from about 0.10 to 0.20.

A wide range of signal strength values improves the consistency of the baseline from day to day, and allows one to differentiate between changes in attention and visual function (Bushnell et al., 1997). That is, poor attending to the signals should cause an increase in P(fa) and a decrease in P(hit) at all signal strengths where P(hit) exceeds P(fa). In other words, the P(hit) by signal strength gradient should shift downward. In contrast, a change in the ability of the rat to see the signal should produce a horizontal shift in the P(hit) by signal strength gradient, such that P(hit) is altered only for signals of intermediate intensity; in addition, P(fa) should not change.

P(hit) and P(fa) can be used to calculate signal detection indices of sensitivity and bias by any of a number of methods (e.g., Grier, 1971; Frey and Colliver, 1973; Green and Swets, 1974). However, interpretation of these derived measures depends upon the particular assumptions upon which their calculation is based, and explanation of their meaning invariably requires reference to the values of P(hit) and P(fa) from which they were derived. Thus the advantages of deriving signal detection indices — which involves trading one pair of measures [P(hit) and P(fa)] for another pair of more derived measures (sensitivity and bias) — are generally outweighed by the effort required to calculate and explain these derived measures.

Response time may also be measured as the latency between insertion of the lever and the rat's response. This value provides an index of motor function similar to a simple reaction time, because rats typically choose which lever to press during the time interval after the signal, by positioning themselves in front of one of the levers, and pressing it during its insertion into the chamber. Latency typically does not vary with signal intensity, but does tend to be shorter for hits and false alarms than for misses and correct rejections (Bushnell et al., 1994).

III.   The 5-Choice Serial Reaction-Time Task

A.   Introduction

The five-choice serial reaction-time task (5-CSRTT) was adapted by Carli et al. (1983) from a test for human subjects originally ascribed to Leonard (Wilkinson, 1963), and still in use (e.g., Sahakian et al., 1993). In a typical application of this test, a rat faces five openings in a horizontal array along one curved wall of a test chamber; a food cup with a clear plastic door is located on the rear wall. The rat initiates a trial by opening the food cup door. After a brief delay, a visual signal is presented, consisting of a brief illumination of one of the five openings. If the rat then breaks a photobeam crossing that opening, a food pellet is delivered into the food cup.

The method posits that the rat must attend to the broad array of openings in order to detect the signal and respond correctly to it. Accurate responding thus requires attention both in the temporal and spatial domains, which provides for a high degree of parametric flexibility and the potential for independent assessments of the spatial and temporal components of attending. In theory, then, the method may be "tuned" for independent assessment of selective and sustained attentional processes.

An impressive series of studies over the last ten years employed a variety of drug and lesion challenges to characterize the roles of the ascending monoaminergic arousal systems in performance of this task (reviewed by Robbins and Everitt, 1995). In addition, it has been proposed as a means to model attention deficit hyperactivity disorder in rats (Puumala et al., 1996) and to evaluate the effects of lesions (Muir et al., 1996) and systemic drugs (Jones and Higgins, 1995; Jäkälä et al., 1992a; 1992b) on performance of the task. In general, accuracy of responding on the basic task appears to depend upon cortical acetylcholine and speed of responding on mesolimbic dopamine. Auditory distractors are particularly disruptive to rats with loss of ceruleocortical norepinephrine, and adequate forebrain serotonin appears to be necessary to suppress premature responding (Robbins and Everitt, 1995).

B.   Materials

Subjects — Same as for the sustained attention task, above.

Apparatus — The 5-CSRT task is typically run in specialized equipment originally built in the laboratory of Experimental Psychology, Cambridge, UK (Carli et al., 1983). It is now commercially available, either from CeNeS Ltd. or from Med Associates, Inc. The test chambers measure 25 cm by 25 cm horizontally, and roughly 30 cm in height. In place of response levers, a series of 9 openings (CeNeS) or 5 openings (Med Associates), each 2.5 cm by 2.5 cm in opening, are arranged along a curved rear wall of the chamber, 2 cm from the floor. Each opening is bisected by a photobeam which is used to detect entry of the rat's nose into the opening. A light mounted inside each opening is illuminated briefly on each trial to serve as a signal, and the rat is trained to poke its nose into the illuminated opening to receive a food pellet, which is delivered into a food cup mounted in the wall opposite to the openings. The food cup must either be accessible via a hinged door with a microswitch, or bisected by a photobeam, so that nosepokes to retrieve food pellets can be detected. A single pellet dispenser is needed to deliver reinforcers to the rat. A houselight is also needed for general illumination of the chamber. The computer system, interface, software, and calibration equipment described above for the sustained attention task are also necessary.

C.   Preparation of the Subjects

Same as for the sustained attention task, above.

D.   Training Steps

1.   Cover the response openings. Place the hungry rat in the chamber and turn on the houselight. Provide food pellets in the food cup. On the first day, begin with about 20 pellets in the cup, and allow the rat 30 min to explore the chamber and collect the food. On 3 following days, deliver the pellets singly at 30 sec intervals and ensure that the rat retrieves the pellets and consumes them.

2.   Remove the covers from the openings. Turn on the houselight and deliver a single food
     pellet to start the session and begin the first trial when the rat retrieves the pellet. A
     trial involves illuminating the signal light in one opening (selected at random) after an
     inter-trial interval (ITI) of 2 sec and recording the nosepokes made by the rat into the
     5 openings. When the rat pokes its nose into the illuminated opening, turn off the signal
     light and deliver a single food pellet. (An auditory cue may be helpful in informing
     the rat that a pellet has been delivered, if the action of the dispenser is quiet.) If the
     rat pokes its nose into a dark opening, turn off the houselight for a 2 sec timeout period
     and do not deliver food. Reset the timeout each time the rat pokes its nose into a dark
     opening. Begin a new trial when the rat pokes its nose into the food cup after either
     timeout or food delivery. Present signals in each opening an equal number of times in
     each session, and select at random the opening to illuminate on each trial. Criterion:
     100 reinforced nosepoke responses in a 30-min test session.

3.   Repeat Step 2, but place a time limit on the signal light and the response period (called
     a limited hold). Allow the rat to initiate each trial as above, and use a 2-sec ITI before
     illuminating a signal. Illuminate each opening for 60 sec and set a 60-sec limited hold
     after the signal period. If the rat pokes into the illuminated opening during this 2-min
     period, deliver a food pellet and count a *correct response*. If the rat pokes into an unlit
     opening during this period, turn off the signal light and houselight and count an *error
     of commission*. If the rat fails to make a nosepoke response in this period, turn off the
     signal light and the houselight and count an *error of omission*. If the rat makes a
     nosepoke into any opening during the ITI, turn off the houselight, count a *premature
     response*, and restart the same trial. Criterion: 100 correct responses in a 30-min test
     session, or a 100-trial session with 80% correct responding (see below).

4.   Repeat Step 3, progressively shortening the signal duration and limited hold, ending
     with a signal duration of 0.5 sec and a limited hold of 5 sec. Lengthen the ITI to 5 sec
     and the timeout period to 3 sec during these steps. A stable baseline of about 80%
     correct responding with about 15% omissions should be achieved in about 30 training
     sessions.

E.   Data Analysis and Notes

Correct responses, errors of commission (nosepokes into openings not illuminated
by a signal), errors of omission (failures to make a nosepoke during the limited
hold), and premature responses (nosepokes into any opening prior to presentation
of a signal) are used to quantify the accuracy of responding in the task. One measure
of accuracy may be calculated as ratio of the number of correct responses divided
by the total number of responses (correct plus incorrect, ignoring premature
responses). In addition, response speed may be quantified in two ways: *response
latency* (the time between onset of the signal and a correct response) and *food
retrieval latency* (the time between delivery of a food pellet and the rat's entry into
the food cup).

    Schedule parameters may be manipulated to challenge attentive and visual
processing of information in this task. Temporal challenges to attending include
lengthening and/or shortening the ITI, or making it variable across trials rather than
constant. In addition, shortening the duration of the signal has been claimed to

increase the attentional load of the task (Muir et al., 1994). In contrast, dimmer signals have been used to challenge visual detection of the signals, a manipulation assumed to differ in nature from reducing duration (Muir et al., 1996). However, this distinction has not been systematically examined in control animals.

Distracting auditory cues may be interpolated at various times during a trial, and are most effective when delivered immediately before onset of the visual signal, when they induce a high frequency of premature responses (Carli et al., 1983). The spatial distribution of signals has not to my knowledge been manipulated systematically.

References

Ali J.S., Olszyk V.B., Dunn D.D., Lee K.-A., Rhoderick R.R., Bushnell P.J. A LOTUS 1-2-3-based animal weighing system with a weight maintenance algorithm. *Behavior Research Methods, Instrumentation, and Computers* 24:82–87, 1992.

Ator N.A. Subjects and instrumentation. In: I.H. Iversen and K.A. Lattal, Eds., *Experimental Analysis of Behavior, Part 1*, Elsevier Science Publishers, Amsterdam, pp. 1–62, 1991.

Bushnell P.J. Behavioral effects of acute *p*-xylene inhalation in rats: Autoshaping, motor activity, and reversal learning. *Neurotoxicol. Teratol.* 10:569–577, 1988.

Bushnell P.J. Concentration-time relationships for the effects of inhaled trichloroethylene on detection of visual signals in rats. *Fundam. Appl. Toxicol.* 36:30–38, 1997.

Bushnell P.J. Detection of visual signals by rats: Effects of signal intensity, event rate and task type. *Behav. Proc.* 46(2):141–150, 1999.

Bushnell P.J., Kelly K.L., Crofton K.M. Effects of toluene inhalation on detection of auditory signals in rats. *Neurotoxicol. Teratol.* 16:149–160, 1994.

Bushnell P.J., Oshiro W.M., Padnos B.K. Effects of chlordiazepoxide and cholinergic and adrenergic drugs on sustained attention in rats. *Psychopharmacology* 134:242–257, 1997.

Carli M., Robbins T.W., Evenden J.L., Everitt B.J. Effects of lesions to ascenting noradrenergic neurones on performance of a 5-choice serial reaction task in rats: Implications for theories of dorsal noradrenergic bundle function based on selective attention and arousal. *Behav. Br. Res.* 9:361–380, 1983.

Craig A., Davies D.R. Vigilance: Sustained visual monitoring and attention. In: J.A. Roufs (Ed) *Vision and Visual Dysfunction, Vol 15: The Man-Machine Interface.* MacMillan, Basingstoke, pp. 83–98, 1991.

Davenport J.W. Combined autoshaping-operant (AO) training: CS-UCS interval effects in the rat. *Bull. Psychon. Sci.* 3:383–385, 1974.

Fleschler M., Hoffman H.S. A progression for generating variable-interval schedules. *J. Exp. Anal. Behav.* 5:529–530, 1963.

Frey P.W., Colliver J.A. Sensitivity and responsivity measures for discrimination learning. *Learn. Motiv.* 4:327–342, 1973.

Green D.M., Swets J.A. *Signal Detection Theory and Psychophysics.* R.E. Krieger Publishing, Huntington, NY, 1974.

Grier J.B. Nonparametric indexes for sensitivity and bias: Computing formulas. *Psych. Bull.* 75:424–429, 1971.

Jäkälä P., Sirviö J., Jolkkonen J., Riekkinen P. Jr., Acsady L., Riekkinen P. The effects of
    p-chlorophenylalanine-induced serotonin synthesis inhibition and muscarinic blockade
    on the performance of rats in a 5-choice serial reaction time task. *Behav. Brain. Res.*
    51:29–40, 1992a.

Jäkälä P., Sirviö J., Riekkinen P. Jr., Haapalinna A., Riekkinen P. Effects of atipamezole, an
    α_2-adrenoreceptor antagonist, on the performance of rats in a five-choice serial reaction
    time task. *Pharmacol. Biochem. Behav.* 42:903–907, 1992b.

Jones D.N.C., Higgins G.A. Effect of scopolamine on visual attention in rats. *Psychophar-
    macology* 120:142–149, 1995.

Koelega H.S. Benzodiazepines and vigilance performance: A review. *Psychopharmacology*
    98:145–165, 1989.

Koelega H.S. Stimulant drugs and vigilance performance: A review. *Psychopharmacology*
    111:1–16, 1993.

Koelega H.S. Alcohol and vigilance performance: A review. *Psychopharmacology*
    118:233–249, 1995.

McGaughy J., Decker M.W., Sarter M. Enhancement of sustained attention performance by
    the nicotinic acetylcholine receptor agonist ABT-418 in intact but not basal forebrain-
    lesioned rats. *Psychopharmacology* 144:175–182, 1999.

McGaughy J., Kaiser T., Sarter M. Behavioral vigilance following infusions of 192 IgG-
    saporin into the basal forebrain: Selectivity of the behavioral impairment and relation to
    cortical AChE-positive fiber density. *Behav. Neurosci.* 110:247–265, 1996.

McGaughy J., Sarter M. Behavioral vigilance in rats: Task validation and effects of age, amphet-
    amine, and benzodiazepine receptor ligands. *Psychopharmacology* 117:340–357, 1995.

Muir J.L., Everitt B.J., Robbins T.W. AMPA-induced excitotoxic lesions of the basal forebrain:
    A significant role for the cortical cholinergic system in attentional function. *J. Neurosci.*
    14:2313–2326, 1994.

Muir J.L., Everitt B.J., Robbins T.W. The cerebral cortex of the rat and visual attentional
    function: Dissociable effects of mediofrontal, cingulate, anterior dorsolateral, and parietal
    cortex lesions on a five-choice serial reaction time task. *Cerebral Cortex* 6:470–481,
    1996.

Nachreiner F., Hänecke K. Vigilance. In: Smith A.P. and Jones D.M. (Eds) *Handbook of
    Human Performance, Vol 3: State and Trait.* Academic Press, London, pp. 261–288, 1992.

Parasuraman R. The psychobiology of sustained attention. In: Warm J.S. (Ed) *Sustained
    attention in human performance.* Wiley, New York, pp. 61–101, 1984.

Parasuraman R., Warm J.S., Dember W.N. Vigilance: Taxonomy and utility. In: Mark L.S.,
    Warm J.S., Huston R.L. (Eds) *Ergonomics and Human Factors* Springer-Verlag, New
    York, pp. 11–32, 1987.

Puumala T., Ruotsalainen S., Jäkälä P., Koivisto E., Riekkinen P. Jr., Sirviö J. Behavioral and
    pharmacological studies on the validation of a new animals model for attention deficit
    hyperactivity disorder. *Neurobiol. Learn. Memory* 66:198–211, 1996.

Robbins T.W., Everitt B.J. Arousal systems and attention. In: M.S. Gazzaniga (Ed) *The
    Cognitive Neurosciences.* MIT Press, Cambridge, MA, pp 703–720, 1995.

Sahakian B.J., Owen A.M., Morant M.J., Eagger S.A., Boddington S., Crayton L., Crockford
    H.A., Crooks M., Hill K., Levy R. Further analysis of the cognitive effects of tetrahy-
    droaminoacridine (THA) in Alzheimer's disease: Assessment of attentional and mne-
    monic function using CANTAB. *Psychopharmacoldogy* 110:395–401, 1993.

Turchi J., Holley L.A., Sarter M. Effects of nicotinic acetylcholine receptor ligands on behavioral vigilance in rats *Psychopharmacology* 118:195–205, 1995.

Wilkinson R.T. Interaction of noise with knowledge of results and sleep deprivation. *J. Exp. Psychol.* 66:332–337, 1963.

V. Names and Addresses of Vendors Discussed in the Text

Behavioral Test Systems

CeNeS Limited
Compass House, Vision Park
Chivers Way, Histon
Cambridge CB4 4ZR
England, UK
cenes@cenes.co.uk
www.cenes.com

Columbus Instruments
950 N. Hague Ave.
Columbus, OH 43204
www.colinst.com

Coulbourn Instruments, LLC
7462 Penn Dr.
Allentown, PA 18106
www.coulbourninst.com

MED Associates, Inc.
Box 2089
Georgia, VT 05468
www.med-associates.com

San Diego Instruments
7758 Arjons Dr.
San Diego, CA 92126

State Systems, Inc.
P.O. Box 2215
Kalamazoo, MI 49003

Calibration

Audiometric
Brüel & Kjær Instruments, Inc.
185 Forest St.
Marlborough, MA 10752

Photometric
EG&G Gamma Scientific
8581 Aero Drive
San Diego, CA 92123

Food pellets
Bio-Serv
One 8th St.
Suite 1
Frenchtown, NJ 08825
www.bio-serv.com

P.J. Noyes Company, Inc.
P.O. Box 381
Bridge St.
Lancaster, NH 03584

Chapter

8

Assessment of Distractibility in Non-Human Primates Performing a Delayed Matching-to-Sample Task

Mark A. Prendergast

Contents

I.   Introduction

The cognitive processes of both working and reference memory are closely linked to, and indeed dependent upon, adequate function of selective attention and memory consolidation in primates. While it is often difficult from a methodological perspective to identify specific behaviors reflective of selective attention, two general components of this process are the abilities to select desired items for conscious processing and to ignore, or filter out, less relevant and distracting information which may compete for cognitive resources.[1] This is perhaps most eloquently demonstrated in the "Cocktail Party Phenomenon" described by Cherry.[2] This work described the ability of human subjects to monitor and evaluate both relevant and irrelevant verbal cues in a crowded room with several different simultaneous conversations taking place. More specifically, an irrelevant conversation not being attended to can immediately become a focus of cognitive processing if a salient word or phrase (i.e., the subject's name) is heard. Thus, the conversation was filtered out of conscious processing until it became relevant. The most significant implications of this and other similar perceptual phenomenon are: 1) selective attention is an active process that involves a regular perception and assessment of both relevant and potentially distracting information; and 2) the results of these processes have significant influences on the allotment of conscious cognitive resources.

It is the effect of reducing distraction and focusing cognitive resources that is perhaps the most beneficial result of selective attention. Impairment of selective attention or, more specifically, the ability to ignore potentially distracting stimuli is seen with normal aging and is a prominent symptom of several clinical conditions. Aged humans with no diagnoses of cognitive impairment, when compared to younger adults, demonstrate susceptibility to distraction by extraneous stimuli (e.g., irrelevant numbers seen during a card sorting task) while performing working memory tasks.[3,4] Further, several investigators have demonstrated that increased distractibility contributes to impairment of sustained attention and working memory seen in patients with AD.[5,6] Increased distractibility is perhaps most evident in those with Attention Deficit Disorder (ADD).[7] In these conditions, and in normal aging, increased susceptibility to distraction contributes to significant difficulty in general cognitive function, as well as in daily living skill competence. This is illustrated best in findings of a link between increased distractibility in children and difficulties in academic and social settings,[8] difficulties which persist throughout adulthood in as many as 60% of those adolescents with ADD.[9]

The relevance of developing animal models to examine and potentially reduce distractibility may be seen in medicine's difficulty in the clinical management of impaired attention. Currently, standard treatment regimens for both childhood and adult ADD most often include a psychomotor stimulant, typically methylphenidate or d-amphetamine.[10,11] The pharmacologic action of these and related compounds fits well with evidence of prefrontal catecholamine dysregulation in disorders of attention, modeled in both human and animal models.[12] Indeed, function of the prefrontal association cortex has been demonstrated to be essential for working memory.[13,14] While many patients with ADD undoubtedly benefit from treatment with these agents, a significant portion may be either refractive to this treatment or suffer adverse consequences warranting discontinuation.[10] Further, methylphenidate has no demonstrated efficacy to improve sustained visual attention or working memory in patients with AD[7] and clinically available acetylcholinesterase inhibitors offer only minimal benefit to cognitive function in these patients.[15] Thus, development of behavioral models to assess distractibility in non-human primates is of significant benefit in not only understanding the relationship between selective attention and higher cognitive function but in the development of novel pharmacological strategies which may enhance cognitive function in these patient populations.

II.   Non-Human Primate Models of Distractibility: Task-Irrelevant Stimuli

Models of brief memory, attention, and distractibility in non-human primates are methodological descendants of work pioneered by American psychologist Walter S. Hunter. In his experiments on mental representation in lower animals, Hunter[16] described decrements in the memories of rats, dogs, and raccoons with increasing delay intervals between presentation of a distinct cue and presentation of choice stimuli, only one of which was previously paired with the initial cue, to obtain a food reward. This early demonstration of delay-dependent recall has provided non-human primate researchers with a valuable methodological foundation to use in assessing memory function in higher species. More specifically, several researchers have, in recent decades, elaborated on the work of Hunter and his contemporaries to develop both automated or manually operated delayed response tasks for use with non-human primates.[17,18] In these paradigms, monkeys were presented with an initial stimulus, a brief illumination of one of nine visible panels or placement of food reward behind one of two panels, and required after variable durations of delay to recall which panel was illuminated or behind which panel food reward was placed. Age- and delay-dependent decrements in performance have been well documented using both of these paradigms,[19,20] with longer delay intervals and older age being associated with impaired recall.

The products of selective attention and distractibility were assessed in these studies by exposing animals to additional stimuli *during* delay intervals, the time at which selective attention and consolidation operate in concert to produce working memory. Thus, presentation of stimuli during delays may best be characterized as disrupting formation of the product of selective attention (i.e., the interaction of these two cognitive processes). It is this product whose characteristics are grouped or consolidated to represent a single salient stimulus, the stimulus that is to be remembered (this process is illustrated in Figure 8.1).

In both paradigms described above, presentation of an irrelevant stimulus during delay intervals produced impairment of subsequent recall in some groups of animals. Several very significant, specific results of this work merit attention. Bartus and Dean[17] compared young and aged monkeys using a distracting stimulus that was presented for the duration of delay intervals. Their data demonstrated that young

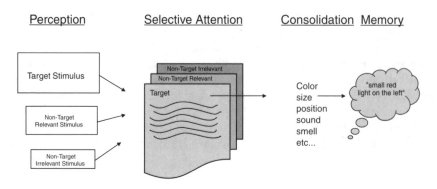

FIGURE 8.1

Schematic representation of working memory formation. At each stage of the process, cognitive resources are allotted to attend to both relevant and non-relevant extraneous stimuli and target stimulus encoding can be diminished by increasing the salience of these non-target stimuli.

monkeys were resistant to the distracting effects of the stimulus. However, delayed recall in aged monkeys was markedly impaired by visual interference during delay intervals. A second key aspect of the methodologies described above is the salience of the stimuli presented during delay intervals. The stimuli in each of these paradigms can be characterized as irrelevant or task-irrelevant in relation to the target stimulus to be remembered. The issue, then, is recognition by the subjects of the salience or lack thereof of the stimulus they are exposed to during variable delay intervals, an issue which is likely to impact the severity of recall impairment. More specifically, in the position memory paradigm using lit panels, the distracting stimulus was a random illumination of all of the panels by a light of a different hue than the target light for the *duration of the delay*. Thus, monkeys were not exposed to stimuli during delays that may be perceived as potentially matching the target sample.[17] Similarly, in the food reward paradigm, animals were exposed to stimuli during delays that were semantically and spatially distinct from the food reward placement.[18]

The latter study included an examination of the temporal relationship between distracting stimuli and cognitive processing during delay intervals. These authors assessed the relative distracting effects of the same stimuli presented either immediately after presentation of the target stimulus (start of delay interval), during the middle of the delay interval, or immediately before presentation of a response opportunity (end of delay interval). Using the task-irrelevant distracting stimuli in these studies, temporal position of the stimulus during the delay interval did not alter the distracting effect of the stimulus. This stands in contrast to what would be predicted based on the hypothesis that distractibility involves disruption of the transition from selective attention to consolidation, which would be reflected in distractors placed early in the delay interval producing markedly greater impairment of recall than those placed elsewhere temporally. However, only cognitively impaired aged animals were employed in this study and it remains to be seen if younger animals respond similarly.

III. Delayed Matching-to-Sample (DMTS): Task-Relevant Stimuli

The DMTS model employed in our laboratory is an adaptation and elaboration of that developed by Bartus and colleagues.[21] Regarding the mnemonic functions that may be assessed using this model, some uncertainty likely exists. Clearly, standard DMTS or delayed matching-to-location tasks require the function of working memory, defined as retention of information needed to complete a given task at a given time. A large body of work conducted by Goldman-Rakic and others has identified many of the anatomical and neurochemical underpinnings of working memory[12-14] and performance on these tasks is, indeed, altered by manipulation of these systems. However, the distinction between working and other complex forms of memory in completing these tasks is unclear. The complexity of most delayed recall task rules, as well as the extensive baseline training regimen in both rodents[22] and non-human primates[19,23] implies the involvement of reference memory in completion of each individual task. This is reflected in the requirement that, to complete a given trial, animals must retain information concerning the format of the task for periods encompassing years. Thus, this suggests then that the tasks described in this section may be characterized as more complex mixed-memory tasks.[24]

A. DMTS Paradigm

Animal Housing and Training — All monkeys (*macaca mulatta* and *macaca nemestrima*) used in our studies are housed at the Medical College of Georgia Animal Behavior Center in stainless steel cages composed of multiple 130 by 75 by 75 cm units. Toys and foraging tubes are provided and all monkeys are allowed to observe television programs each afternoon after testing to promote psychological well-being. During a test week, monkeys are maintained on a feeding schedule that allows approximately 15% of their normal daily food intake to be derived from banana-flavored reinforcement pellets awarded for correct responses. Testing is typically conducted on five days of each week. Standard laboratory monkey chow, fresh fruits, and vegetables compose the remainder of an animal's daily food intake following testing each day. Water is available *ad libitum*.

It is important to note that even with the continuous daily testing that monkeys are subjected to over the course of several months or years, no reinforcer devaluation has been observed in our animals. Previous animal studies have documented reinforcer devaluation in rodents performing various instrumental tasks following over-training,[25] however, it is likely that the dietary modification employed with our monkeys aids in maintaining reinforcer salience.

For initial training in use of the test apparatus and to develop competence in performing the delayed matching task, all monkeys are gradually presented with a program of acclimation, habituation, and basic rule learning in the test environment, their home cages. Other investigators have suggested that behavioral testing of non-human primates in home cages may prove distracting.[18] Indeed, others typically perform delayed recall testing in non-human primates only following removal from the home cage, in a dedicated test chamber. However, we have not observed evidence that animals fail to attend to test panels. Further, monkeys can be trained to a very high degree of accuracy on shorter delay trials and recall accuracy (following different delays) across months and even years is quite stable. Perhaps most significant in this regard are our observations that latencies to respond to sample and choice stimuli are stable and typically less than 3 to 4 seconds in all animals.

Test panels were attached to the front of home cages such that animals had free access to the entire face plate of the panel. After initially placing panels on the front of home cages, animals were trained by shaping to approach the panel for a banana pellet reward delivered by a remotely operated feeder bin attached to the panel exterior. Following habituation to the panel and mechanical feeding apparatus, which typically requires several short sessions over the course of several days, monkeys were exposed to the first of several fully automated training programs incorporating the test stimuli. Stimuli used during all tasks were 2.54 cm diameter colored disks (red, yellow, or green) presented by light-emitting diodes located behind clear plastic push-keys positioned in a pyramidal shape on the face plate of the test panel. We have found the duration of training with each program to vary markedly between animals and some acceleration or reversion between programs may be necessary for individual animals. For each of the initial training programs described, no delay intervals between sample and choice push-key illumination are included. At each stage of training, it is essential to monitor patterns of responding for the presence of several different forms of strategic or reflexive behaviors that may be elicited by animals, particularly during transition to more demanding cognitive tasks. During the early stages of training, prior to attaining competence in the DMTS paradigm, we have observed limited numbers of monkeys, particularly aged monkeys, to employ strategies of side or color perseveration, or perseverative behavior with regard to sample key pressing. In many instances, the behavior appears to be a trial and error attempt to apply a basic, assumed rule to the task to obtain reward (i.e., red = reward). However, we have found that with counterbalancing for all side and color combinations for each delay interval, these forms of experimentation by the monkey can be quickly extinguished. Representative training programs and the order in which they may be employed are described below.

Matching with Delays — Completion of each of the training programs above, in succession, is followed by gradual, titrated imposition of variable delay intervals. Four possible delay intervals between a monkey's response to the sample light and the presentation of the two choice lights are employed: Zero seconds delay or a Short, Medium, and Long delays. The duration of these delay intervals should be gradually titrated to ensure continued reinforcement at a level above chance levels for delays other than long delays. Short, medium, and long delay intervals are individually adjusted to produce stable performance levels approximating the following levels of accuracy: short (75 to 85% correct); medium (65 to 75% correct); and long (55 to 65% correct). Monkeys performance for zero seconds delay trials typically averages 85 to 100% correct. Monkeys complete 96 trials on each day of

Training program	Criterion for completion
A. Stimulus push-key (top) is illuminated by one of the three colored lights in random order. Animals receive a food reward for depression of the stimulus push-key after illumination and another trial begins.	100 presses of sample key for reward each day (during one hour) for four consecutive days
B. Stimulus push-key is illuminated. After depressing this key, no reward is given and the top light remains illuminated. One of two bottom push-keys is illuminated with the matching color. The other is not illuminated. Monkeys are rewarded for depressing; the error results in the start of a new trial.	80% stable accuracy in depressing correct, illuminated choice key for reward over 1 to 2 weeks *Color of sample is constant until monkey reaches 80% accuracy over several days. One of the other three colors is then substituted using the same criterion (*color titration*)
C. *Simultaneous Matching-to-Sample*: Same as Program B, with both choice lights illuminated. Monkey is rewarded for depressing the matching key. Press of the non-matching key extinguishes the trial and a new trial begins.	80% stable accuracy over 3 to 4 weeks *color titration
D. *Simultaneous Matching-to-Sample with Correction*: Same as above with the exception that incorrect choices result in presentation of the same color problem until it is completed correctly. Colors are varied for each trial.	80% stable accuracy over 3 to 4 weeks
E. *Matching-to-Sample without Delay*: Sample key is extinguished after pressed and choice keys are both lit. Colors are varied for each trial.	85% stable accuracy over 3 to 4 weeks

testing with trials of each delay interval (including zero delay) presented an equal number of times. Typical delay intervals for aged and non-aged macaques, and their accuracy of recall following these intervals are illustrated in Figure 8.2.

The progression of a standard DMTS trial is as follows:

1. A trial begins with the illumination of the sample key by one of the colored disks. The sample light remained lit until the sample key was depressed by an animal, initiating one of four pre-programmed delay intervals, during which no disks were illuminated (i.e. no distractors present).

2. Following the delay interval, the two choice lights located below the sample key are illuminated. One of the choice lights matches the color of the sample light. These disks remain illuminated until a monkey presses one of the two lit keys.

3. Key-presses of choice stimuli that matched the color of the sample stimulus are rewarded by a 300 mg banana-flavored pellet. Non-matching choices are neither rewarded nor punished. A new trial is initiated 5 seconds after the second key-press on a preceding trial.

As may be seen in Figure 8.2, use of this method yields age-dependent differences in not only baseline DMTS performance (i.e., recall without delays), but also in the ability to accurately recall target stimuli after increasingly longer delay periods.

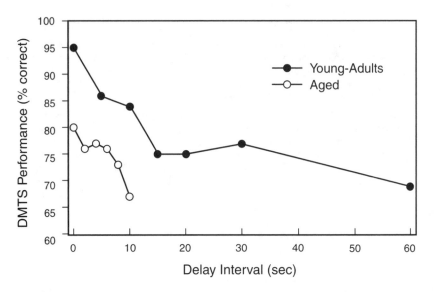

FIGURE 8.2
Representative durations of delay intervals and accuracy of recall following these intervals in aged and young-adult macaques. Aged, relative to younger, animals are impaired in accuracy of both immediate and delayed recall.

As would be predicted, aged monkeys are markedly impaired in both aspects of memory in this paradigm. Using this paradigm, we have demonstrated the beneficial effects of many different pharmacologic agents including acetylcholinesterase inhibitors used in the treatment of Alzheimer's Disease;[26] adrenergic agents;[27] and agonists of CNS nicotinic acetylcholine receptors.[28,29] While selective attention clearly functions in the performance of this DMTS paradigm and contributes to successful completion of the task, this model was not specifically designed to allow for a specific empirical assessment of attention and/or distractibility. However, the adaptability of this and similar models of delayed recall make them ideally suited for the modifications described below.

B.   DMTS with Distractor Stimuli

The methodology described here was developed by the author and colleagues at the Medical College of Georgia Alzheimer's Research Center (MCGARC) and represents a modification of the automated version of the DMTS task developed at the MCGARC.[23,30] A primary aim of this methodology was to produce impairment of the consolidation process or the transition from selective attention to consolidation in many instances, rather than the process of initial visual attention. This is reflected in the placement of all distractor stimuli discussed below *after* animals had attended to and depressed the sample stimulus key. A modification of this paradigm could be made to more selectively examine selective attention by presenting animals with both target stimuli and distracting stimuli concurrently.

The distraction paradigms discussed above[17,18] employed what may be characterized as task-irrelevant distracting stimuli. This is based on the observation that, with each of these methods, the stimuli presented during delay intervals were distinct from those presented initially (the target stimuli). In contrast, the method described here was designed to examine effects of exposure to a task-relevant distracting stimulus (described below) on delayed recall. The use of a task-relevant stimuli in this model provides an assessment of the animal's ability to discern when a familiar stimulus is and is not salient to completion of the immediate task, a process that we theorized to be of greater complexity than assessing a task-irrelevant distractor.

Test sessions with different distractor paradigms (termed interference sessions[18]) were conducted twice each week with a minimum of three days of standard DMTS testing conducted between each session. A schematic representation of a trial completed during interference sessions is shown in Figure 8.3. On 18 of the 96 trials completed during test sessions, distractor stimuli were presented during delay intervals. The stimuli were presented simultaneously on the sample and choice keys for one or three seconds and consist of a random pattern of the three colored lights flashing in an alternating manner. The distractor lights were comprised of the same three colors used for sample and choice stimuli presentation. The total duration of illumination for a given colored light was 0.33 seconds. Immediately, as one colored light was extinguished, a different colored light was presented in random order. Each color is presented in random order on each key once during a one second distractor and three times during a three second distractor. Distractor stimuli were present an equal number of times on trials with short, medium, and long delay intervals. The remaining trials were completed with no delay interval or distractor and were randomly placed throughout the test session. The onset of the distractor stimuli described here (either one or three seconds in duration) began one second after depression of the sample key. We have previously used a variation of this, placing the distractor stimuli one or three seconds prior to the end of each delay interval.[30] Though many age-dependent effects of distractor exposure were observed, the experimenter is not always assured that animals will attend to test panels after several seconds or minutes of a delay interval. Predictably, we have found that use of late-onset distractors produced marked variability of responding.

IV. Age- and Time-Dependent Effects of Visual Distractor Presentation on Delayed Recall in Aged and Young Macaques (*macaca mulatta* and *macaca nemestrima*)

Data illustrated in Figure 8.4 demonstrate the effects of either a one or three second visual distractor presented immediately after depression of the sample key on delayed recall accuracy in six aged *macaca mulatta* (>21.5 yrs). The mean duration of short, medium, and long delay intervals in these animals were 4.38 ± 0.71, 12.25 ± 4.18, and 23.38 ± 8.58 seconds, respectively. Exposure to the visual distractor lasting one

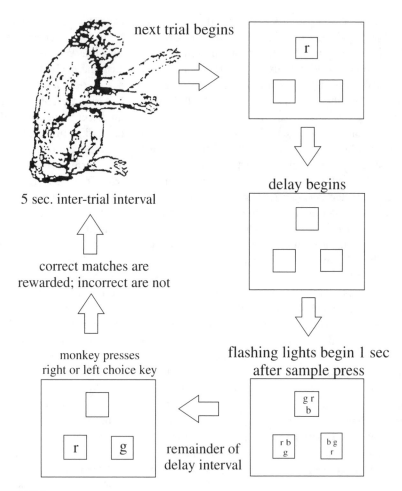

FIGURE 8.3
Schematic representation of a monkey performing an automated DMTS trial during which distractors are presented at the beginning of a given delay interval.

second markedly impaired accuracy on trials with the shortest delay intervals. Performance on trials with longer delay intervals was not impaired by this distractor stimulus. It is interesting to note the accuracy of recall on trials that did not include exposure to the distractor stimulus. Though these trials did not include the visual distractor, they were completed during the same 96 trial sessions and, temporally, were very close to those which did. As can be seen in the left panel of Figure 8.4, accuracy of recall on these non-distractor trials was impaired following the shortest delay interval. However, it is critical to emphasize that short delay non-distractor trials were interspersed randomly throughout the test sessions, as were all delay intervals, and were not presented immediately after short delays with distractors. In most instances, medium or long delay trials (with or without distractors) were presented immediately after short delay trials with distractors. Thus, impairment of non-distractor trials with short delays does not readily imply the presence of a

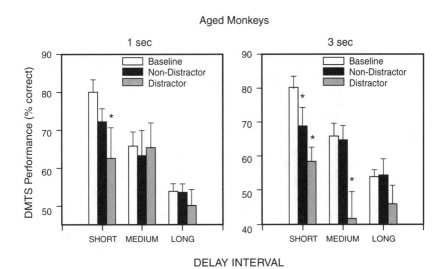

FIGURE 8.4
DMTS accuracy on distractor and non-distractor trials in aged rhesus monkeys exposed to a one or three second distractor at the beginning of different delay intervals. * = P < 0.05 vs. baseline performance in DMTS.

long-lasting proactive interference, as one would predict impairments on non-distractor trials of longer durations. Rather, it implies the presence of a brief proactive interference that, with longer delay intervals, decays in salience.

Presentation of the longer duration distractor (three seconds) produced an even greater impairment of delayed recall accuracy, extending to include trials with medium delay intervals. As with the briefer distractor stimulus, accuracy of recall on non-distractor trials (short delay) was markedly reduced. The relative degree of impairment produced by this distractor is predictable given that short delay durations averaged 4.38 seconds; thus, the visual distractor encompassed nearly the entire average delay interval. It is of interest that accuracy on medium delay trials (averaging 12.25 seconds in duration) was also impaired by exposure to this longer distracting stimulus. Clearly, then, these data demonstrate a time-dependent disruption of the early phases of stimulus characteristic consolidation and short-term memory formation.

Young-adult macaques (both *m. mulatta* and *m. nemestrima*; 10 to 13 yrs) displayed a resistance to the deleterious effects of distractor exposure, relative to aged monkeys. A one second distractor, presented immediately after sample key press, did not markedly reduce recall accuracy after short or longer delay intervals (Figure 8.5). However, presentation of a distractor stimulus lasting three seconds did produce a significant reduction in accuracy of recall on trials with short delay intervals. Decrements of greater than 20% in accuracy were observed in the short delay recall of all monkeys. However, trials with longer delay intervals were unaffected by presentation of the distractor stimulus. Further, recall accuracy on non-distractor trials with short delays, completed during the same 96 trial test session,

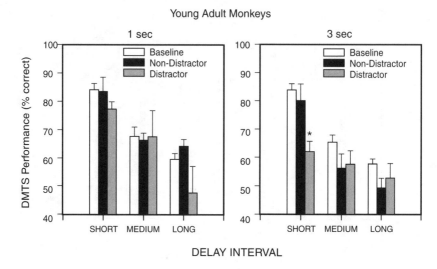

FIGURE 8.5

DMTS accuracy on distractor and non-distractor trials in young-adult rhesus and pigtail macaques exposed to a one or three second distractor at the beginning of different delay intervals. * = P < 0.05 vs. baseline performance in DMTS.

was unaffected. This stands in contrast to the effects of distractor exposure on aged animals, which demonstrated evidence of brief proactive interference. These data demonstrate the widely documented resistance of younger non-human primates to recall impairment by extraneous stimuli, either task-relevant or irrelevant, compared to the aged monkeys. Further, younger animals were not subject to a putative effect of proactive interference or expectancy.

V. Effects of Methylphenidate Administration on Distractibility in Young-Adult Monkeys

A second series of studies was conducted to examine the sensitivity of our distractibility paradigm to potentially beneficial effects of methylphenidate, the primary pharmacologic treatment option for disorders of attention in both children[7] and young adults.[11] We have previously reported on the lack of benefit provided by methylphenidate to aged monkeys exposed to the distractor stimulus, negative findings which are similar to those observed by others in aged humans with AD.[31]

Eight young-adult macaques, those described above, were administered saline or methylphenidate (0.125–1.0 mg/kg) in the gastrocnemius muscle in a volume of 0.035 ml/kg 10 min prior to distractor testing, using a three second early-onset distractor as described above. The dose order was chosen randomly for each animal and a minimum of five days was alotted between administration of successive doses of drug. As described above, accuracy of recall on all trials/delay with distractors was compared to mean performance during a test session when no distractors were

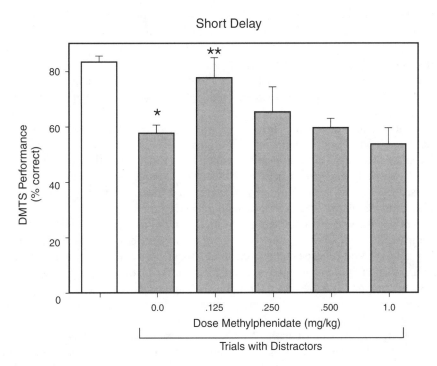

FIGURE 8.6
Acute methylphenidate administration (0.125 mg/kg) in young-adult macaques significantly attenuates
the impairment of recall on short delay trials produced by exposure to a distractor lasting three seconds.
* = P < 0.05 vs. baseline performance in DMTS (open bar at left). ** = P < 0.05 vs. distractor.

present. Figure 8.6 illustrates the beneficial effects of acute methylphenidate exposure on distractor-impaired recall accuracy following trials with short delays. Exposure to the distractor produced greater than a 20% reduction in short delay DMTS
performance, to near-chance levels. Methylphenidate administration significantly
attenuated the recall impairment produced by distractor exposure. No effects of
methylphenidate on other performance for other delay intervals or on latencies to
respond to sample or choice stimuli were noted.

VI. Discussion and Interpretation of DMTS Performance

Habituation to Distractors — An additional aspect of these findings that warrants
consideration is the potential for development of habituation to distracting stimuli.
Arnsten and Contant[18] assessed delayed recall with and without distracting stimuli
and did observe marked distractibility in their sample of aged monkeys, an effect that
extended to non-distractor trials. However, they reported that all animals habituated
to the distracting stimulus employed. They attempted to address this consideration

by alternating between two types of task-irrelevant distracting stimuli, a brief illumination of a light and a brief auditory stimulus, when habituation to one stimulus
was observed. However, it is difficult to assess the comparability of salience of these
distractors, given their distinct properties. Further, it is likely that animals would
habituate to a second, novel distracting stimulus with repeated exposures. These
findings underscore the importance of regularly monitoring for potential devaluation
of the distractor.

Employing a distracting stimulus we characterize as task-relevant, we have not
observed evidence of habituation to the distractor in either aged or young-adult
macaques during the course of several months of testing. While the exact reason for
this is not fully understood, it is likely that the continued salience and disruption
caused by the distracting stimulus over repeated exposures is a function of the
distractor's similarity or relevance to the given DMTS trial at hand. More specifically,
it is our contention that employing a distracting stimulus that may be perceived by
subjects as potentially relevant to obtaining reinforcement (i.e., selecting a choice
key color) should be resistant to habituation. Despite this difference in susceptibility
to habituation, it seems likely that examination of the distracting properties of both
task-relevant and task-irrelevant stimuli are of significant relevance to the applica-
bility of these behavioral paradigms in understanding cognitive pathology such as
that observed in patients with ADD or AD.

Matching or Rule Learning? — Many behavioral researchers have, for years,
questioned the premise that most species of non-human primates are capable of
learning the general, complex rule of matching-to-sameness.[32] Rather, it is possible
that monkeys, macaques in this instance, learn a less complex, specific rule of
choosing the same color. Rather than mere curiosity of methodology, the distinction
between learning the complex sameness rule or learning a specific, unimodal rule
such as match-to-color is of great significance in understanding the complexity of
cognitive function in non-human primates and in relevance of this function in
modeling the human condition. This is perhaps most significant in regard to use of
non-human primates to model human cognitive pathology in the course of therapeutics discovery.

From a methodological standpoint, the relative degree of general or specific rule
learning observed in the monkey can be assessed by employing techniques that
assess transfer of training. In the context of delayed matching-to-sample, transfer
of training can be assessed quite readily by training animals to proficiency in a
paradigm such as that described above and then by switching the sample and choice
stimuli characteristics. For example, monkeys can be trained initially using colored
stimuli as above, and then exposed to the same general paradigm wherein the
different colors are replaced by different shapes. If the less complex specific rule of
matching-to-color is indeed the rule governing an animal's performance, then accuracy of recall with no or even brief delays will be near chance levels. However,
evidence of a more complex general rule learning would be seen if animals perform
significantly above chance levels of performance with the novel stimuli. It is likely
that some decrement of performance would be seen even if a general rule of sameness
is learned as the novel stimuli may prove distracting. Once familiarity with the new

stimuli is achieved, immediate and delayed recall accuracy should rapidly return to levels of accuracy observed with the initial stimuli.

Clearly, there are distinct differences between lower monkeys' ability to learn the more complex general rule. Chimpanzees (*pan troglodytes*) readily transfer general rule learning, demonstrating transfer of training in matching-to-sample paradigms of greater than 80%.[33] This was observed using complex, three-dimensional training stimuli, a factor that may have increased the salience of sample and choice stimuli. In contrast, some have suggested that long-tailed macaques (*m. fascicularis*) are incapable of demonstrating transfer using highly complex training stimuli, human facial expression.[34] In contrast, several researchers have provided evidence that rhesus monkeys (*m. mulatta*) readily transfer delayed matching-to-sample training using relatively simple stimuli not only within sensory modalities, but between modalities (i.e., visual stimuli-tactile stimuli).[35] Further, rhesus monkeys have demonstrated evidence of transfer of training from a manually operated delayed recall task to a video-generated recall task.[36] As a whole, these and other findings do suggest that macaques are capable of the complex, general rule learning necessary to assess sameness relatively independent of specific stimulus characteristics, though greater stimulus complexity may make transfer more difficult in macaques.

References

1. Treisman, A. M., Riley, J. G. A., Is selective attention perception or selective response? A further test, *J. Exp. Psych.*, 79, 27, 1969.
2. Cherry, E. C., Some experiments upon the recognition of speech, with one and with two ears, *J. Acoust. Soc. Am.*, 32, 884, 1960.
3. Davis, H. P., Cohen, A., Gandy, M., Colombo, P., VanDusseldorp, G., Simokle, N., Romano, J., Lexical priming deficits as a function of age, *Behav. Neurosci.*, 104, 288, 1990.
4. Hoyer, W. J., Rebok, G. W., Marx Sved, S., Effects of varying irrelevant information on adult age differences in problem solving, *J. Gerontol.*, 34, 553, 1992.
5. Broks, P., Preston, C., Traub, P., Poppleton, P., Ward, C., Stahl, S. M., Modeling dementia: effects of scopolamine on memory and attention, *Neuropsychologia*, 28, 685, 1988.
6. Jones, G. M. M., Sahakian, B. J., Levy, R., Warburton, D. M., Gray, J. A., Effects of acute subcutaneous nicotine on attention, information processing and short-term memory in Alzheimer's disease, *Psychopharmacology*, 108, 485, 1992.
7. Elia, J., Drug treatment for hyperactive children. Therapeutic guidelines, *Drugs*, 46, 863, 1993.
8. Barkley, R. A., Attention-deficit hyperactivity disorder, *Sci. Am.*, 279, 66, 1998.
9. Wender, E. H., Attention-deficit hyperactivity disorders in adolescence, *J. Dev. Behav. Pediatr.*, 16, 192, 1995.
10. Greenhill, L. L., Pharmacologic treatment of attention deficit hyperactivity disorder, *Pediatr. Pharmacol.*, 15, 1, 1992.

11. Spencer, T., Wilens, T., Biderman, J., Faraone, S. V., Ablon, J. S., Lapey, K., A double-blind, crossover comparison of methylphenidate and placebo in adults with childhood-onset attention-deficit hyperactivity disorder, *Arch. Gen. Psychiatry*, 52, 434, 1995.

12. Goldman-Rakic, P. S., Brown, R. M., Regional changes in monoamines in cerebral cortex and subcortical structures of aging rhesus monkeys, *Neuroscience*, 6,177, 1981.

13. Goldman, P. S., Rosvold, H. E., Localization of function within the dorsoloateral prefrontal cortex of the rhesus monkey, *Exp. Neurol.*, 27, 291, 1970.

14. Goldman-Rakic, P. S., Circuitry of the primate prefrontal cortex and the regulation of behavior by representational memory, in *Handbook of Physiology: the nervous systems, higher functions of the brain*, Plum, F., Ed., American Physiological Society, Bethesda, 1987, 373.

15. Nordberg, A., Svensson, A. L., Cholinesterase inhibitors in the treatment of Alzheimer's disease: a comparison of tolerability and pharmacology, *Drug Saf.*, 19, 465, 1998.

16. Hunter, W. S., The delayed reaction in animals and children, *Behav. Mon.*, 2, 6, 1913.

17. Bartus, R. T., Dean, R. L., Recent memory in aged non-human primates: hypersensitivity to visual interference during retention, *Exp. Aging Res.*, 5, 385, 1979.

18. Arnsten, A. F. T., Contant, T. A., Alpha-2 adrenergic agonists decrease distractibility in aged monkeys performing the delayed response task, *Psychopharmacology*, 108, 159, 1992.

19. Bartus, R. T., Flemming, D., Johnson, H. R., Aging in the rhesus monkey: effects on short-term memory, *J. Ger.*, 33, 858, 1978.

20. Arnsten, A. F. T., Goldman-Rakic, P. S., Catecholamines and cognitive decline in aged nonhuman primates, *Ann. NY Acad. Sci.*, 444, 218, 1984.

21. Bartus, R. T., Johnson, H. R., Short-term memory in the rhesus monkey: disruption from the anticholinergic scopolamine, *Pharm. Biochem. Behav.*, 5, 31, 1976.

22. Terry, A. V. Jr., Buccafusco, J. J., Decker, M. A., Cholinergic channel activator, ABT-418, enhances delayed-response accuracy in rats, *Drug Dev. Res.*, 41, 304, 1997.

23. Buccafusco, J. J., Jackson, W. J., Beneficial effects of nicotine administered prior to a delayed matching-to-sample task in young and aged monkeys, *Neurobio. Aging*, 12, 233, 1980.

24. Prendergast, M. A., Terry, A. V. Jr., Buccafusco, J. J., Effects of chronic, low-level organophosphate exposure on delayed recall, discrimination, and spatial learning in monkeys and rats, *Neurotox. Teratol.*, 20, 115, 1998.

25. Adams, C. D., Variations in the sensitivity of instrumental responding to reinforcer devaluation, *Quart. J. Exp. Psych.*, 34B, 77, 1982.

26. Buccafusco, J. J., Prendergast, M. A., Terry, A. V. Jr., Jackson, W. J., Cognitive effects of nicotinic cholinergic receptor agonists in nonhuman primates, *Drug Dev. Res.*, 38, 196, 1996.

27. Terry, A. V. Jr., Jackson, W. J., Buccafusco, J. J., Effects of concomitant cholinergic and adrenergic stimulation on learning and memory performance by young and aged monkeys, *Cereb. Cortex*, 3, 304, 1993.

28. Buccafusco, J. J., Jackson, W. J., Terry, A. V. Jr., Marsh, K. C., Decker, M. W., Arneric, S. P., Improvement in performance of a delayed matching-to-sample task by monkeys following ABT-418: a novel cholinergic channel activator for memory enhancement, *Psychopharmacology*, 120, 256, 1995.

29. Prendergast, M. A., Terry, A. V., Jr., Jackson, W. J., Marsh, K. C., Decker, M. A., Arneric, S. P., Buccafusco, J. J., Improvement in accuracy of delayed recall in aged and non-aged mature monkeys after intramuscular or transdermal administration of the CNS nicotinic receptor agonist ABT-418, *Psychopharmacology*, 130, 276, 1997.

30. Prendergast, M. A., Jackson, W. J., Terry, A. V. Jr., Kille, N. J., Arneric, S. P., Decker, M. A., Buccafusco, J. J., Age-related differences in distractibilty and response to methylphenidate in monkeys, *Cereb. Cortex*, 8, 164, 1998.

31. Goodnick, P., Gershon, S., Chemotherapy of cognitive disorders in geriatric subjects, *J. Clin. Psychiat.*, 45, 196, 1984.

32. Hewett, T. D., Ettlinger, G., Cross-modal performance: the nature of the failure at "transfer" in non-human primates capable of "recognition," *Neuropsychologia*, 17, 511, 1979.

33. Oden, D. L., Thompson, R. K. R., Premack, D., Spontaneous transfer of matching by infant chimpanzees (Pan troglodytes), *J. Exp. Psych.*, 14, 140, 1988.

34. Kummer, H., Anzenberger, G., Hemelrijk, C. K., Hiding and perspective taking in long-tailed macaques (Macaca fascicularis), *J. Comp. Psychol.*, 110, 97, 1996.

35. DiMattia, B. V., Posley, K. A., Fuster, J. M., Crossmodal short-term memory of haptic and visual information, *Neuropsychologia*, 28, 17, 1990.

36. Washburn, D. A., Hopkins, W. D., Rumbaugh, D. M., Video-task assessment of learning and memory in macaques (Macaca mulatta): effects of stimulus movement on performance, *J. Exp. Psych. Anim. Behav. Process*, 15, 393, 1989.

Chapter

9

Inhibitory Avoidance Behavior and Memory Assessment

John H. Graham and Jerry J. Buccafusco

Contents

I.   Introduction

Avoidance paradigms require the subject to initiate a specific type of behavior, usually escape or avoidance, in order to preclude the administration of an aversive event, or negative reinforcer. Inhibitory avoidance, also frequently referred to as passive avoidance, requires that the subjects (typically rodents) to behave in a manner contrary to their normal inclination or predilection. In most cases, subjects must not act if they are to avoid the consequences of a negative reinforcer. During the training component of the task, the animal is punished for making an instinctive response, such as moving from a brightly lit chamber into a darkened one (a highly probable event for rodents). After a specified delay interval (from a few minutes to one or two days later), the animal is tested under the same circumstances, and the length of time the animal refrains from repeating the punished response is recorded. The latency to repeat the punished response serves as an index of memory. Presumably the greater the avoidance latency, the more efficacious the memory process. When multiple consecutive training trials are used (short inter-trial intervals), the number of trials required for the animal to achieve a predetermined latency represents a better estimate of task acquisition efficiency — an index of general cognition. Because inhibitory avoidance paradigms are technically relatively simple procedures, they are almost certainly the most widely used methods for cognition screening efforts in lower animals. They are routinely used to ascertain the effects of drug administration, brain lesions, strain differences, etc. on aspects of learning and memory.[1] Another appealing aspect of inhibitory avoidance paradigms is the rapidity with which they can be performed — usually only 2 to 3 days to train and test animals.

Certain potential drawbacks of the inhibitory avoidance paradigms include the often large inter-subject variability, and the differing tolerance to shock reported to occur among subjects (which is probably one reason for the overall subject response variability). Accordingly, relatively large numbers of subjects are often required when detecting modest changes in task performance. One other disadvantage of the paradigm (at least with respect to other types of delayed response tasks that use changeable discriminable stimuli) is that in inhibitory avoidance a subject can only be used one time, and he cannot serve as his own control. However, inhibitory avoidance is a classic paradigm that is well traveled. As of this writing, *Medline's* most recent database (1996–1999) listed over 600 articles with passive avoidance or inhibitory avoidance as a main subject. Thus, in addition to its relative longevity, the method continues to be used extensively.

This task may be said to provide a rather rudimentary estimate of cognitive function. There is a component of working memory during the acquisition phase of the task since the animal is presented with a novel environment when training begins. The multi-trial repeated acquisition method allows for some quantification of this cognitive component, and, therefore, it is particularly useful for measuring task acquisition.[2]

The single trial version of the task can be used to measure the effect of experimental manipulations on either acquisition, consolidation (establishing the memory trace) or recall components of memory. Such specificity is possible when the experimental manipulation is administered either before training, just after the training trial, or just prior to testing, respectively. For an accurate estimate of task consolidation, however, the experimental manipulation should have its CNS action within the first few minutes after the training trial.[2] Rapidly acting pharmacological agents are one of the few types of experimental manipulation that fit this criterion. The passive avoidance paradigm is useful for assessing memory decrements in response to certain surgical procedures and to many pharmacological and toxicological agents. In addition, the memory-enhancing properties attributed to CNS agents can be assessed in animals subjected to an experimentally induced amnesia (e.g., cytotoxic or mechanical lesions), and in animal strains that model specific disease states (e.g.,[2-6]). The argument has been made that this method assesses only learning and memory associated with emotionally salient events, for instance, those that are stressful or anxiogenic.[7-9] If this is in fact the case, the argument goes, then inhibitory avoidance behavior is not a broad spectrum test of memory, but a measure of only very specific types of memory. This chapter will not attempt to resolve this debate, but only point out the potential confounds of interpretation associated with this method. Despite their limitations, however, inhibitory avoidance paradigms have proven to be remarkably predictive, at least in terms of those experimental manipulations that have subsequently been tested in other, perhaps more sophisticated behavioral measures of cognition, and in higher species.

II. Methods

As mentioned above, variability is rather large with this procedure, so that 12 to 20 or more subjects per experimental group is typical for these types of studies.[2-6,11,12] Of the approaches described below, the multi-trial repeated acquisition method would require fewer subjects per group. Anecdotal reports indicate that mice are more difficult to use in this procedure than rats, and typically, studies using mice have used more subjects/group.[13,14] It is important in this type of study that animals are naïve to the test apparatus, until training is to begin. Therefore, miscues in training or testing usually require that the subject be removed from the experiment.

A.   Animal Subjects

Most any standard laboratory rat may be used in this procedure. The only requirements are that the subjects are freely mobile, and that they can easily fit through the door to the next chamber (see below). The animals must also exhibit some response to footshock. For example, drug regimens or other interventions that cause significant analgesia or antinociception in test subjects may render them incapable of performing the task, or in the least, may make the data difficult to interpret.

B.   Equipment

1. **Shuttle cage with retractable door** (e.g., Coulbourn Instruments [Lehigh Valley, PA], — E10-16PA with E10-16D); approximately 16.5 x 35.5 x 20 in. (see Figure 9.1). Typically, the dual compartment shuttle cage is mounted on a rocker that tilts in response to the animal's weight. As the subject passes from one side to the other the cage tilts, and this activates a trigger which closes the door (prevents escape back to the safe side) and activates the shock. One compartment (the safe side) is fitted with a bright house light; the other is covered (black cardboard or other opaque material).

2. **Shocking grid**: The shocking grid forms the floor of the shuttle cage, but only the floor under the dark (unsafe) side is wired to receive the shock.

3. **House light**: light (safe) side.

Passive Avoidance

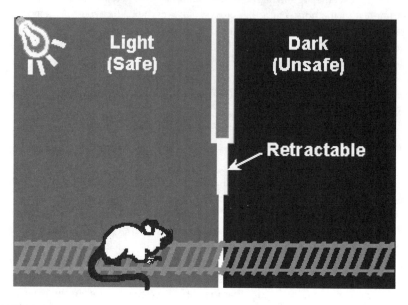

FIGURE 9.1
Schematic representation of a typical shuttle box inhibitory avoidance task.

4. **Shocker and scrambler**: activated by a trigger on the dark side.

5. **Automated triggered timer** (optional) or hand-held stopwatch.

6. **Retractable magnetic door**: divides the cage into two equal compartments and is trigger-activated or manually operated (if building your own chamber, make sure the door has about a 1 cm clearance when closed; this prevents the animal's tail from being caught or pinched.

C. Procedure: Single Trial Retention Method

Each subject is placed in a holding cage and is allowed at least 15 min to adapt to the darkened testing room each day before initiating the trial. After this acclimation period the subject may receive any experimental regimen. During the interval between receiving the experimental regimen and beginning the trial the subject should remain in the holding cage. To begin the test, the subject is placed into the "safe" compartment, which remains darkened for a specified acclimation time. Acclimation periods of 20 sec to 5 min have been used, but times from 1 to 3 min are considered optimum. During this time the retractable door is closed. At the completion of this period the compartment light is illuminated and the door is retracted. Measurement of the training latency (nearest 0.1 sec) also begins at this point.

Note: *Most normal rats will cross to the dark compartment within 30 sec, but certainly no longer than 60 after the light is illuminated and the door retracted. Measurement of training latencies has not always been reported; however, this measure provides some indication of the subject's general ambulatory activity and exploratory motivation. This information may aid interpretation later on.*

After the subject crosses to the dark side (for manual methods, some anatomical criteria should be used to establish a complete crossing, such as hind limb clearance or a point marked on the tail as it clears the door). When the animal crosses over to the dark (unsafe) side, the door is lowered and an inescapable footshock is delivered through the grid floor.

Note: *A scrambled footshock of 0.5 to 1.0 mA intensity, and 2 to 5 sec duration should be delivered.[2-6,15,16] What works best will depend on the strain and size of the subjects. Also, shock intensity and duration can be manipulated according to the particular experiment. For example, when testing potential amnestic agents, it is often desirable to ensure a high probability (higher intensity shocks) of control subjects avoiding those used during the testing session. In contrast, when testing potential memory-enhancing agents, lower levels of shock may be used so that a much lower percentage of controls avoid during testing. An alternative (or concomitant) procedure that reduces avoidance probability during testing is to increase the time interval between training and testing (i.e., 24 to 48 hr). Test runs prior to the actual experimental procedure should allow for the determination of optimal shock intensity and duration, and training-testing interval, but once established, the settings should be consistent for all subjects.*

Caution: *The footshock should be aversive, but not incapacitating or damaging. The shock parameters should not be overly stressful for animals. A good evaluation is to place your hand on the grid and deliver the shock. You should feel no more than a mild tingling. This procedure should be performed on a daily basis (and be sure to mention this in your methods section when you publish, etc.). Finally, clean the grid regularly to avoid variations in conductance, which often alters the shock intensity.*

After the shock is delivered, immediately remove the animal and return him to his home cage. The retention interval is the time interval between the training trial and the testing trial. Typically, a retention interval of 24 hr or 48 hr is used. However, longer intervals may be appropriate, particularly when attempting to enhance forgetting e.g., in order to better test a memory-enhancing regimen. In general, it is considered inappropriate to test an animal more than once. Since learning occurs in every session, it is difficult to make inferences about subject performance on subsequent testing sessions, since the first testing session informed the animal that shock was no longer available.[12] After the completion of the retention interval the subject is returned to the testing room and treated in the same manner as during the training trial except that no experimental regimen is delivered.

Note: *As indicated above, the inhibitory avoidance task may be used to determine the effect of an acute experimental manipulation on three aspects of memory: attention, consolidation, and retrieval, depending upon when the experimental manipulation is administered during the phases of the task. Thus, there may be times when a regimen may be administered after the retention interval and prior to the testing trial. In such cases the regimen should be given in the holding cage as described above. In some cases, the inhibitory avoidance task may be used to determine whether a particular regimen causes state-dependent learning.[2] It might be necessary to determine whether the presence of some drug- or regimen-induced physical or mental state induced during the training trial might have to be repeated during the testing trial to allow for some alteration in task performance. For example, it has been established that individuals who smoke tobacco products while studying for an exam, when examined later, often perform the exam more accurately if they are allowed to smoke during the examination period than if they are not allowed to smoke. Thus, it may be necessary to compare the task performance levels of subjects administered a drug regimen and then placebo, vs. drug regimen and drug regimen prior to the training and testing trials, respectively (also see reference[2]).*

The testing trial is initiated by placing the subject in the safe compartment. The procedure is identical to that for the training trial except that when the subject crosses to the dark compartment, no footshock is delivered. This time the latency between light-on and door open to the animal entering the dark compartment is termed the Step-through latency (STL). Again the STL is measured to the nearest 0.1 sec. In most cases the paradigm includes a cut-off latency while the subject is in the safe compartment. We have used 300 sec, but other time periods may be used. Some authors have used a value of 3 times the average training latency. Once a subject reaches the cut-off latency it is considered that he has learned the task and the session is ended.

D.   Procedure: Multi-Trial Repeated Acquisition Method

The multi-trial method is used most often when the animals are subjected to some experimental manipulation that interferes with memory processes. This is because, with sufficient foot shock parameters, most normal rats will perform the inhibitory avoidance task successfully in the first session. However, for rats treated with 0.8 mg/kg of scopolamine prior to training in a standard passive avoidance task, it required 5 consecutive trials to approach normal STLs.[2] The acclimation period during the first trial of training is the same as that for the single trial method (e.g., 1 to 3 min.). For the remaining trials the acclimation period is usually reduced to 30 sec. Each trial is followed by an inter-trial interval of 5 min, during which time the subject is returned to his holding cage. During each consecutive trial the animal receives a foot shock until he either completely avoids (remains in the safe compartment until cutoff) or undergoes a predetermined maximal number of trials (usually 5). The maximum number of trials and the length of the inter-trial interval should be determined with a consideration of the duration of action of the drug or regimen being examined. For this paradigm it is best to require subjects to reach the maximal STL on two consecutive trials to ensure that the behavior has been learned.[2,12,15] This version of the task also may be performed with longer, more standard, inter-trial intervals, e.g., 24 hr or 48 hr. In this case, the drug or regimen would have to be re-administered after each inter-trial interval.

Note: *For those with a sufficient budget we recommend a new system that we recently purchased — San Diego Instruments' GEMINI Avoidance System. This is a computer-automated system that requires no experimenter intervention during testing. The equipment is expandable to allow up to 8 stations. The equipment includes the two-compartment enclosure with separate grid floors for rat or mouse studies. The system uses banks of photobeam detectors to follow the subject's ambulations, and to control the movement of compartment door and the status of the cue lamps. One disadvantage of this system is that the subject cannot be observed during trials.*

Other variations of the inhibitory avoidance task are replete in the literature. One example uses a narrow runway connected to the chamber containing a shock grid.[12,17] During testing, the subject is required to avoid the chamber by remaining on the runway. Also, warm and cool chambers have been used wherein cool air (unpleasant) is blown into the safe side. During training sessions the animal migrates to the warm chamber where it is subsequently met by the aversive stimulus (footshock).[18] In situations where electrical shocks cannot be used as the aversive stimulus, a brief air puff has been used effectively.

Finally, it is possible to alter the discriminability of the shock source by modifying an open grid floor so that a specific section of the floor could be either not discriminated from the safe floor, or variably discriminable from the safe floor according to visual differences and/or texture.[11]

E.   Procedure: Inhibitory Avoidance in Mice

The light vs. dark inhibitory avoidance procedure can be used for testing mice.[13,14] Shuttle cages and retractable doors are available for the mouse (Coulbourn Instruments E10-15PA, E10-15D), and the procedure can be run in the same manner as described above. Shock intensities as high as 4 mA and shock durations as long as 3 sec have been used,[14] but it would be appropriate to first screen for the effects of lower intensities and shorter durations. As mentioned above, this may necessitate larger subject groups than are required with rats, but both the single trial retention method and the multi-trial repeated acquisition method have been used successfully.

III.   Typical Applications

Inhibitory avoidance most commonly has been used to assess memory decrements or disruptions in animal models of CNS disease or trauma.[4-6,15,18] This paradigm also has been used frequently in those studies seeking to elucidate the neuroanatomical pathways or neural mechanisms responsible for short-term memory, memory consolidation, and memory recall.[2,3,10,15] Because of its utility in these various types of experiments, it also has proven to be valuable in studies examining the neuroprotective properties or the memory-enhancing potential of various pharmacological agents. Rats and mice can be assessed for strain-specific deficits in cognitive or nmemonic abilities that are related to known cellular or molecular deficits that have been previously isolated.[5,6,13]

IV.   Analysis and Interpretation

The results of inhibitory avoidance are typically subjected to non-parametric statistical tests because of the large variability in the data, and the fact that cut-off points are generally used. The Mann-Whitney U test is frequently used for the test-day STL, especially if a comparison of the means of two groups is the main interest.[2,12,14] When multiple groups are being compared, analysis of variance procedures, such as the Kruskal-Wallace test, are commonly employed for *post-hoc* analysis.[2,5,6,10,12,14] In addition to measurement of average STL, if experimental groups are large enough, it may be possible to compare the percentage of subjects having learned the task (avoided for the complete duration or cut-off period) after a particular trial. These population or frequency statistics can be applied to a Chi-squared procedure. This is particularly helpful for the multi-trial repeated acquisition version, wherein task acquisition is determined either by the percentage of subjects having learned the task, or by the mean number of trials required to learn the task.[10,12] Because of the non-parametric nature of the data, we also have expressed the STL in terms of the median STL, and have provided the range between the 25th and 75th percentiles as a representation of the error.[16]

As with all animal behavior paradigms, data derived from inhibitory avoidance tasks should be carefully and cautiously interpreted. When an animal avoids (fails to cross to the darkened chamber) during the testing trial, consideration should always be given to the possibility that this lack of behavior may occur for reasons other than the recall of the previously aversive experience. Certain drugs, disease states, or even strain effects may impair the subject's general locomotor activity, natural exploratory behavior, or motivation to perform the task. Alternatively, drugs or conditions that appear to produce an amnestic action (i.e., enhance the probability of crossing) may act simply to enhance exploratory activity, to reduce visual acuity, or to reduce fear. It may be necessary (and often is) to use one or more of the other tasks (such as the measurement of locomotor activity in an open field) described in this text to rule out these possibilities.

V.   Representative Data

The data presented in Figure 9.2 exemplify the versatility of the inhibitory avoidance task as it relates to evaluating components of memory. Scopolamine, a cholinergic muscarinic receptor antagonist and a classical amnestic agent, was administered to rats during various phases of the classical one-trial 24-hr foot shock avoidance task. Using the equipment and the methodology described above, we found that scopolamine was only effective as an apparent amnestic agent if the compound was administered prior to the training phase of the task. This suggested that the drug's action on memory was relegated primarily, if not exclusively, to the acquisition component of memory. Note that irrespective of the time of injection, training steps through latencies were similar for each group. This result indicated that the drug did not interfere with the subjects' ability to discriminate between the two environments, nor did it affect general ambulatory activity.

Effect of Scopolamine on Inhibitory Avoidance Performance

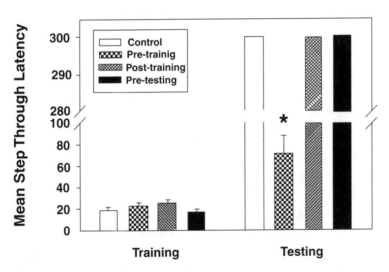

FIGURE 9.2
Effect of differential scopolamine (0.8 mg/kg, sc.) treatment on one-trial 24-hr passive avoidance performance. Each bar represents the mean ± S.E.M. * = p<0.01 vs. the other groups. Control rats (30) were administered saline (vehicle) 30 min prior to training. Pre-training rats (12) received scopolamine injection 30 min prior to training (acquisition). Post-training rats (9) received scopolamine immediately after training (retention). Pre-testing rats (14) received scopolamine 30 min prior to testing (recall).

The data presented in Figure 9.3 provide an example of an experiment that utilized a multi-trial version of the inhibitory avoidance task for which the inter-trial interval was 5 min. In this case rats received either vehicle injection, or ibotenic acid (glutamic acid receptor agonist) into the nucleus basalis two weeks before behavioral testing. The regimen used produces a rather selective lesion of the forebrain cholinergic system, and a decrease in cognitive function. Note that Trial 1

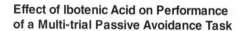

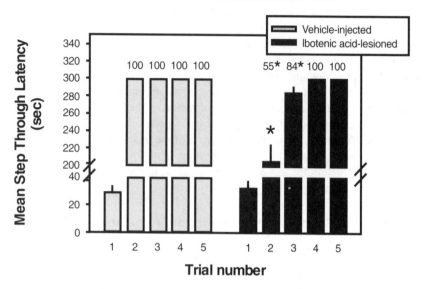

FIGURE 9.3

Effect of infusion of vehicle or ibotenic acid in rats on their performance of a multi-trial passive avoidance task. Rats received either 5 μl of vehicle (5 μM PBS, pH 7.5, n = 18), or 25 μg of ibotenic acid (n = 38) by direct bilateral injection into the nucleus basalis magnocellularis 2 weeks prior to behavioral testing. Each bar represents the mean ± S.E.M. The numbers above the bars refer to the % of subjects to have learned the task (reached cut-off). * = p<0.05 vs. vehicle-injected group (Mann-Whitney, χ^2 test). Note: Trial 1 essentially represents the training step through latency.

essentially represents a standard inhibitory avoidance training trial, and that the average step through latencies (obtained during training) were similar between the two groups. However, the ibotenic acid-lesioned group exhibited a reduced rate of task acquisition as indicated by the significant decrease in STL measured during Trial 2, and the decrease in the percentage of rats having learned the task (reached cut-off) during Trials 2 and 3. During Trial 4 all subjects avoided, and this was confirmed during Trial 5.

References

1. Porsolt, R. D., McArthur, R. A., and Lenègre, A., Psychotropic screening procedures, in *Methods in Behavioral Pharmacology*, van Haaran, H., Ed., Elsevier, New York, 1993, chap. 2.

2. Elrod, K. and Buccafusco, J. J., An evaluation of the mechanism of scopolamine-induced impairment in two passive avoidance protocols. *J. Pharmacol. Biochem. Behav.* 29, 15, 1988.

3. Elrod, K. and Buccafusco, J. J., Correlation of the amnestic effects of nicotinic antagonists with inhibition of regional brain acetylcholine synthesis in rats. *J. Pharmacol. Exp. Ther.* 258, 403, 1991.

4. Buccafusco, J. J., Heithold, D. L., and Chon, S. H., Long-term behavioral and learning abnormalities produced by the irreversible cholinesterase inhibitor soman: Effect of a standard pretreatment regimen and clonidine. *Toxicol. Let.* 52, 319, 1990.

5. Gattu, M., Pauly, J. R., Boss, K., Summers, J, B., and Buccafusco, J. J., Cognitive impairment in hypertensive rats: role of central nicotinic receptors. Part I. *Brain Res.* 771, 89, 1997.

6. Gattu, M., Terry, A. V. Jr., Pauly, J. R., and Buccafusco, J. J., Cognitive impairment in hypertensive rats: role of central nicotinic receptors. Part II. *Brain Res.* 771, 104, 1997.

7. LeDoux, J. E., Emotional Memory Systems of the Brain. *Behav. Brain Res.* 58, 69, 1993.

8. Tomaz, C., Dickinson-Anson, H., McGaugh, J. L., Souza-Silva, M. A., Viana, M. B., and Graeff, F. G., Localization in the amygdala of the amnestic action of diazepam on emotional memory. *Behav. Brain Res.* 58, 99, 1993.

9. Feldman, S. F., Meyer, S. M., and Quenzer, L. F., *Principles of Neuropsychopharmacology.* Sinaur Assoc., Inc., Sunderland, MA, 1997.

10. Elrod, K. and Buccafusco, J. J., Microinjection of vehicle into the nucleus basalis magnocellularis results in task-specific impairment of passive avoidance responding. *Res. Com. Psychol. Psych. Behav.* 13, 271, 1988.

11. Knardahl, S. and Karlsen, K., Passive-avoidance behavior of spontaneously hypertensive rats. *Behav. Neural Biol.* 42, 9, 1984.

12. Sahgal, A., Passive avoidance procedures. In *Behavioural Neuroscience. A Practical Approach,* Sahgal, A., Ed., Oxford University Press, New York, 1993, vol. 1, chap. 4.

13. Picciotto, M. R., Zoll, M., Lena, C., Bessls, A., Lallemand, Y., LeNovere, N., Vincent, P., Pich, E. M., Brulet, P., and Changeux, J. P., Abnormal avoidance learning in mice lacking functional high-affinity nicotine receptor in the brain. *Nature* 374, 65, 1995.

14. Calamandrei, G., Venerosi, A., Branchi, I., Chiarotti, F., Verdina, A., Bucci, F., and Alleva, E., Effects of prenatal AZT on mouse neurobehavioral development and passive avoidance learning. *Neurotoxicol. Teratol.*, 21, 29, 1999.

15. Fine, A., Dunnett, S. B., Bjorklund, A., and Iversen, S.D., Cholinergic ventral forebrain grafts into the neocortex improve passive avoidance memory in a rat model of Alzheimer disease. *Proc. Natl. Acad. Sci. U.S.A.*, 82, 5227, 1985.

16. Terry, A. V., Jr., Gattu, M., Buccafusco, J. J., Sowell, J. W., and Kosh, J. W., Ranitidine analog, JWS-USC-75IX, enhances memory-related task performance in rats. *Drug Devel. Res.* 47, 97, 1999.

17. Bohus, B., Ader, R., and de Wied, D., Effects of vasopressin on active and passive avoidance behavior. *Hormones & Behavior*, 3, 191, 1972.

18. Blozovski, D., Deficits in passive avoidance learning in young rats following mecamylamine injections in the hippocampo-entorhinal area. *Exper. Brain Res.*, 50, 442, 1983.

Chapter

Spatial Navigation (Water Maze) Tasks

Alvin V. Terry Jr., Ph.D.

Contents

I.   Introduction

Since the early part of the 20th century, a variety of experimental procedures have been developed for animals that employ the escape from water as a means to motivate learning and memory processes.[1-4] Water maze tasks primarily designed to measure spatial learning and recall have become quite useful for evaluating the effects of aging, experimental lesions, and drug effects, especially in rodents. For nearly two decades the Morris Water Maze (MWM)[5] has been the task most extensively used and accepted by behavioral physiologists and pharmacologists. A cursory literature search revealed that well over 500 journal articles have been published since 1992 in which this model was used to assess and compare learning and memory in rodents.

The MWM, while simple at first glance, is a challenging task for rodents that employs a variety of sophisticated mnemonic processes. These processes encompass the acquisition and spatial localization of relevant visual cues which are subsequently processed, consolidated, retained, and then retrieved, in order to successfully navigate and thereby locate a hidden platform to escape the water[5] (see also review[6]). The general processes utilized for visuospatial navigation in rats also contribute considerably to human day to day cognitive processes. Importantly, several lines of evidence confirm the utility of the model for investigations relevant to the study of neurodegenerative diseases such as Alzheimer's Disease (AD) and Parkinson's Disease (PD) which feature cognitive decline. While one would readily acknowledge the differences in complexity between human and rodent behaviors, several salient observations regarding the utility of the MWM are notable: (1) The functional integrity of forebrain cholinergic systems that are critical for efficient performance of the MWM appears to be consistently disrupted in patients who suffered AD. This disruption correlates well with the degree of dementia (see reviews[7,8]) and is also present in many PD patients who suffer cognitive decline.[9,10] (2) Cortical and hippocampal projections from the nucleus basalis magnocellularis (NBM) and medial septum (MS), respectively, are reproducibly devastated in AD (reviewed[7]) and, accordingly, reductions in central cholinergic activity in rodents resulting from brain lesions (e.g., NBM, MS, etc.) and age reproducibly impair spatial learning in the MWM (reviewed[6]). (3) Other data implicate the hippocampus as an essential structure for place learning[13] which, incidentally, is commonly atrophic in patients with AD.[14,15] It is interesting to note that the hippocampal formation (in particular the hippocampal-dentate complex and the adjacent entorhinal cortex), which undergoes significant degeneration with age (and particularly so in the setting of dementia), is intimately involved in cognitive mapping and the facilitation of context dependent behavior in a changing spatio-temporal setting (reviewed[16]). (4) Anticholinergic agents (e.g., scopolamine) that are used routinely to impair performance in the MWM also impair memory in humans and worsen the dementia in those with AD[11] (see also review[12]). (5) Finally, it is also important to note that spatial orientation, navigation, learning, and recall (which are utilized extensively in the MWM) are quite commonly disrupted in AD patients. Visuospatial and visuoperceptual deficits and topographic disorientation are detectable very early in the course of AD and become more pronounced as the disease progresses.[17-19] The common observations of spatial and visual agnosia in AD patients also indicate the disruption of complex processes which involve both visual pathways and mnemonic processing.[20,21]

The MWM procedure offers a number of advantages as a means of assessing cognitive function in rodents when compared to other methods: (1) It requires no pre-training period and can be accomplished in a short period of time with a relatively large number of animals. For example, young adult, unimpaired (control) rats can accomplish the most commonly employed versions of the task with asymptotic levels of performance achieved in 10 to 20 trials, generally requiring no more than a few days of testing. (2) Through the use of training as well as probe or transfer trials, learning as well as retrieval processes (spatial bias)[5] can be analyzed and compared between groups. (3) The confounding nature of olfactory trails or cues is eliminated. (4) Through the use of video tracking devices and the measure of swim speeds, non-mnemonic behaviors or strategies (i.e., taxon, praxis, thygmotaxis, etc.) can be delineated and motoric or motivational deficits can be identified. (5) Visible platform tests can identify gross visual deficits that might confound interpretation of results obtained from standard MWM testing. (6) By changing the platform location, both learning and re-learning experiments can be accomplished. Accordingly, several doses of experimental drugs can be tested in the same group of animals. (7) While immersion into water may be somewhat unpleasant, more aversive procedures such as food deprivation or exposure to electric shock are circumvented. (8) Through the use of curtains, partitions, etc., operation of the video tracking system by the experimenter out of sight of the test subjects also reduces distraction. (9) Finally, the MWM is quite easy to set up in a relatively small laboratory, is comparatively less expensive to operate than many types of behavioral tasks, and is easy to master by research and technical personnel. We have found the method quite useful in drug development studies for screening compounds for potential cognitive enhancing effects,[22] as well as delineating deleterious effects of neurotoxicants on cognition.[23] For a more extensive discussion of the advantages of the MWM, see Morris[5] and reviews.[6,24]

II.   Standard Procedures

The MWM generally consists of a large circular pool of water maintained at room temperature (or slightly above) with a fixed platform hidden just below (i.e., ~ 1.0 cm) the surface of the water. The platform is rendered invisible by one of several means: (1) adding an agent (i.e., powdered milk) to render the water opaque; (2) having a clear plexiglass platform in clear water; or (3) having the platform painted the same color as the pool wall and floor (e.g., black on black). Rats are tested individually and placed into various quadrants of the pool and the time elapsed and/or the distance traversed to reach the hidden platform is recorded. Various objects or geometric images (e.g., circles, squares, triangles) are often placed in the testing room or hung on the wall in order that the rats can use these visual cues as a means of navigating in the maze. With each subsequent entry into the maze the rats progressively become more efficient at locating the platform, thus escaping the water

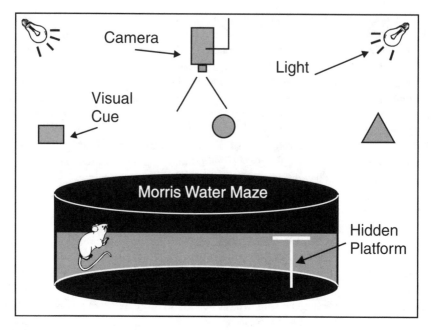

FIGURE 10.1
Diagrammatic illustration of the Morris Water Maze (MWM) testing room and apparatus.

by learning the location of the platform relative to the distal visual cues. The learning curves are thus compared between groups. An illustration of a typical Morris Water Maze setup (as used in our laboratory) appears in Figure 10.1.

A.   Methodology

1.   Testing Apparatus

1. Maze testing should be conducted in a large circular pool (e.g., diameter: 180 cm, height: 76 cm) made of plastic (e.g., Bonar Plastics, Noonan, GA) with the inner surface painted black.

2. Fill the pool to a depth of 35 cm of water (maintained at 25 ± 1.0°C) to cover an invisible (black) 10 cm square platform. The platform should be submerged approximately 1.0 cm below the surface of the water and placed in the center of the northeast quadrant.

Note: *We use a black platform in a pool with the sides and floor painted black to obviate the need for addition of agents to render the water opaque. If the experimenter is unsure whether or not the platform is still visible, closing the curtains to eliminate spatial cues and subsequent testing of a few rats will resolve this question. Rats will not become more successful with each entry into the pool if the platform is invisible and room lighting is diffuse.*

3.   The pool should be located in a large room with a number of extramaze visual cues including highly visible (reflective) geometric images (squares, triangles, circles, etc.) hung on the wall, with diffuse lighting and black curtains used to hide the experimenter and the awaiting rats. Swimming activity of each rat may be monitored via a ccTV camera mounted overhead, which relays information including latency to find the platform, total distance traveled, time and distance spent in each quadrant, etc. to a video tracking system. Tracking may be accomplished via a white rat on a black background.

Note: *We have found the Poly-Track® Video Tracking System, San Diego Instruments, San Diego, CA to be a very reliable system which is also easy to set up. Several other vendors market similar systems.*

2.   Hidden Platform Test
We employ a method in which each rat is given four trials per day for four consecutive days.

1.   Each day, a trial is initiated by placing each rat in the water facing the pool wall in one of the four quadrants (designated NE, NW, SE, SW) which are set up on the computer software such that each quadrant is equal in area and color coded. The daily order of entry into individual quadrants is randomized such that all 4 quadrants are used once every day.

Note: *Do not place the rat in adjacent quadrants sequentially since the rat may adopt positional or other non-mnemonic strategies (e.g., all right turns) to locate the platform. Further, the order should be changed on each subsequent day of testing.*

2.   For each trial, the rat is allowed to swim a maximum of 90 sec in order to find the hidden platform. When successful, the rat is allowed a 30 sec rest period on the platform. If unsuccessful within the allotted time period, the rat is given a score of 90 seconds and then physically placed on the platform and also allowed the 30 sec rest period. In either case the rat is immediately given the next trial (ITI = 30 sec) after the rest period.

Note: *In some cases the rat may fall or jump off of the platform and resume swimming before the elapsed 30 sec interval. When this occurs, the stopwatch should be immediately stopped, the rat retrieved and placed on the platform again. The stopwatch should be reactivated such that the remainder of the time interval (30 sec) is elapsed. This assures that each rat has equal time to observe spatial cues after each trial.*

3.   Transfer Tests (Probe Trials)
On Day 5, two trials are given in which the platform is removed from the pool to measure spatial bias.[5] This is accomplished by measuring the time and distance traveled in each of the four quadrants. The important measure will be the percentage

of the total time elapsed and distance swam in which the rat is in the boundaries of the previous target quadrant.

1. Place each rat in the pool and track the animal for 90 sec. This procedure is repeated one time, since in some cases an unusual level of variance in performance will be observed in this first trial. It is assumed that some of the rats are in some way disoriented after the change in testing conditions. Two trials performed identically one after the other (and averaged) generally reduce the variance and provide a good measure of the overall accuracy and mastery of original learning.

Note: *More than two trials will result in extinction effects with less time spent in the target quadrant and is thus undesirable for a measure of spatial bias.*

2. The time elapsed and distance swam in the previous target quadrant is recorded. An annulus ring can be circumscribed around the previous target location (on the computer screen) to localize more closely the previous target location. The number of crossings through this region may be recorded. Alternatively, crossings of the actual 10-cm square platform target outlined in the previous trials can be recorded and compared between groups.

Note: *In our hands both of these measures are associated with an unacceptable level of variance and we have had much more success measuring and comparing the time spent and the distance traveled in the entire quadrant which previously held the target.*

4.   Visible Platform Test

A visible platform test may be performed to identify if a drug or other experimental manipulation results in crude changes in visual acuity which would thus confound the analyses of data that depend on the use of distal visual cues for task performance. One must be aware, however, of certain behaviors that might be interpreted as impaired visual acuity. For example, the absence of search behaviors or thymgotaxis (swimming constantly along the perimeter of the pool) might be misinterpreted as visual deficits since the animal does not locate the platform in a reasonable period of time. Thus, animals must make attempts to cross the pool and then be impaired at locating the platform in order for an interpretation of visual deficits to be made.

1. Immediately following the transfer test on Day 5, place the platform into the pool in the quadrant located diametrically opposite the original position (SW quadrant).
2. A cover (available from San Diego Instruments, Inc.) which is rendered highly visible (i.e., with light-reflective glossy or neon paint) is attached to the platform to raise the surface above the water level (approximately 1.5 cm).
3. Change the lighting such that extra-maze cues are no longer visible and a spotlight illuminates the visible platform only.

Note: *The video tracker is not used for this procedure and only a stopwatch is needed.*

4. Allow each rat one trial in order to acclimate to the new set of conditions and locate the platform visually. This is accomplished by lowering the rat into the water in the NE quadrant and allowing location of the platform. No time limit is placed on this first trial. Once the platform is located, allow the rat 30 sec on the platform. The rat should then immediately be given a second trial in the same manner and the latency to find the platform measured as a comparison of visual acuity.

Note: *This procedure may be repeated additional times; however, the platform location should be changed on each subsequent trial to ensure that visual location of the platform is actually made from a distance and the rat is not first using nearby stationary visual cues.*

5. Relearning Phases

After completion of the first five days of water maze testing and a rest period (generally at least one week and often longer), a second series of trials (Phase 2) may be conducted as described above (hidden platform test) except that the location of the platform is changed to a different quadrant. Daily performances (average of four trials/day/rat) are then compared as described above. This method may be used in order to compare different drug doses or other additional manipulations with the same groups of animals.

Note: *It must be realized that learning curves will generally be steeper than in the first phase of testing since a number of factors not associated with the actual platform location will have been previously learned (e.g., use of visual cues to navigate, the fact that escape is not associated with the pool wall, etc.).*

B. Statistical Analyses

1. Hidden Platform Test

For the hidden platform test, we average the latencies and the distances swam across the four trials for each rat each day. These means are then analyzed across the four days of testing. A two-repeated measures analyses of variance (ANOVA) is used with day as the repeated measure and latency or distance swam as the dependent variables.

2. Probe Trials

For probe trials, each of the two trials for each rat are averaged and the means compared between groups via a one-way analyses of variance (ANOVA).

3. Visible Platform Tests

For the visible platform test, the second trial for each rat is recorded and compared between groups via one-way analyses of variance (ANOVA).

4.   Relearning Phase

The relearning phase is analyzed identically to the hidden platform test described above.

C.   Representative Data

A representative water maze study (hidden platform test) from our laboratory appears in Figure 10.2. We evaluated commonly used doses of two muscarinic-cholinergic antagonists (scopolamine and atropine) and one nicotinic-cholinergic antagonist (mecamylamine) for their ability to inhibit learning in this study. As expected, under saline conditions the rats learned to locate the hidden platform with progressively shorter latencies and distances swam until the end of the study. Asymptotic levels of performance were approached by Day 3 under control conditions. In contrast, each of the cholinergic antagonists evaluated significantly impaired performance of the task and asymptotic levels of performance were not achieved by the end of the 4-day study. The curves, however, do indicate a significant day effect (i.e., the platform was located more quickly and with less distance traveled each day) and thus learning (although impaired) did occur after administration of each of the cholinergic antagonists. This learning by day effect is important when selecting doses of amnestic compounds for drug studies. When screening potential cognitive enhancing compounds for their ability to reverse the effects of cholinergic antagonists, it is important that some level of learning be accomplished in the presence of the antagonists alone. This assures that the rats are in fact memory-impaired, not simply disoriented, thygmotaxic, or motivationally impaired, etc. Swim speeds and visual acuity, etc. (described below) may also be evaluated in order to further ensure memory impairment.

Figure 10.3 illustrates the evaluation of the cholinergic antagonists in typical transfer tests (probe trials). Each cholinergic antagonist (compared to saline) reduced both the total time and distance swam in the previous target quadrant. These experiments are performed on the day immediately following the last day of the hidden platform tests and reflect a spatial bias of animals toward the previous location of the hidden platform. The results are analyzed separately from the hidden platform tests and offer a second, easily performed method of estimating the strength and accuracy of original learning processes.[6] It is important to note that since the pool is divided into 4 quadrants of equal area, a chance level of performance would mean that the % of time or distance swam (of the total) in the previous target quadrant would generally approximate 25%. Once again, it is generally desirable to use doses of amnestic compounds which allow the animal to perform at a level somewhat above 25% for the reasons described above.

Figure 10.4 illustrates the effects of the cholinergic antagonists in a visible platform test (A) and on swim speeds both by day (B) and as an average across the study (C). These data are useful to identify gross deficits or changes in visual acuity and motoric or motivational changes. In this study, no significant drug effects were observed by either measure. It is important to note that rats will often increase swim speeds somewhat as the learning of the platform location occurs and this increase

Water Maze
Hidden Platform Test

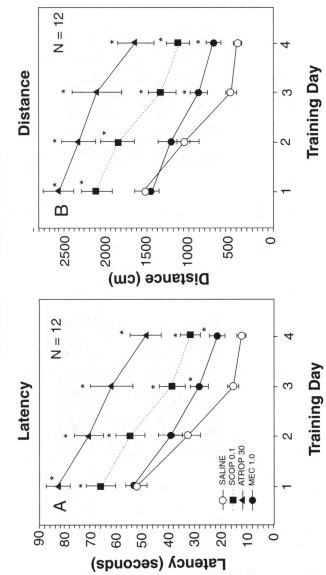

FIGURE 10.2

Hidden Platform Test in the Morris Water Maze (Testing Days 1 to 4). Effects of saline, scopolamine 0.1 mg/kg (SCOP 0.1), atropine 30 mg/kg (ATROP 30), or mecamylamine 1.0 mg/kg (MEC 1.0) administered IP, 30 min before testing in the maze. The latency in seconds (A) and the distance swam in centimeters (B) to find the hidden platform over four consecutive days of testing are presented for each group. Each point represents the mean ± SEM of 4 trials per day. N = 12 rats per group. * = significantly different from saline control group (Two-way repeated measures ANOVA p<0.05).

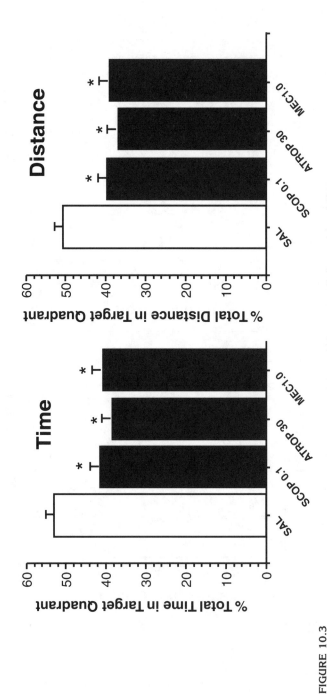

FIGURE 10.3

Transfer Test (Probe Trials) in the MWM (Testing Day 5). Effects of saline, scopolamine 0.1 mg/kg (SCOP 0.1), atropine 30 mg/kg (ATROP 30), or mecamylamine 1.0 mg/kg (MEC 1.0) administered IP, 30 min before testing. The % total time spent in seconds (**A**) and the % total distance swam in centimeters (**B**) in the previous target quadrant is presented. Each bar represents the mean ± SEM of 2 trials per rat in each group. N = 12 rats per group. * = significantly different from saline control group (One-way ANOVA p<0.05).

will often become statistically significant by the end of the study under control conditions. This may not be the case when animals are impaired with amnestics. Thus, the use of average speeds (for the whole study), using only Day 1 speeds (before significant learning occurs), or using swim speeds only during probe trials, may be compared to obviate the confounding nature of learning effects on swim speeds.

III.   Alternative Procedures

A number of variations of the water maze tasks described above have been employed for the study of memory processes in rats and a full review of these procedures is beyond the scope of this chapter. A short summary of a few of these procedures is outlined below, however. For a more detailed overview of these and additional water maze procedures, see Morris[5] and reviews.[6,24]

A.   Place Recall Test

In this procedure, hidden platform tests are first performed as described above in intact animals such that the location of the platform is well learned. Subsequently, the rats are experimentally manipulated (i.e., given brain lesions, drugs, or other physiological manipulations, etc.) and then retested with either additional hidden platform tests or probe trials. Thus, the effects of the experimental manipulations on all processes used to solve the task with the exception of learning and memory formation may be studied. Namely, processes such as memory retrieval, spatial bias, as well as motoric, sensory, and motivational effects of the manipulations may be delineated.

B.   Platform Discrimination Procedures

These methods require rats to discriminate between two visible platforms in order to successfully escape the water. One of the platforms is rigid and able to sustain the weight of the rat while the other platform is floating (often made of styrofoam) and not able to sustain the rat's weight. Both spatial and non-spatial versions of this task have been used. In the spatial version of the task the platforms appear identical (visually) and rats are required to discern the viable platform by learning its location relative to distal visual cues in the room. In the non-spatial version of the task, the rats learn to visually discriminate between two platforms of different appearance. For example, discrimination between platforms may be engendered via a difference in shape, brightness, or painted pattern. Curtains are drawn to exclude the influence of extra-maze cues.

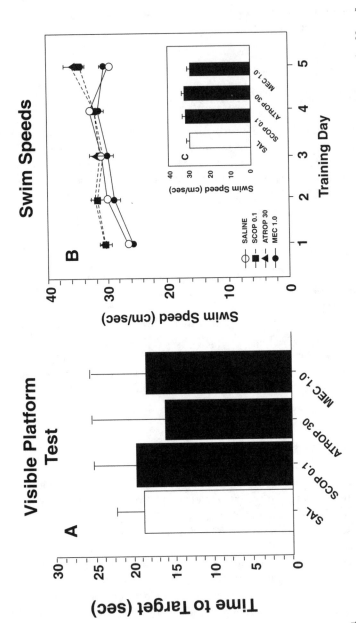

FIGURE 10.4

A. Latency (sec) to locate a highly visible platform by each group of rats tested. The effects of saline, scopolamine 0.1 mg/kg (SCOP 0.1), atropine 30 mg/kg (ATROP 30), or mecamylamine 1.0 mg/kg (MEC 1.0) administered IP, 30 min before testing are presented. Each bar represents the mean ± SEM. **B.** Swim speeds in cm/sec for each of the five days of testing. Swim speeds averaged across the five days of testing. Each bar represents the mean ± SEM (N = 12 rats per group). No significant differences between groups were found (one-way ANOVA and two-way repeated measures ANOVA, respectively, p>0.05).

## C.	Working Memory Procedures

Working memory procedures in the MWM (sometimes referred to as spatial matching to sample procedures) generally involve a two trials/day paradigm in which a hidden platform is located in one of four quadrants and randomly relocated on each of several subsequent days of testing. The assumption drawn is that each rat will obtain information regarding the location of the platform on the first trial which will be of benefit for discerning its position on Trial 2. The intertrial interval can be manipulated in order to alter the difficulty of the task.

# IV.	Summary and Conclusions

The Morris Water Maze (MWM) equipped with a video tracking system has become a commonly used and well-accepted behavioral task for rodents. It is quite easy to set up in a relatively small laboratory, is comparatively less expensive to operate than many types of behavioral tasks, and is easy to master by research and technical personnel. It utilizes a number of mnemonic processes in rats that are relevant to the study of human learning and memory and disorders thereof. In addition, it is a very versatile paradigm, which can be used to study both spatial and non-spatial (discriminative) learning as well as working memory processes, and offers several means of delineating and dissociating confounding non-mnemonic processes.

References

1.	Glaser, O.C. The formation of habits at high speed. *J. Comp. Neurol.* 20, 165, 1910.
2.	Wever, E.G. Water temperature as an incentive to swimming activity in the rat. *J. Comp. Psychol.* 14, 219, 1932.
3.	Waller, M.B., Waller, P.F., and Brewster, L.A. A water maze for use in studies of drive and learning. *Psychol. Rep.* 7, 99, 1960.
4.	Woods, P.J., Davidson, E.H., and Peters, R.J., Instrumental escape conditioning in a water tank: effects of variation in drive stimulus intensity and reinforcement magnitude. *J. Comp. Psychol.* 57, 466, 1964.
5.	Morris, R.G.M. Development of a water-maze procedure for studying spatial learning in the rat. *J. Neurosci. Meth.* 11, 47, 1984.
6.	McNamara, R.K. and Skelton, R.W., The neuropharmacological and neurochemical basis of place learning in the Morris water maze. *Brain Res. Rev.* 18, 33, 1993.
7.	Perry, E., Walker, M., Grace, J., and Perry, R. Acetylcholine in mind: a neurotransmitter correlate of consciousness? *Trends Neurosci.* 22, 273, 1999.
8.	Francis, P.T., Palmer, A.M., Snape, M., and Wilcock, G.K. The cholinergic hypothesis of Alzheimer's disease: a review of progress. *J. Neurol. Neurosurg. Psychiatry* 66, 137, 1999.
9.	Whitehouse, P.J., Hedreen, J.C., White, C.L. 3d, and Price, D.L. Basal forebrain neurons in the dementia of Parkinson disease. *Ann. Neurol.* 13, 243, 1983.

10. Perry, E.K., Curtis, M., Dick, D.J., Candy, J.M., Atack, J.R., Bloxham, C.A., Blessed, G., Fairbairn, A., Tomlinson, B.E., and Perry, R.H. Cholinergic correlates of cognitive impairment in Parkinson's disease: comparisons with Alzheimer's disease. *J. Neurol. Neurosurg. Psychiatry* 48, 413, 1985.

11. Sunderland, T., Tariot, P.N., and Newhouse, P.A. Differential responsivity of mood, behavior, and cognition to cholinergic agents in elderly neuropsychiatric populations *Brain Res.* 472, 371, 1988.

12. Ebert, U. and Kirch, W. Scopolamine model of dementia: electroencephalogram findings and cognitive performance. *Eur. J. Clin. Invest.* 28, 944, 1998.

13. McDonald, R.J. and White, N.M. Hippocampal and nonhippocampal contributions to place learning in rats. *Behavi. Neurosci.* 109, 579, 1995.

14. Terry, R.D. and Katzman, R. Senile dementia of the Alzheimer type. *Ann. Neurol.* 14, 497, 1983.

15. Mann, D.M. The topographic distribution of brain atrophy in Alzheimer's disease. *Acta. Neuropathol. (Berl.)* 83, 81, 1991.

16. Scheibel, A.B. The hippocampus: organizational patterns in health and senescence. *Mech. Ageing Dev.* 9, 89, 1979.

17. Eslinger, P.J. and Benton, A.L. Visuoperceptual performances in aging and dementia: clinical and theoretical implications. *J. Clin. Neuropsychol.* 5, 213, 1983.

18. Huber, S.J., Shuttleworth, E.C., and Freidenberg, D.L. Neuropsychological differences between the dementias of Alzheimer's and Parkinson's diseases. *Arch. Neurol.* 46, 1287, 1989.

19. Morris, J.C., McKeel, D.W. Jr., Storandt, M., Rubin, E.H., Price, J.L., Grant, E.A., Ball, M.J., and Berg, L. Very mild Alzheimer's disease: informant-based clinical, psychometric, and pathologic distinction from normal aging. *Neurology* 41, 469, 1991.

20. Henderson, V.W., Mack, W., and Williams, B.W. Spatial disorientation in Alzheimer's disease. *Arch. Neurol.* 46, 391, 1989.

21. Mendez, M.F., Tomsak, R.L., and Remler, B. Disorders of the visual system in Alzheimer's disease. *J. Clin. Neuroophthalmol.* 10, 62, 1990.

22. Terry, A.V., Jr., Gattu, M., Buccafusco, J.J., Sowell, J.W., and Kosh, J.W. Ranitidine Analog, JWS-USC-75IX, Enhances Memory-Related Task Performance in Rats. *Drug Dev. Res.* 47, 97, 1999.

23. Prendergast, M.A., Terry, A.V., Jr., and Buccafusco, J.J. Chronic, low-level exposure to diisopropylfluorophosphate causes protracted impairment of spatial navigation learning. *Psychopharmacol.* 129, 183, 1997.

24. Brandeis, R., Brandys, Y., and Yehuda, S. 1. The use of the Morris Water Maze in the study of memory and learning. *Int. J. Neurosci.* 48, 29, 1989.

The Delayed Non-Match-to-Sample Radial Arm Maze Task: Application to Models of Alzheimer's Disease

Carl A. Boast, Thomas J. Walsh,
and Adam Bartolomeo

Contents

I.   Introduction/Rationale

Spatial memory provides animals the ability to learn about the location of food or predators and to utilize precise spatial information to guide their seasonal migrations. The history of animal learning is firmly rooted in the use of mazes to study the behavioral properties of spatial memory and its neurobiological substrates (see Olton, 1979[1] for a historical review). Mazes provide a rigidly structured spatial environment that allows the investigator to study the dynamics of choice behavior and the processes that guide spatial learning and memory. One of the most commonly used mazes has been the radial-arm maze (RAM), first introduced by David Olton and for over two decades proven to be a useful tool for analyzing cognitive processes in rodents.[2] The task is based on the natural foraging tendencies of rats. Simply, several straight alleys, or arms, extend from a central hub and food can be found at the end of each arm. Once the food has been retrieved from a given arm in a particular session, food is no longer available in that arm. Therefore, a foraging strategy of visiting each arm once is optimal. Many strategies can be used to achieve this optimum, but strategies based on the use of extramaze spatial cues requiring memory for previous arms visited have been most studied.[3,4] The win-shift strategy is commonly used by rodents to solve appetitive spatial problems. Rodents and many avian species learn to avoid recently depleted patches of food. In contrast, win-stay strategies are used to contend with aversive spatial problems.[5]

Assessment of the effects of neurosurgical,[6] pharmacological,[7] or molecular[8] manipulations, as well as age,[9,10] gender,[11,12] or strain[13,14] related differences have been conducted using the RAM. These studies have implicated various neuroanatomical structures and neurochemical pathways in the acquisition, retention, and/or recall of information relevant to successful RAM performance.[3,4] Particularly, the hippocampal formation and its associated cholinergic pathways have been well studied in these RAM experiments.[15,6] Performance in the RAM task can be impaired by a variety of treatments including: administration of cholinergic antagonists (e.g., scopolamine,[16,17] lesions of brain areas such as the hippocampus[6] or cholinergic nuclei (e.g., NBM),[18] or selective cholinergic depletion following central administration of toxins (e.g., AF64A[7] or 192-IgG-saporin[19]). This combination of hippocampal pathology, cholinergic hypofunction, and cognitive deficits may provide an animal model of Alzheimer's Disease (AD).[20]

Although this task has been used extensively to evaluate the effects of a variety of experimental manipulations, there are important methodological considerations that can be overlooked. For instance, though often referred to as a spatial task, certain commonly used versions of the procedure may allow other non-spatial strategies to be used. Specifically, the common practice of hand-baiting maze arms may allow maze behavior to be guided by food related cues such as odor or visibility and should be avoided if possible; in addition, response patterning may provide non-spatial alternatives and should be prevented. These and other methodological issues are addressed in more detail below, following a detailed description of the apparatus and procedure. Examples of data obtained using the procedure and pertaining to models of AD are provided and some advantages and disadvantages of using the procedure are briefly discussed.

The authors of this chapter represent the diverse backgrounds of both industry and academe and, as is common in behavioral research, have utilized different specifics in apparatus construction, variable parameters, and experimental design. Despite these differences, several consistencies in general procedures and especially in outcomes have emerged. Similarly, shared experiences have resulted in the identification of some common issues related to the use of the RAM. These major similarities form the substance of this chapter.

A.   Working and Reference Memory

Working memory is involved in temporarily maintaining representations of previously experienced events or episodes, whereas reference memory contributes to the performance of well-learned responses in the presence of an appropriate discriminative stimulus.[15] This component theory of memory serves to provide a conceptual framework for interpreting the dissociation in memory processes observed in both humans and animals following damage to the hippocampus (HPC) or its cholinergic innervation. While several nomenclatures have been introduced to describe this behavioral dissociation, the working memory-reference memory distinction is attractive since each of these processes can be operationally defined within given experimental contexts. We will use this terminology throughout this chapter.

II.   Detailed Methods

A.   Subjects

Mostly Sprague-Dawley rats have been used, but studies with other rat strains,[13] Mongolian gerbils,[16] and different mouse strains[14] have also been conducted. Food restriction must be undertaken to sufficiently motivate subjects because this is a food-motivated task. Typically, subjects are reduced to 85% of their free-feeding body weight; aged animals may have to be reduced to 75–80%. Food is then provided daily in measured portions during the late afternoon. This food restriction regimen may significantly increase the life span of rats.[21] Water is available *ad libitum* when the animals are in the colony room. The care of the animals and testing procedures should all be approved by internal animal care and use committees.

B.   Apparatus

The RAM apparatus (Figure 11.1) typically consists of eight maze arms (50 to 58 cm long, 10 cm wide) extending outward from an octagonal/circular central arena (20 to 23 cm across) which is elevated (58 to 76 cm) above the floor. Arms can be fitted with a removable clear plastic cover 35 to 43 cm long, 9 cm wide, and 13 cm high, to prevent inappropriate climbing from one arm to another. Removable barriers (8.5 to 12 cm wide by 9 to 25 cm high) made of clear Plexiglas attached to a black wooden base, or alternatively made entirely of black Plexiglas, can be used to block selected arms during the pre-delay session of the RAM task. Alternatively, automated doors may be installed on the apparatus (see below). Food cups are recessed holes located 4 to 5 cm from the end of each arm.

   Automated RAMs are available commercially. However, if good support personnel are available, a RAM can be constructed, fitted for the appropriate hardware, and computer interfaces/programs generated in-house. The automated RAM that we use has three sets of photocells in each arm (located 6, 27, and 52.5 cm from the opening of the arm). The photocells are used to track the movement of the animal in the maze. The computer program (labVIEW, National Instruments) controls the dispensation of food when the outermost photocell of an arm is passed for the first time. The program also records arm entries/exits; an entered arm is considered exited when the innermost photobeam of another arm is interrupted. Errors are also automatically recorded; an error is assigned when either the middle or the outermost photocell is triggered upon a re-entry into a previously rewarded arm. Two 45mg chocolate pellets (Bio-Serv) are dispensed by Coulbourn Pellet Feeders (E14 to 24), located above the food cups at the end of each arm. Pellet feeders provide a means of removing food odors which might guide the animal through the task (see below). Another advantage of automation is the control of the barriers to the arms. However, barrier selection is critical since the animal's view of the room must not be blocked, so clear plastic barriers should be used (we had once attempted to install opaque guillotine doors that were raised above the arm openings and the animals were unable to reach what we had previously established as learning criteria, probably due to blocking the visual cues located around the test room). The maze is located in a testing room with black and white geometric posters on each wall to serve as visual cues. White noise is audible (~70 db) during all training and testing procedures.

FIGURE 11.1
Drawing depicting an 8-arm radial maze. Shaded arms represent one option for initial blockade by the experimenter when using a delayed-non-match-to-sample paradigm.

C. Habituation/Training Procedure (All Arms Accessible Design)

Animals are subjected to 5 days of acclimation trials in which they are allowed to explore the maze for a 5 to 10 minute period each day. During these trials, chocolate-flavored food pellets are scattered throughout the maze. Shaping procedures can be used, by confining the scattered pellets closer to the food cups on successive days (e.g., Day 1: food pellets are scattered throughout the maze; Day 2: food pellets are located in the arms only; Day 3: food pellets are located on the outer half of the arm; Day 4: food pellets are located in the outermost third of the arm; Day 5: a few food pellets are located around the outside of the food cup). During acclimation the pellet dispensers release food whenever the outermost photobeam is disrupted, to adapt subjects to the sound and action of the pellet dispensers. The E14-24 Coulbourn pellet dispenser is quieter than previous models and we have not had animals drop out due to the sound of this dispenser.

Following the acclimation sessions, the animals are given one session per day with all arms accessible. In this procedure an animal is placed in the center of the RAM and allowed to visit the eight arms; food pellet rewards are available upon the first visit to each arm. Re-entry into an arm previously visited within any daily trial is not rewarded and is scored as an error. After each arm has been visited once, the session ends; a default time limit can be 5 to 10 minutes. Training criteria can be based on either a predetermined number of trials, such as 15, or asymptotic performance of ≤ 2 total errors for 3 consecutive days, which could occur after as few as 5 trials.

D.   Delay Non-Match-to-Sample (DNMTS) Procedure

Following acquisition of the all arms open RAM task, animals are trained to perform the task with a one-hour delay imposed between the fourth and fifth arm choices. Plexiglas barriers are used to force the animals to select a predetermined set of four arms during the pre-delay session (Figure 11.1). This eliminates potential algorithmic response patterns, which could result from free-choice of the four arms and skew the error scores from these sessions (see below). There were two patterns of blocked arms that were never used: four adjacent arms (e.g., 1,2,3,4) and every other arm (e.g., 1,3,5,7). The daily pattern of blocked arms is randomly selected and is different for sequentially tested animals. During the pre-delay session, animals are allowed to choose freely until all four accessible arms have been chosen. Upon completion of the pre-delay session, animals are returned to their home cages for a one-hour delay interval. Following the delay, animals are returned to the maze and allowed to choose freely among all eight arms. Entry into an arm visited during the pre-delay session is not rewarded and constitutes an error, as does repeated entry into post-delay choices. The pre-delay and the post-delay sessions are limited to five minutes each. If an animal does not complete the pre-delay session in the ascribed time, this subject is not tested in a post-delay session, since not all of the pre-delay information was obtained. If an animal does not make at least four choices in five minutes during the post-delay session, the data from that animal are not used and the animal is scored as a time-out. The maze is wiped with a 70% ethanol solution between the pre-delay and the post-delay to minimize odor cues. To also minimize intra-maze cues, the maze can be rotated between the pre-delay and the post-delay, while maintaining a consistent orientation of the arms of the maze relative to the landmarks in the test room. For instance, to rotate the maze two arm positions, what was arm 1 in the pre-delay would be either arm 3 or 7 during the post-delay, depending on clockwise or counterclockwise rotation. With our in-house built maze, we had a lazy Susan-type device serve as the base of the maze; it was slotted with an adjustable pin that would fix the maze into position following rotation. Of course, this rotation manipulation to reduce the role of intra-maze cues also required additional code in the software to track the transformed arm numbers, ensuring that correct arms and previously visited arms were appropriately designated.

This procedure allows for greater experimental control over the animal's performance in that the set of arms accessible during the pre-delay session are under the control of the experimenter. Therefore, particular response patterns that the animal might use are not beneficial for accurate responding. In addition, this version of the RAM task represents a delayed non-match-to-sample task which can be used to specifically assess working memory processes.

E.   Different Types of Errors in the RAM

The dependent measures in the DNMTS procedure are the number of correct choices (CC) during the first four arm entries following the delay and the number of errors. It is also possible to define two types of post delay errors; retroactive errors (RE) and proactive errors (PE). For example, if arms 1,3,5, and 6 are baited during the pre-delay (training) session and arms 2,4,7, and 8 are baited during the post-delay (testing) session, then RE are defined as re-entries into those arms (1,3,5,6) that were entered during the training session. An increased number of RE represents retrograde amnesia. In contrast, PE are repeated entries into the correct set of arms (2,4,7,8) during the post-delay session. An increased number of PE can reflect anterograde amnesia or alterations in non-memorial processes such as attention. Latency per choice provides an index of non-memorial factors such as motor ability and/or motivation. The microanalysis of error types can provide information about the nature of cognitive impairments (retrograde or anterograde amnesia), the temporal dynamics of an amnestic treatment, and the potential importance of proactive drug effects on performance.[22]

In this task, the rat must remember a specific array of cues within an episode to correctly perform the task. The pattern of open and closed arms varies in a random manner from day to day. Therefore, the animal confronts a new working memory problem each day. This repeated acquisition feature promotes the use of within-subjects crossover designs in which a single rat can receive different doses of a given drug and/or vehicle and serve as its own control. Furthermore, performance on non-injection days gives an index of proactive and/or persistent drug effects on performance. This task maximizes the data obtained from a single rat. The task is sensitive to age-related cognitive impairments[9] and to damage to the HPC,[19] areas of the temporal cortex[23] and to the cholinergic septohippocampal pathway.[19]

Another important aspect of this task is that the memory demands can be parametrically manipulated by varying the duration of the delay. Within any study, performance over four or five different delay conditions can be examined. A memory deficit should be sensitive to task demands and subtle deficits might only be observed when the delay is long and demands are stringent. This variable-delay feature maximizes the sensitivity of the task by minimizing the possibility of missing a subtle memory deficit due to marginal task demands or missing a deficit if the task is too difficult for control subjects.

III. Major Variations of Ram Tasks

A. Non-Match-to-Sample (Free Choice)

Perhaps the most commonly used RAM paradigm is one in which the animal is free to select any arm, initial arm entries are rewarded, and no delay is imposed between any choices. As detailed elsewhere in this chapter, this paradigm suffers from the potential development of a constant sequence of arm visitation, resulting in the use of reference memory rather than working memory to perform the task. Introducing a forced choice component to this task would make it trial dependent and therefore working memory would be required to perform the task.

B. DNMTS Free Choice

This task[24] is similar to the one that we have used, the DNMTS forced choice. Animals, in a preliminary training phase, have learned the win-shift rule that re-entries into an arm are not rewarded; during the DNMTS task the animal visits a subset of the total arms (e.g., 4 of 8 arms), and then a delay is interposed (seconds to hours). Subsequently, the animal is permitted to continue selecting arms but is rewarded only for first entries into arms that were not selected in the pre-delay. The difference from the forced choice paradigm described above is that the animal is free to select the arms to visit during the pre-delay portion of the task; no barriers or doors are utilized to determine the arm selection for the animal. A problem with this task can occur if an individual animal consistently chooses either the same arms or the same pattern of arms during the pre-delay session. This consistent pattern of choices would suggest that the animal is not using spatial working memory, but rather by selecting the same arms during each pre-delay session, a reference memory task is evolving. Since rewarded and unrewarded arms are consistent on each trial, the task has become trial-independent, which defines reference memory. If the animal consistently selects the same pattern of arms across trials, the animal is employing an egocentric strategy, that is, the subject is guiding its responses based on its location in the RAM as opposed to utilizing the spatial, extra-maze cues to navigate to obtain reward, i.e., an allocentric strategy.

C. Delayed Match-to-Sample (DMTS)

This task[25] is similar to the DNMTS versions of RAM paradigms except that a win-stay strategy must be used. Subjects undergo training in the all arms accessible task. Then, during the predelay session a subset of arms are accessible and are rewarded upon entry. A delay is interposed and, during the post-delay session, arms that were rewarded in the predelay are rewarded during the first visit during the post-delay. The issues with free or forced choice arm selections during the pre-delay session are the same as mentioned in the previous section. Again the accessible arms could vary from trial to trial, which would involve trial-dependent or working memory.

D.   Reference/Working Memory Tasks

This design often utilizes mazes with more than eight arms.[11,26,27] Typically, at least four of the arms are never rewarded across trials (the reference memory component) while the remaining (often eight) arms are rewarded for initial entries. Thus, within a given session the animal must remember which arms are never baited (reference memory) and those arms which are rewarded but have been previously entered during that session (working memory). Please refer to previous sections regarding these designs and also to the Additional Methodological Issues section below.

E.   Serial Position

This paradigm was developed by Kesner.[28] Arm barriers/doors must be controllable for this task. Following maze familiarization, the initial training consists of opening one door, the animal enters that arm, obtains the reward, returns to the central platform, the visited arm door closes, a different arm door opens, and the cycle repeats with a different arm accessible each time. This continues until each arm is visited once. This continues using the same arm sequence for several days. The next phase involves presenting a variable sequence of arms accessible one at a time; the arms are a subset of the total arms on the maze. The test phase utilizes this variable sequence of accessible arms, followed by two arms being made available. One of those arms is a previously visited arm, while the other arm is not. The animal is rewarded for returning to the previously rewarded arm; this is a win-stay rule. The trial ends after choosing one arm, rewarded or not. The two arms presented should be either adjacent or one arm apart and the correct arm should be balanced to be 50% on the left or on the right. The serial order of the arm should be balanced as well; arms earlier in the serial order may be more difficult to choose due to the interference effects of subsequent arm visits. Multiple test trials can be given during a single day, but these test trials should be separated (e.g., by an hour). This task has been used to measure the neurobiological correlates of temporal memory, that is, the relative order of events such as arm entries.

F.   Submerged RAM

By placing walls into a tank of water, creating a RAM,[29] animals must learn to escape from the water, and must find one or more of the hidden, slightly submerged platforms at the end of one or more arms. Any of the above mentioned dry RAM paradigms could be used in the submerged RAM, but a DMTS task would probably be the easiest to learn since rodents tend to use win-stay strategies to solve aversive spatial tasks. The difference with this version of the RAM is that the motivation to complete the task is not appetitive.

It should be noted that any of these variations on RAM tasks could be worthwhile to investigate different aspects of learning, memory, or performance. The specific variation selected is a function of the experimental questions being addressed. Our laboratories have focused upon the forced choice DNMTS task due to its implications as a model for Alzheimer's disease.

IV.   Additional Methodological Issues

A.   Working Memory vs. Reference Memory

Impairments in RAM performance have been interpreted as deficits in working memory since, it is argued, subjects must retain an updated list of visited and non-visited arms during a given session.[2] Variations of the basic procedure have included procedures in which food is never available in certain arms of the maze.[11,26,27] In these paradigms, animals learn not to enter these unreinforced maze arms. This invariant aspect of the procedure is interpreted to reflect reference memory. It is sometimes overlooked that in procedures where reinforcement is available in all maze arms, there are aspects of the task that can still be ascribed to reference memory. It should also be recognized that although it is commonly interpreted that working memory is selectively impaired in many conventional RAM paradigms, this is in part due to the relative ease of the working memory component of the task. When the working memory component of a different task (e.g., split-stem T-maze) is made more difficult relative to the reference memory component, selective disruption of reference memory can be observed.[16]

B.   Pre-Baited Arms

When food is present in some arms and not in others (i.e., after some choices have been made and food consumed), it is possible that maze behavior is guided by food-related cues (odor, visibility) and not by spatial cues. We have examined this by testing animals in the dark. We reasoned that in the dark, visual spatial cues are not available to guide behavior and therefore maze performance should be impaired. When we used pre-baited arms, maze performance in the dark was not impaired and we concluded that food odor cues might be guiding behavior. Therefore, pellet feeders were installed at the ends of each arm and photocells were used to detect the presence of the rat at the end of each arm. When the photobeam was disrupted, food was delivered. Thus, food odors were always present in all arms, but could not be used to guide behavior. Under these conditions, maze performance in the dark was impaired.

Another approach to ensuring that food odor cues are not guiding behavior in the maze is to demonstrate delay-dependent impairments in the task. Specifically, when a delay is imposed between the fourth and fifth choices in the maze, longer delays will result in increased errors during the post-delay session. If odor cues were guiding behavior, no delay-dependent change in errors would be expected. If delay-dependent maze performance can be demonstrated with prebaiting of arms, then its use is appropriate.

C.   Response Patterning

In versions of the RAM task that involve a design in which all arms are accessible and rewarded, the subject must visit each arm once to obtain a reward; subsequent visits are not rewarded and are scored as errors. Results obtained using this design may be confounded due to the subject not employing a spatial strategy to optimize its search for reward. Subjects may use an algorithmic pattern of arm selections, such as always visiting the immediately neighboring arm to the left; thus the subject would not make any errors. (Actually, in our experience, a common strategy used by rats in the RAM is to choose every second arm [moving clockwise or anti-clockwise] until an error [i.e., no food reward] occurs, then to select the immediately adjacent arm and continue to visit every second arm.) However, the adoption of such an egocentric behavioral strategy resembles what is defined as skill or reference memory and minimizes or eliminates the working memory aspect of the task. These egocentric strategies are based upon how the animal responds, not on spatial, extra-maze, or allocentric cues. Behavior guided by spatial cues is thought to be mediated by the hippocampus.[30] Rats with hippocampal or caudate-putamen lesions,[31] or aged rats[9,10] have been reported to engage in egocentric response strategies. Aged rats, for example, tend to select more distal arms than proximal ones, whereas younger rats tend to select proximal over distal arms.[32,33] Obviously, other treatments of interest may influence egocentric strategy and thus confound interpretations of treatment effects on working memory, unless procedures are adopted to prevent the use of these egocentric strategies. These procedures, detailed above, include the use of a DNMTS task which precludes the use of egocentric response strategies.

V.   Example Applications

A.   Chlordiazepoxide Modulation

In a series of studies in one of our laboratories, we examined whether an acute disruption of the septo-hippocampal pathway would produce a profile of errors in the DNMTS task that was different from that produced by a chronic disruption of this system. In these studies, rats were injected with chlordiazepoxide (CDP) into the medial septum at different times relative to training or testing. Intraseptal injections of benzodiazepines such as CDP, or GABA-A or GABA-B agonists, decrease high affinity choline transport (HAChT) in the hippocampus, and reduce acetylcholine turnover and release.[34-36]

These studies examined the time-dependence, site-specificity, and potential mechanism of action of CDP. Rats injected with CDP into the medial septum, but not the lateral septum, amygdala, or nucleus basalis, exhibited dose-dependent deficits in DNMTS performance. Furthermore, there was a distinct time-dependent profile with deficits observed when CDP was injected immediately, but not 15, 30, or 45 min after the pre-delay session. Rats injected with CDP immediately after training were able to acquire and use information during the post-delay session, but they were unable to utilize information from the pre-delay session to guide performance. They exhibited a significant decrease in CC and an increase in RE, but no alterations in PE. These rats exhibited impaired memory of the arms entered in the pre-delay session but they could acquire and retain new information during the post-delay session. CDP disrupted working memory (i.e., retrograde amnesia) but did not produce an anterograde amnesia or performance deficit. The lack of PE is important since it demonstrates that working memory is intact during the post-delay session in this group.

Deficits were also observed when CDP was injected immediately prior to the post-delay session but there was a qualitatively different pattern of errors, with a selective increase in PE, but not RE. Rats were able to use information acquired prior to the delay but their acquisition of information during the test session was impaired. These data demonstrate that intraseptal CDP impairs working memory, and the degree to which different phases of working memory are disrupted depends upon the time of injection. Post-training injection of CDP impairs the encoding or maintenance of working memory, while pretest infusion impairs the ability to use or encode information to guide performance during the test session. In contrast, rats injected with CDP immediately before the post-delay test session exhibited a significant increase only in PE. An increase in PE suggests an impaired ability to store or maintain current arm choices into working memory.

B.   Cholinergic Modulation

Cholinergic antagonists also impair spatial memory in a wide variety of cognitive paradigms across several species, including humans.[15] Systemic and intra-hippocampal administration of the muscarinic cholinergic antagonist, scopolamine, prior to training impairs working memory performance of rats in the RAM.[37,38] Using a RAM paradigm similar to the DNMTS task described above,[39] it was found that scopolamine: (1) injected prior to *training* increased the number of RE but not PE; and (2) injected immediately prior to *testing* increased both RE and PE. The effects of pre-training and pre-testing scopolamine are further support for the involvement of cholinergic-dependent processes in the encoding and retrieval of working memory.

C.   Attenuation of Scopolamine Impairment

Cholinergic hypofunction has been associated with AD[40] and can be mimicked to some extent by blockade of muscarinic cholinergic receptors by scopolamine.[41] RAM

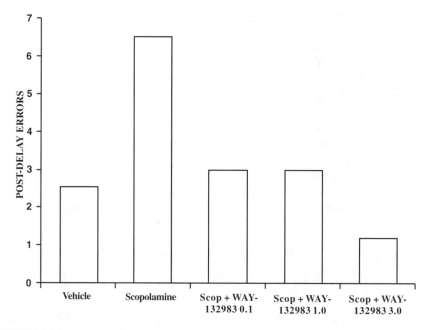

FIGURE 11.2

Stylized representation of the effects of the M1 preferring agonist, WAY-132983, on post-delay errors in radial maze following scopolamine treatment. Details of the experiment can be found in Bartolomeo et al.[46]

impairments can be reliably produced by administration of scopolamine.[42,43] Agents that mimic the effects of acetylcholine (ACh) or modulate endogenous ACh can attenuate the scopolamine-induced impairment[44,45] Specifically, we have previously reported[46] that a selective M1 muscarinic agonist, WAY-132983, can attenuate a scopolamine-induced impairment of RAM performance (Figure 11.2). Other cholinergic modulators have also been reported to show similar effects.[47] Interestingly, a variety of agents with other noncholinergic mechanisms have also been shown to attenuate scopolamine-induced impairments.[48] Unfortunately, many of these agents have been tested in AD patients with little or no apparent benefit. The only drugs that have shown activity in this model and which have benefited AD patients are acetylcholinesterase inhibitors[45] or m1 agonists.[49] Thus, although attenuation of a scopolamine impairment of RAM would be expected from a drug that would be beneficial in treating AD, this effect is not sufficient to warrant the initiation of clinical trials.

D. AF64A Impairment

One reason that attenuation of scopolamine-induced deficits may not be a very good predictor of efficacy in AD is that in AD, cholinergic neurons actually die,[50] whereas after scopolamine treatment, receptors are blocked but cholinergic neurons are intact.

Therefore, following scopolamine treatment, agents that modulate intact cholinergic neurons may result in attenuation of the impairment, but be false positives for treating AD. One way to establish a more robust model of AD would be to selectively destroy cholinergic neurons by treatment with the ethylcholine mustard azirdinium ion known as AF64A,[51] a treatment which results in RAM impairment. AF64A is a choline analog that has been used to study the function of central cholinergic neurons. This compound targets the critical biochemical events that regulate the synthesis of acetylcholine. AF64A inhibits the activity of the high affinity choline transport system (HAChT) that is the rate-limiting step in the synthesis of acetylcholine. Choline is taken into the cholinergic nerve terminal by HAChT and, once inside, it is acetylated by choline acetyltransferase (ChAT) to form acetylcholine. AF64A is a cytotoxic analog of choline that combines a choline-like structure (i.e., ethylcholine), that is recognized by the HAChT system, with a highly reactive cytotoxic aziridinium ring. Due to its structural similarity to choline, AF64A is taken into the terminal by the HAChT system and, once inside the terminal, the highly reactive aziridinium induces cholinergic hypofunction and the death of the cell (reviewed in[52,53]). Injection of AF64A into the lateral cerebroventricles reduces all indices of presynaptic cholinergic function, including regional concentrations of acetylcholine, the activity of choline acetyltransferase, HAChT, K+ – and ouabain-stimulated release of acetylcholine from hippocampal slices. AF64A appears to produce a series of toxic events in cholinergic neurons that culminate in chronic disability of the neuron or cell death. The mechanisms of cell death are still under investigation but they probably involve the generation of oxidative stress and its impact on nucleic acid function. The time course and specificity of AF64A are reviewed by Hanin[53] and by Walsh and Opello.[20] AF64A is a useful tool to selectively compromise the cholinergic innervation of the hippocampus.

A number of laboratories have used the radial-arm maze to examine the effects of AF64A on cognitive function. An early question that guided these studies was whether AF64A produced a profile of cognitive changes that resembled those observed in Alzheimer's disease. Tests that evaluate spatial learning are commonly used since they share formal similarities with the most common tests of human cognition and they can be manipulated to examine the components of cognitive behavior.[54,55]

One of our laboratories initiated a series of studies to determine whether AF64A produced a dissociation of memory in which working memory was compromised and reference memory was spared. In the first set of studies rats were trained in the DNMTS task prior to surgery. After acquisition they were bilaterally injected icv with either CSF or 3.0 nmoles of AF64A and given a 14-day recovery period before being retested. Following recovery, rats were returned to the DNMTS task and tested in 30 sessions with random delays of 0, 0.5, 1, 5, 15, or 30 mins imposed between the fourth and fifth arm choices. Rats injected with vehicle (CSF) exhibited excellent performance at each of the delay intervals (80 to 85% correct). However, the AF64A-injected group was performing at approximately chance levels (50 to 60%) of accuracy regardless of delay condition.

Despite the impairments in the DNMTS task, the AF64A group was able to acquire a reference memory task (i.e., position discrimination) in which they learned to perform an invariant response of always going to the same location in a T-maze. All animals required a comparable number of trials to achieve a performance criterion of 70, 80, or 90% correct in any 10-trial block. A similar dissociation between working memory and reference memory was also observed in a T-maze task that independently assessed both types of memory within the same trial.[56] In this task, each trial was initiated by placing the rat into either the left or the right start/goal box of the T-maze for 15 sec. The rat was then rewarded with a single food pellet for turning into and traversing the stem of the maze and for subsequently returning to the alternate start/goal location. This task involves: (1) a trial-independent component, which involves an invariant response of running down the stem of the maze (reference memory); and (2) a trial-dependent response (working memory) in which the animal had to remember the start location in order to perform accurately.[56] Rats injected icv with CSF readily reacquired both components of the task. The AF64A group, however, exhibited a transient decrease in reference memory that recovered within 30 trials and a persistent decrease in working memory that was evident throughout the 75 post-operative trials. Working memory was impaired in this group even when their reference memory performance was comparable to controls and close to 100% accurate.

AF64A produced a deficit in working memory without affecting reference memory in these appetitive-motivated spatial tasks. Further studies demonstrated that the cognitive deficits induced by AF64A were dose dependent and related to the cognitive demands of the task.[57] Rats were bilaterally injected icv with 0, 0.75, or 1.5 nmoles AF64A and tested with either 0, 1, or 2 hr delays in the DNMTS task. While none of the groups were impaired at the 0 delay condition, the 1.5 nmole group was impaired at both the 1 and the 2 hr delay. However, the 0.75 nmole group was impaired only at the 2 hr delay. Therefore, the memory impairments were related to the dose of AF64A and to the demand placed upon the working memory system by the task. Furthermore, the deficits depended upon the cholinergic hypofunction induced by AF64A since preventing the cholinilytic effects of AF64A with hemi-cholinium-3, a more potent inhibitor of HAChT than AF64A,[58] prevented the AF64A-induced deficits in the DNMTS task and the decreases in HAChT.[59]

Rats treated with AF64A were also deficient in the DNMTS task. The deficit exhibited by these subjects, in addition to an impairment of encoding/maintenance of working memory, included deficits in the ability to utilize current arm selections made during the post-delay test session. That is, AF64A-treated rats made fewer CC and more RE and PE than CSF-treated controls. The AF64A group exhibited no recovery with time or repeated testing. The chronic compromise of the SHC induced by AF64A appeared to impair the encoding, maintenance, and perhaps retrieval of working memory. This dose of AF64A has been shown to consistently decrease measures of pre-synaptic cholinergic function in the HPC without affecting cholinergic parameters in the cortex, striatum, or other brain areas.[60,61] The number of both PE and RE during the last five days of testing was negatively correlated with HAChT in the HPC. A greater number of either type of error was related to lower HAChT values.

E.   Attenuation of AF64A Impairment

We have previously reported that muscarinic agonists (e.g., Arecoline, xanomeline, WAY-132983) can attenuate the AF64A impairment when delivered by osmotic minipump over a two-week period (Figure 11.3).[46] Interestingly, assessment of a small number of AD patients has shown that the nonselective muscarinic agonist arecoline can have beneficial effects[62] and xanomeline has been reported to be beneficial in AD,[49] but side effects and pharmacokinetic issues have limited their development. WAY-132983 has not been tested in AD patients. With admittedly limited data, so far the AF64A model appears to have improved the prediction of potential efficacy in AD.

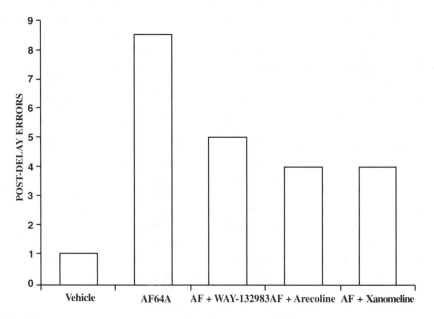

FIGURE 11.3
Stylized representation of the effects of the muscarinic agonists WAY-132983, arecoline, and xanomeline on post-delay errors in radial maze following AF64A treatment. Details of the experiment can be found in Bartolomeo et al.[46]

VI.   Advantages and Disadvantages

The RAM task offers a variety of experimental advantages for the study of cognitive function in rodents. For example, its behavioral substrates and its temporal dynamics are now well characterized. The task assesses working memory and is sensitive to parametric manipulations which vary the demand on memory processes. Furthermore, there is an extensive literature regarding the brain areas and neurotransmitter systems that are critical for its performance. The RAM has been successfully used to study normal cognition in rodents and the changes in spatial memory that result from age, circumscribed lesions, and pharmacological manipulation of transmitter systems.

Like any useful tool, the RAM is best suited to address specific questions and problems. The co-authors of this chapter utilize the RAM to elaborate the neural substrates of spatial memory and to develop new therapies to treat the cognitive deficits associated with AD. To maximize the utility of the RAM, investigators must be aware of the inherent advantages and limitations of using this task within the context of specific paradigms. Some of these are enumerated below.

Advantages — Performance in radial mazes is based on natural foraging behaviors of rats and therefore generalization of results to the behavioral repertoire of the animal may be more accurate than those based on non-natural behaviors. Once animals have been trained to a baseline criterion, the task requirements can be changed for each session, thereby allowing the use of within-subject designs. This reduces the number of animals required for experiments. As has been indicated above, working memory can be selectively studied using the radial maze. Due to the relatively short testing duration which includes both an information acquisition and an information retrieval session, state-dependent effects are minimized when using radial maze tasks. Finally, the task is sensitive to cognitive alterations induced by drugs, lesions, genetic manipulations, and aging.

Disadvantages — Since the task is appetitively motivated, food restriction regimens are required that contrast with cognitive tasks which are aversively motivated. Some practical limitations include the need for intensive labor, extensive time, and sufficient space. For example, the majority of studies using the radial maze are done on well-trained animals and extensive training is required to establish stable baseline performance.

Overall, the radial maze is an excellent choice for providing insights regarding memory processing. It can be used: (1) to examine the anatomical, biochemical, and molecular substrates of memory; (2) to develop models of cognitive disorders such as AD; and (3) to evaluate new and innovative therapies for the treatment of cognitive dysfunction. It is a powerful behavioral tool that can address a multitude of hypothesis-driven questions about the complexity and dynamics of memory, its underlying biology, and its attendant diseases.

References

1. Olton, D.S. Mazes, maps, and memory. *American Psychologist*, 34, 583, 1979.
2. Olton, D.S. and Samuelson, R.J. Remembrance of places passed: Spatial memory in rats. *Journal of Experimental Psychology: Animal Behavior Processes*, 2, 97, 1976.
3. Hodges, H. Maze procedures: the radial-arm and water maze compared. *Cognitive Brain Research*, 3, 167, 1996.
4. Rawlins, J.N.P. and Deacon, R.M.J. Further developments of maze procedures. In *Behavioural Neuroscience Volume I A practical approach*, Saghal, A., Ed., IRL Press, Oxford, 1993, chap 8.
5. Morris, R.G., Garrud, P., Rawlins, J.N., and O'Keefe, J. Place navigation impaired in rats with hippocampal lesions. *Nature*, 297, 681, 1982.

6. Cassel, J.C., Cassel, S., Galani, R., Kelche, C., Will, B., and Jarrard, L. Fimbria-fornix vs selective hippocampal lesions in rats: effects on locomotor activity and spatial learning and memory. *Neurobiology of Learning & Memory*, 69, 22, 1998.

7. Bartolomeo, A.C., Morris, H., and Boast, C.A. Arecoline via miniosmotic pump improves AF64A-impaired radial maze performance in rats: a possible model of Alzheimer's Disease. *Neurobiology of Learning and Memory*, 68, 333, 1997.

8. Hsiao, K., Chapman, P., Nilsen, S., Eckman, C., Harigaya, Y., Younkin, S., Yang, F., and Cole, G. Correlative memory deficits, Abeta elevation, and amyloid plaques in transgenic mice. *Science*, 274, 99, 1996.

9. Chrobak, J.J., Hanin, I., Lorens, S.A., and Napier, T.C. Within-subject decline in delayed-non-match-to-sample radial arm maze performance in aging Sprague-Dawley rats. *Behavioral Neuroscience*, 109, 241, 1995.

10. Tanila, H., Shapiro, M., Gallagher, M., and Eichenbaum, H. Brain aging: changes in the nature of information coding by the hippocampus. *Journal of Neuroscience*, 17, 5155, 1997.

11. Williams, C.L., Meck, W.H., Heyer, D.D., and Loy, R. Hypertrophy of basal forebrain neurons and enhanced visuospatial memory in perinatally choline-supplemented rats. *Brain Research*, 794, 225, 1998.

12. Forgie, M.L. and Kolb, B. Sex differences in the effects of frontal cortex injury: role of differential hormonal experience in early development. *Behavioral Neuroscience*, 112, 141, 1998.

13. Nakamura-Palacios, E.M., Caldas, C.K., Fiorini, A., Chagas, K.D., Chagas, K.N., and Vasquez, E.C. Deficits of spatial learning and working memory in spontaneously hypertensive rats. *Behavioural Brain Research*, 74, 217, 1996.

14. Hyde, L.A., Hoplight, B.J., and Denenberg, V.H. Water version of the radial-arm maze: learning in three inbred strains of mice. *Brain Research*, 785, 236, 1998.

15. Walsh, T.J. and Chrobak, J.J. Animal models of Alzheimer's disease: Role of hippocampal cholinergic systems in working memory. In: *Current Topics in Animal Learning: Brain, Emotion and Cognition*, Dachowski, L. and Flaherty, C. Eds. Lawrence, Erlbaum, NJ, 347, 1991.

16. Maurer, S.A., Storch, F.E., LaForge, R.R., and Boast, C.A. Task difficulty determines the differential memory-impairing effects of EAA antagonists in gerbils. *Pharmacology, Biochemistry & Behavior*, 51, 345, 1995.

17. Sessions, G.R., Pilcher, J.J., and Elsmore, T.F. Scopolamine-induced impairment in concurrent fixed-interval responding in a radial maze task. *Pharmacology, Biochemistry & Behavior*, 59, 641, 1998.

18. Stackman, R.W. and Walsh, T.J. Anatomical specificity and time-dependence of chlordiazepoxide-induced spatial memory impairments. *Behavioral Neuroscience*, 109, 436, 1995.

19. Walsh, T.J., Herzog, C.D., Gandhi, C., Stackman, R.W., and Wiley, R.G. Injection of IgG 192-saporin into the medial septum produces cholinergic hypofunction and dose-dependent working memory deficits. *Brain Research*, 726, 69, 1996.

20. Walsh, T.J. and Opello, K.D. The use of AF64A to model Alzheimer's disease. In *Animal Models of Toxin-Induced Neurological Disorders*, M. Woodruff and A. Nonneman, Eds. Plenum Press, N.Y., 259, 1994.

21. Novelli, M., Masiello, P., Bombara, M., and Bergamini, E. Protein glycation in the aging male Sprague-Dawley rat: effects of antiaging diet restrictions. *Journals of Gerontology. Series A, Biological Sciences & Medical Sciences*, 53, B94, 1998.

22. Stackman, R.W. and Walsh, T.J. Distinct profile of working memory errors following acute or chronic disruption of the cholinergic septohippocampal pathway. *Behavioral Neural Biology*, 64, 226, 1995.

23. Otto, T., Wolf, D., and Walsh, T.J. Combined lesions of perirhinal and entorhinal cortex impair rats' performance in two versions of the spatially-guided radial arm maze. *Neurobiology of Learning and Memory*, 68, 21, 1997.

24. Porter, M.C. and Mair, R.G. The effects of frontal cortical lesions on remembering depend on the procedural demands of tasks performed in the radial arm maze. *Behavioural Brain Research*, 87, 115, 1997.

25. Young, H.L., Stevens, A.A., Converse, E., and Mair, R.G. A comparison of temporal decay in place memory tasks in rats (Rattus norvegicus) with lesions affecting thalamus, frontal cortex, or the hippocampal system. *Behavioral Neuroscience*, 110, 1244, 1996.

26. Demas, G.E., Nelson, R.J., Krueger, B.K., and Yarowsky, P.J. Spatial memory deficits in segmental trisomic Ts65Dn mice. *Behavioural Brain Research*, 82, 85, 1996.

27. Arendash, G.W., Sanberg, P.R., and Sengstock, G.J. Nicotine enhances the learning and memory of aged rats. *Pharmacology, Biochemistry & Behavior*, 52, 517, 1995.

28. Kesner, R.P., Crutcher, K., and Beers, D.R. Serial position curves for item (spatial location) information: Role of the dorsal hippocampal formation and medial septum. *Brain Research*, 454, 219, 1988.

29. Nelson, A., Lebessi, A., Sowinski, P., and Hodges, H. Comparison of effects of global cerebral ischaemia on spatial learning in the standard and radial water maze: relationship of hippocampal damage to performance. *Behavioural Brain Research*, 85, 93, 1997.

30. Eichenbaum, H. Is the rodent hippocampus just for 'place'? *Current Opinion in Neurobiology*, 6, 187, 1996.

31. Oliveira, M.G., Bueno, O.F., Pomarico, A.C., and Gugliano, E.B. Strategies used by hippocampal- and caudate-putamen-lesioned rats in a learning task. *Neurobiology of Learning & Memory*, 68, 32, 1997.

32. Kobayashi, S., Kametani, H., Ugawa, Y., and Osanai, M. Age difference of response strategy in radial maze performance of Fischer-344 rats. *Physiology & Behavior*, 42, 277, 1988.

33. Caprioli, A., Ghirardi, O., Giuliani, A., Ramacci, M.T., and Angelucci, L. Spatial learning and memory in the radial maze: a longitudinal study in rats from 4 to 25 months of age. *Neurobiology of Aging*, 12, 605, 1991.

34. Blaker, W.D., Cheney, D.L., Gandofi, O., and Costa, E. Simultaneous modulation of hippocampal cholinergic activity and extinction by intraseptal muscimol. *Journal of Pharmacology & Experimental Therapeutics*, 225, 361, 1983.

35. Walsh, T.J., Stackman, R.W., Emerich, D.F., and Taylor, L.A. Intraseptal injection of GABA and benzodiazepine receptor ligands alters high affinity choline uptake in the hippocampus. *Brain Research Bulletin* 31, 267, 1993.

36. Imperato, A., Dazzi, L., Obinu, M.C., Gessa, G.L., and Biggio, G. Inhibition of hippocampal acetylcholine release by benzodiazepines: antagonism by flumazenil. *European Journal of Pharmacology*, 238, 135, 1993.

37. Okaichi, H. and Jarrard, L.E. Scopolamine impairs performance of a place and cue task in rats. *Behavioral & Neural Biology*, 35, 319, 1982.

38. Brito, G.N., Davis, B.J., Stopp, L.C., and Stanton, M.E. Memory and the septo-hippocampal cholinergic system in the rat. *Psychopharmacology*. 81, 315, 1983.

39. Beatty, W.W. and Bierley, R.A. Scopolamine impairs encoding and retrieval of spatial working memory in rats. *Physiological Psychology*, 14, 82–86, 1986.

40. McKinney, M. and Coyle, J.T. The potential for muscarinic receptor subtype-specific pharmacotherapy for Alzheimer's disease. *Mayo Clinic Proceedings*, 66, 1225, 1991.

41. Barker, A., Jones, R., Prior, J., and Wesnes, K. Scopolamine-induced cognitive impairment as a predictor of cognitive decline in healthy elderly volunteers: a 6-year follow-up. *International Journal of Geriatric Psychiatry*, 13, 244, 1998.

42. Peele, D.B. and Baron, S.D. Effects of selection delays on radial maze performance: Acquisition and effects of scopolamine. *Pharmacology Biochemistry and Behavior*, 29, 143, 1988.

43. Decker, M.W. and Gallagher, M. Scopolamine-disruption of radial arm maze performance: Modification by noradrenergic depletion. *Brain Research*, 417, 59, 1987.

44. Levin, E.D., Kaplan, S., and Boardman, A. Acute nicotine interactions with nicotinic and muscarinic antagonists: Working and reference memory effects in the 16-arm radial maze. *Behavioural Pharmacology*, 8, 236, 1997.

45. Wang, T. and Tang, X.C. Reversal of scopolamine-induced deficits in radial maze performance by (-)-huperzine A: Comparison with E2020 and tacrine. *European Journal of Pharmacology*, 349, 137, 1998.

46. Bartolomeo, A.C., Morris, H., Buccafusco, J.J., Kille, N., Rosenzweig-Lipson, S., Husbands, M.G., Sabb, A.L., Abou-Gharbia, M., Moyer, J.A., and Boast, C.A. The preclinical pharmacological profile of WAY-132983, a potent m1 preferring agonist. *Journal of Pharmacology & Experimental Therapeutics*, 292, 584, 2000.

47. Fisher, A., Heldman, E., Gurwitz, D., Haring, R., Karton, Y., Meshulam, H., Pittel, Z., Marciano, D., Brandeis, R., Sadot, E., Barg, Y., Pinkas-Kramarski, R., Vogel, Z., Ginzburg, I., Treves, T.A., Verchovsky, R., Klimowsky, S., and Korczyn, A.D. M1 agonists for the treatment of Alzheimer's disease. Novel properties and clinical update. *Annals of the New York Academy of Sciences*, 777, 189, 1996.

48. Sarter, M. Animal models of brain ageing and dementia. *Comprehensive Gerontology. Section A, Clinical & Laboratory Sciences*, 1, 4, 1987.

49. Bodick, N.C., Offen, W.W., Levey, A.I., Cutler, N.R., Gauthier, S.G., Satlin, A., Shannon, H.E., Tollefson, G.D., Rasmussen, K., Bymaster, F.P., Hurley, D.J., Potter, W.Z., and Paul, S.M. Effects of xanomeline, a selective muscarinic receptor agonist, on cognitive function and behavioral symptoms in Alzheimer disease. *Archives of Neurology*, 54, 465, 1997.

50. Boissiere, F., Hunot, S., Faucheux, B., Hersh, L.B., Agid, Y., and Hirsch, E.C. Trk neurotrophin receptors in cholinergic neurons of patients with Alzheimer's disease. *Dementia & Geriatric Cognitive Disorders*, 8, 1, 1997.

51. Walsh, T.J., Tilson, H.A., DeHaven, D.L., Mailman, R.B., Fisher, A., and Hanin, I. AF64A, a cholinergic neurotoxin, selectively depletes acetylcholine in hippocampus and cortex, and produces long-term passive avoidance and radial-arm maze deficits in the rat. *Brain Research*, 321, 91, 1984.

52. Hortnagl, H. and Hanin, I. Toxins affecting the cholinergic system. In *Handbook of Experimental Pharmacology: Selective Neurotoxicity* Vol. 102, H. Herken and F. Hucho, Eds. Springer-Verlag, Berlin, 293, 1992.

53. Hanin, I. The AF64A model of cholinergic hypofunction: an update. *Life Sciences.* 58, 1955, 1996.

54. Olton, D.S. The use of animal models to evaluate the effects of neurotoxins on cognitive processes. *Neurobehavioral Toxicology & Teratology.* 5, 635, 1983.

55. Walsh, T.J. and Chrobak, J.J. The use of the radial-arm maze in neurotoxicology. *Physiology and Behavior*, 40, 799, 1987.

56. Chrobak, J.J., Hanin, I., and Walsh, T.J. AF64A (ethylcholine aziridinium ion), a cholinergic neurotoxin, selectively impairs working memory in a multiple component T-maze task. *Brain Research* 414, 15, 1987.

57. Chrobak, J.J. and Walsh, T.J. Dose and delay dependent working/episodic memory impairments following intraventricular administration of ethycholine aziridinium ion (AF64A). *Behavioral & Neural Biology* 56, 200, 1991.

58. Barker, L.A. and Mittag, T.W. Comparative studies of substrates and inhibitors of choline transport and choline acetyltransferase. *Journal of Pharmacology & Experimental Therapeutics*, 192, 86, 1975.

59. Chrobak, J.J., Spates, M.J., Stackman, R.W., and Walsh, T.J. Hemicholinium-3 prevents the working memory impairments and the cholinergic hypofunction induced by ethycholine aziridimium ion (AF64A). *Brain Research,* 504, 269, 1989.

60. Chrobak, J.J., Hanin, I., Schmechel, D.E., and Walsh, T.J. AF64A-induced working memory impairment: Behavioral, neurochemical and histological correlates. *Brain Research*, 463, 107, 1988.

61. Hanin I. AF64A-induced cholinergic hypofunction. *Progress in Brain Research*, 84, 289, 1990.

62. Soncrant, T.T., Raffaele, K.C., Asthana, S., Berardi, A., Morris, P.P., and Haxby, J.V. Memory improvement without toxicity during chronic, low dose intravenous arecoline in Alzheimer's disease. *Psychopharmacology*, 112, 421, 1993.

12

Use of the Radial-Arm Maze to Assess Learning and Memory in Rodents

Edward D. Levin

Contents

Abstract

The radial-arm maze has proven to be a very useful technique for assessing spatial learning and memory in rodents. Many different sizes of radial-arm mazes have been used, with the most common being the 8-arm maze. The radial maze takes advantage of rodents' natural tendency to explore new places for food reinforcement. They

0-8493-0704-X/01/$0.00+$.50
© 2001 by CRC Press LLC

quickly learn the maze to asymptotic levels of choice accuracy performance against which drug, lesion, or behavioral manipulations can be made to assess the neurobehavioral bases of cognitive function.

I.   Introduction

The radial-arm maze has proven to be very useful for assessment of spatial learning and memory in rodents. It was originally developed for rats but has also been used for other rodents such as mice[1-3] and gerbils.[4-6] The radial-arm maze takes advantage of the natural tendency of rodents to learn and remember different spatial locations for food. It has been used in a variety of configurations to assess the neurobehavioral bases for learning and memory, the adverse effects of toxic chemicals, and the beneficial effects of novel therapeutic treatments.

The radial-arm maze in its present configuration was developed by the late David Olton and his colleagues.[7] Similar types of mazes have been used since the early and middle parts of the 20th century.[8-11] Olton and co-workers provided extensive information concerning the neurobehavioral processes involved in radial-arm maze performance. Rats use extra-maze visual cues to navigate about on the maze.[12,13] Brain areas such as the hippocampus and the frontal cortex, as well as cholinergic transmitter systems, are of critical importance for solving the radial-arm maze.[14-18] Several reviews are available which describe the behavioral methods for and neural systems involved in radial-arm maze performance.[16,19-24] The extensive information concerning the neurobehavioral aspects of radial-arm maze is an important advantage for its use.

II.   Radial-Arm Maze Design

The radial-arm maze is flexible. The numbers of arms can vary from three on up. The most common configuration is eight arms. Larger mazes from 12 to 17 arms have often been used to provide a greater challenge to the subject, with sufficient numbers of arms to test both working and reference memory. There are also a variety of ways to run the radial-arm maze to access different types of information concerning cognitive function.

Radial-arm mazes can be purchased prefabricated or made in the laboratory. Automated radial-arm mazes can be purchased from a variety of vendors. The advantage of such mazes include the ability to easily add and remove arms, record arm entries, and dispense reinforcements in an automated fashion. Disadvantages include high cost, mechanical breakdowns, superstructure attenuating extra-maze visual cues, and the noise of the mechanics of the maze distracting the subject. Many labs have used to good effect "homemade" radial-arm mazes made of wood or plastic.

A.   8-Arm Maze

We have used radial 8-arm mazes, constructed of wood and painted black. The octagonal central arena is 50 cm across with eight 10 by 60 cm arms extended radially. Food cups are located 2 cm from the end of each arm. There are rails 2 cm high along the sides of the arms to keep the rat from slipping off the maze. The maze is raised 30 cm from the floor. There is a removable opaque plastic ring that fits on the central platform, and serves as the starting point for each session.

The maze needs to be located in a testing room which contains many extra-maze visual cues which normally should be kept in the same position during testing. A behavioral challenge can be used by altering the configuration of visual cues in the test room or by switching the rat to a maze in a different room. Typically upon such a switch, performance accuracy immediately drops off and then over the course of several sessions returns to the previously attained baseline. This technique can be used to identify subtle impairments caused by small neural insult. The testing environment needs to a quiet room without noise and distraction, unless of course one would like to provide these as distractors.

B.   16-Arm Maze

Adding additional arms (e.g., 16-arm maze) is useful for increased cognitive load and for providing sufficient capability for simultaneous determination of working and reference memory.[25-27] An 8-arm maze can be easily converted into a 16-arm maze by adding an additional arm between each of the eight existing arms. With a large number of arms going out of a standard-size central platform, clear plastic barriers are needed to keep the rat from jumping directly from one arm to the next without re-entering the center platform.

C.   Mouse Radial-Arm Maze

Similar procedures can be used for other rodents such as mice. The principal advantage of using mice is the ability to use knockouts and transgenic over expressers of various genes involved with cognitive function. Mice learn the radial-arm maze quite readily.[2] The maze is made of wood painted black and consists of a center platform 12 cm in diameter, with eight extending arms (24 by 4 cm) with food cups located 1 cm from the end of each arm. There are 1 cm high rails to keep the mouse from slipping off the maze. The maze is elevated 25 cm from the floor.

III.   Procedures

A.   Adaptation

For successful studies using the radial-arm maze it is important to perform preliminary procedures to assure that too much stress or too little motivation interfere with
learning the maze.

It is important to adapt the rats and mice to human handling before the start of
testing in the radial-arm maze. Animals not adapted will show a stress response
when picked up to be put on the maze. This will complicate the interpretation of
the choice accuracy data. Holding and petting the rats or mice for several minutes
2 to 5 times is usually sufficient to adapt them to not show behavioral signs of stress.

Since the rats and mice are to run the maze for food reinforcement it is important
to shape them to eat the food reinforcements. Rewards used in the maze are often
sweetened cereal such as Froot Loops® (Kellogg's, Battle Creek, MI), but can also
be operant food pellets, water, or sweetened milk. The following is a procedure we
have found to be useful for food reinforcements. After the handling procedure
described above, put the number of rewards to be consumed during a session in the
animal's home cage once per day for two days. Then, put the rat in the central arena
start barrel with the number of rewards to be consumed during a session within the
barrel. Record the time it takes for the subject to eat all of the reinforcements up to
a maximum time equal to the maximum session length (5 minutes for the 8-arm
maze). Repeat daily until the rat eats all of the reinforcements within the maximum,
then begin training. We do not spread rewards throughout the arms of the maze as
a part of the shaping procedure, because this conflicts with the task the rat will be
called upon to learn after the shaping is complete. This is especially important in
studies in which the initial rate of acquisition is of interest.

B.   Win-Shift Acquisition

This is the most common way to run the radial-arm maze. The rat is only reinforced
once for an entry into each arm. Before each session, all the arms of the maze are
baited with a 1/3 to 1/2 piece of sugar-coated cereal (Kellogg's Froot Loops®). The
reinforcements are not replaced during the session. A radial-arm maze test session
is started when the rat was placed in a circular plastic ring in the central platform.
After 10 seconds the ring was lifted and the rat is allowed to freely explore the
maze. Arm choices are recorded when the rat had placed all of its paws beyond the
threshold of the arm. The radial-arm maze test session continues until the rat enters
all eight arms or 5 minutes elapse. Wipe the maze surface between rats to remove
fecal boli and urine. Repeat training 3 to 5 times per week. Rats and mice typically
take 18 to 24 sessions to complete the learning phase of training and enter a period
of stable performance during which effects on memory can be tested. Figure 12.1
shows an example of learning curves in the win-shift procedure on the 8-arm radial
maze with juvenile rats having quicker learning than adults. Figure 12.2 shows an

Radial-Arm Maze Age-Effect Function

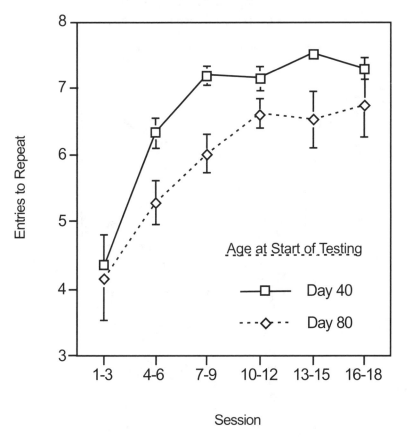

FIGURE 12.1

Age-effect function on learning in the 8-Arm radial maze.[30]

example of learning curves in the win-shift procedure on the 8-arm radial maze with transgenic mice having overexpression of extracellular superoxide dismutase (EC-SOD) or EC-SOD knockouts having impaired acquisition.

C. Working/Reference Memory

The radial-arm maze can be used to distinguish between working memory, defined as memory with changing contents, as opposed to reference memory, defined as memory with fixed contents. The procedure is to leave some arms of the maze without bait. Entries into these arms would be reference memory errors since their status as unbaited arms is unchanging. The task for the animal is to retrieve the baits from the other arms which are baited once at the beginning of the session. Re-entries into these arms are working memory errors since their status has changed from

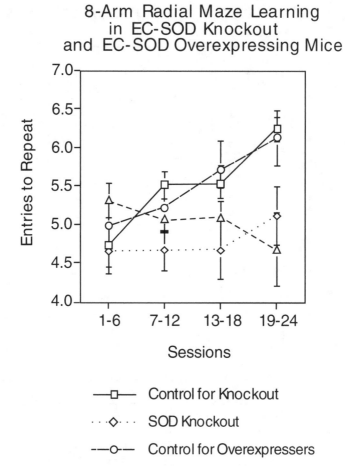

FIGURE 12.2
Acquisition of win-shift 8-Arm radial maze performance by mice.[31]

baited to unbaited. Although some investigators test working and reference memory
on an 8-arm maze with four baited and four unbaited arms, this lessens the sensitivity
of the task as the memory demands for the location of four baits is low. We have
found that use of a 16-arm maze with 12 baited and 4 unbaited arms provides a
useful measure of working vs. reference memory while keeping task demands high.
Figure 12.3 shows an example of working and reference memory performance on
a 16-arm radial maze with the muscarinic antagonist scopolamine causing a selective
impairment in working memory performance.

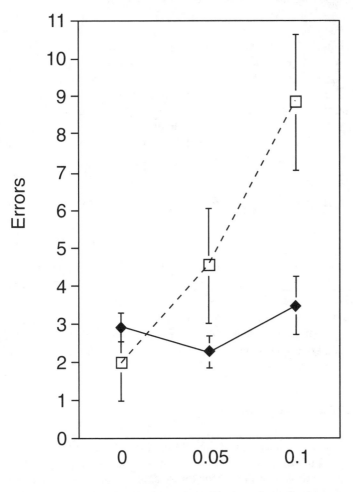

Working and Reference Memory Performance
in the 16-Arm Radial Maze
Effects of Scopolamine Challenge

Scopolamine Dose (mg/kg)

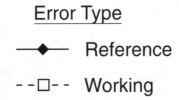

Error Type

♦ Reference

--□-- Working

FIGURE 12.3
Selective effects of the muscarinic cholinergic antagonist scopolamine on working memory in the 16-Arm radial maze.[26]

D.   Repeated Acquisition

This is a method for repeatedly assessing learning in the radial-arm maze. The repeated acquisition procedure for the radial-arm maze was developed by Peele and Baron.[28] The method we have used with success is to bait three arms at the beginning of each trial of a session. Allow the rat to freely enter arms until it has entered all three of the baited arms. Count the number of errors (entries into unbaited arms or re-entries into initially baited arms). For each session, repeat for five trials the same three arms baited, with a maximum of 180 seconds per trial. On each session, change which three arms are baited. We use a rule that there be no more than two adjacent arms baited. After 18 sessions of training the rat should show a reliable learning curve during each session, with a decreasing number of errors across the five trials of the session as the rats learns the new location of the reinforcements. Figure 12.4 shows an example of within session repeated acquisition curves on the 8-arm radial maze with a nicotinic agonist improving performance.

E.   Delayed Matching to Position

In contrast to the conventional win-shift version of the radial-arm maze task, in this task the rewarded strategy was win-stay. In each session, three two-run trials are done. On the first run of each trial a food reward is placed in one randomly selected arm (the other seven arms are blocked off) and the rat is rewarded for entering it. After this sample run the rat is placed in the center ring for a delay of 30 seconds. The blocks are removed and the sample arm is rebaited. The rat is then allowed to freely select arms until it finds the reward. This procedure is repeated two more times each session with other sample arms used. Between trials the rat is placed in a holding cage for 30 seconds.

F.   Non-Spatial Discrimination

For the non-spatial discrimination learning, differentially textured floor inserts can be used as critical stimuli for reward placement in the radial-arm maze.[29] Rewards will be placed in the arms with four of the textures but not the others. The location of the textured floor inserts are shifted each session.

IV.   Data Analysis

There are a variety of ways of measuring choice accuracy in the radial-arm maze. The most common method is to count total errors to complete the task. This includes all of the errors but does not specify when during the session they occurred. In addition, the rats often do not finish the task in the maximum time allotted. For the win-shift task we use the entries to repeat measure. This is the number of correct entries until an error

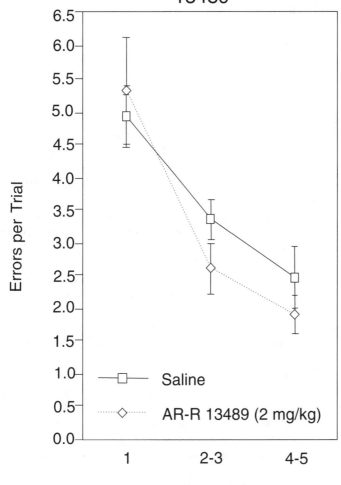

Repeated Acquisition in the 8-Arm Radial Effects of the Nicotinic Agonist AR-R 13489

FIGURE 12.4
Repeated acquisition in the 8-Arm radial maze.[31]

is made. Random chance performance for the entries to repeat measure on an 8-arm maze as determined by computer simulation is 3.25. The response latency measure is the total session duration divided by the number of arms entered (seconds per entry).

It is common to block sessions for analysis. In this way a more stable measure of performance can be attained and sessions in which the rat does not choose enough arms to provide a valid choice accuracy score before the maximum session length will not result in holes in the data set. One can average together the number of sessions in the block that do have valid accuracy scores.

Acknowledgment

Work was supported by National Institute on Drug Abuse grant # DA 11943.

References

1. Ammassari-Teule, M. and Caprioli, A., Spatial learning and memory, maze running strategies and cholinergic mechanisms in two imbred strains of mice. *Behav. Brain Res.*, 17, 9, 1985.

2. Levin, E.D., Brady, T.C., Hochrein, E.C., Oury, T.D., Jonsson, L.M., Marklund, S.L., and Crapo, J., Molecular manipulations of extracellular superoxide dismutase: functional importance for learning. *Behav. Genetics*, 28, 381, 1998.

3. Toumane, A., Durkin, T., and Marighetto, A., The durations of hippocampal and cortical cholinergic activation induced by spatial discrimination testing of mice in an eight-arm radial maze decrease as a function of acquisition. *Behav. Neural Biol.*, 52, 279, 1989.

4. Schwartz-Bloom, R.D., McDonough, K.J., Chase, P.J., Chadwick, L.E., Inglefield, J.R., and Levin, E.D., Long-term neuroprotection by benzodiazepine full vs. partial agonists following transient cerebral ischemia in the gerbil. *J. Cerebral Blood Flow Metabolism*, 18, 548, 1998.

5. Takai, R.M. and Wilkie, D.M., Foraging experience effects gerbils' (Meriones unguiculatus) radial arm maze performance. *J. Comp. Psychol.*, 99, 361, 1985.

6. Wilkie, D.M. and Slobin, P., Gerbils in space: performance on the 17-arm radial maze. *J. Expr. Anal. Behav.*, 40, 301, 1983.

7. Olton, D.S. and Samuelson, R.J., Remembrance of places passed: Spatial memory in rats. *J. Exp. Psychol.*, 2, 97, 1976.

8. Hamilton, G., A study of trial and error reactions in mammals. *J. Anim. Behav.*, 1, 33, 1911.

9. Hamilton, G., A study of perseverance reactions in primates and rodents. *Behav. Monogr.*, 3(2), 1, 1916.

10. Tolman, E.C., Ritchie, B.F., and Kalish, D., Studies in spatial learning. I. Orientation and the short-cut. *J. Exp. Psychol.*, 63, 13, 1946.

11. Lachman, S.J. and Brown, C.R., Behavior in a free choice multiple path elimination problem. *J. Psychol.*, 43, 27, 1957.

12. Foreman, N. and Stevens, R., Visual lesions and radial maze performance in rats. *Behav. Neural Biol.*, 36, 126, 1982.

13. Zoladek, L. and Roberts, W.A., The sensory basis of spatial memory in the rat. *Anim. Learn. Behav.*, 6, 77, 1978.

14. Becker, J.T., Walker, J.A., and Olton, D.S., Neuroanatomical bases of spatial memory. *Brain Res.*, 200, 307, 1980.

15. Kolb, B., Pittman, K., Sutherland, R.J., and Wishaw, I.Q., Dissociation of the contributions of the prefrontal cortex and dorsomedial thalamic nucleus to spatially guided behavior in the rat. *Behav. Brain Res.*, 6, 365, 1982.

16. Levin, E.D., Psychopharmacological effects in the radial-arm maze. *Neurosci. Biobehav. Rev.*, 12, 169, 1988.

17. Levin, E.D. and Rose, J.E., Nicotinic and muscarinic interactions and choice accuracy in the radial-arm maze. *Brain Res. Bull.*, 27, 125, 1991.

18. Olton, D.S. and Werz, M.A., Hippocampal function and behavior: Spatial discrimination and response inhibition. *Physiol. Behav.*, 20, 597, 1978.

19. Olton, D.S., Becker, J.T., and Handelmann, G.E., Hippocampus, space, and memory. *Behav. Brain Sci.*, 2, 313, 1979.

20. Olton, D.S., The use of animal models to evaluate the effects of neurotoxins on cognitive processes. *Neurobehav. Toxicol. Teratol.*, 5, 635, 1983.

21. Olton, D.S., The radial arm maze as a tool in behavioral pharmacology. *Physiol. Behav.*, 40, 793, 1987.

22. Olton, D. and Markowska, A., Mazes: Their uses in delayed conditional discriminations and place discriminations, In F. van Haaren (Ed.), *Methods in Behav. Pharmacol.*, Elsevier, New York, pp. 195, 1993.

23. Rawlins, J. and Deacon, R., Further developments of maze procedures, In A. Sahgal (Ed.), *Behavioral Neuroscience: A Practical Approach, Volume 1*, IRL Press at Oxford University Press, New York, pp. 95, 1994.

24. Walsh, T.J. and Chrobak, J.J., The use of the radial arm maze in neurotoxicology. *Physiol. Behav.*, 40, 799, 1987.

25. Levin, E., Kim, P., and Meray, R., Chronic nicotine effects on working and reference memory in the 16-arm radial maze: Interactions with D_1 agonist and antagonist drugs. *Psychopharmacology*, 127, 25, 1996.

26. Levin, E., Kaplan, S., and Boardman, A., Acute nicotine interactions with nicotinic and muscarinic antagonists: Working and reference memory effects in the 16-arm radial maze. *Behav. Pharmacol.*, 8, 236, 1997.

27. Levin, E.D., Bettegowda, C., Weaver, T., and Christopher, N.C., Nicotine-dizocilpine interactions and working and reference memory performance of rats in the radial-arm maze. *Pharmacol. Biochem. Behav.*, 61, 335, 1998.

28. Peele, D.B. and Baron, S.P., Effects of scopolamine on repeated acquisition on radial-arm maze performance by rats. *J. Exp. Anal. Behav.*, 49, 275, 1988.

29. Gray, J.A., Mitchell, S.N., Joseph, M.H., Grigoryan, G.A., Bawe, S., and Hodges, H., Neurochemical mechanisms mediating the behavioral and cognitive effects of nicotine. *Drug Dev. Res.*, 31, 3, 1994.

30. Chambers, R.A., Moore, J., McEvoy, J.P., and Levin, E.D., Cognitive effects of neonatal hippocampal lesions in a rat model of schizophrenia. *Neuropsychopharmacology*, 15, 587, 1996.

31. Levin, E., Bettegowda, C., Blosser, J., and Gordon, J., AR-R 17779, An $\alpha 7$ nicotinic agonist improves learning and memory in rats. *Behav. Pharmacol.*, 10, 675–680, 1999.

Chapter 13

An Operant Analysis of Fronto-Striatal Function in the Rat

Stephen B. Dunnett and Peter J. Brasted

Contents

I.   Introduction

The basal ganglia were once believed to function as part of an extrapyramidal motor system, operating separately from the pyramidal tract.[1,2] However, this concept has been discarded for two fundamental reasons. First, the basal ganglia have been shown to be an intrinsic part of well-defined anatomical circuits that not only receive cortical input but that also send projections, via the thalamus, back to those cortical areas that control motor output. Second, a wealth of experimental work has shown that the striatum, the main input structure for the basal ganglia, can no longer be regarded purely as a motor structure. The observation that striatal damage could induce deficits in cognitive function led researchers such as H. Enger Rosvold to state that the striatum may reflect the function of those areas of neocortex that project to it. The realisation that the striatum may mediate a wide variety of functions that reflect its diverse cortical innervation has also become evident in studies of patients with basal ganglia disorders. Thus, impairments in cognitive function are now well documented in patients with neurodegenerative diseases such as Huntington's disease (HD) or Parkinson's disease (PD), disorders once regarded as entirely movement related.

Attempts to examine disease states such as HD in experimental animals can provide both insight into normal brain function and a means by which to assess potential therapeutic strategies. In either scenario, an operant analysis of behaviour can prove particularly powerful. The detailed functional analyses that are permitted by operant paradigms not only allow more specific questions to be asked of normal brain function, but can also provide experimental paradigms which are extremely sensitive to brain insults and subsequent recovery.

II.   The Neuropathological and Behavioural Profile of HD

A.   HD Pathology

Originally reported by George Huntington in 1872,[3] HD is a fatal inherited neuro-degenerative disorder, the genetic basis of which has recently been identified.[4] The disease is characterised pathologically by the loss of GABAergic medium spiny projection neurones within the neostriatum (caudate nucleus and putamen). The disease process begins in the caudate nucleus and progresses through the entire striatum in a medial to lateral and dorsal to ventral fashion.[5,6] Striatal degeneration involves in particular loss of the medium spiny projection neurones, with relative sparing of the large aspiny interneurones. There is post mortem evidence that the earliest striatal neurones affected are those in the striosomes, projecting to the substantia nigra pars compacta, and Hedreen has proposed that more diverse striatal projections via this nigral feedback is an essential component in the spread of the disease.[7] As the disease progresses, striatal atrophy, gliosis and cell loss become progressively more marked, which has been characterised by Vonsattel in a widely used 5-point grading system. In advanced disease, not just the striatum but widespread areas of the forebrain — in particular areas such as neocortex or substantia nigra pars reticulata that are sites of afferent and efferent connections with the striatum — also undergo atrophy and cell loss.

Although the mechanism by which the genetic mutation involving an expansion of the polyglutamine repeat in a previously unknown peptide of unknown function (designated "huntingtin") remains unresolved,[8] the first transgenic models of the disease have recently identified the formation of abnormal nuclear inclusions of N-terminal truncated fragments of huntingtin,[9-12] which have now been confirmed as being widely distributed in the brains of affected individuals.[13] Nevertheless, at the time of writing, there remain many unresolved issues regarding the relationship of intranuclear inclusions to disease progression, not least whether inclusion formation is itself a fundamental component in the pathogenesis of Huntington's disease, or is an epiphenomenal marker of some other neurodegenerative process.

B. HD Symptomatology

The uncontrollable movements (or chorea) that characterise HD are now recognised to be only one part of the disease's behavioural profile. In fact HD presents with a triad of motor, cognitive, and affective symptoms, all of which worsen as the disease progresses inexorably in parallel with the progress of the underlying degeneration. Indeed, the introduction of genetic screening in conjunction with more sensitive clinical tests has led to the suggestion that subtle cognitive and psychiatric aspects of the disease are apparent before the onset of chorea.[14]

The most striking aspect of HD is the chorea that originally gave its name to the disorder, until the diverse nature of impairments in HD was acknowledged. However, these unwanted choreic actions often mask an underlying bradykinesia, and deficits in initiating responses in reaction time paradigms,[15,16] are indicative of impairments in initiating and selecting motor programmes. Consequently, it is now recognised that both hypokinetic and hyperkinetic symptoms co-exist in HD.[17-19]

Moreover, it has become obvious that HD patients express a profile of neuro-psychological deficits not dissimilar to those seen in patients with prefrontal cortical damage.[20] This includes impairments in learning[21,22] and working memory,[21,23] as well as deficiencies in executive tasks that assess planning and attentional control.[14,23,24] A number of psychiatric symptoms are also present in HD, such as depression and anxiety.[25]

III. Excitotoxic (and Other) Lesions of the Rat Striatum

The impact of striatal damage on motor and cognitive function and motivational state can be all examined experimentally in the rat. By recreating a *milieu* in experimental animals that resembles the pathology of diseases such as HD, we can attempt to define the precise functional role of particular neural mechanisms more precisely than can be achieved in the human condition. These animal models of human disease can also provide a basis for assessing potential strategies for repair (e.g., neural transplantation, neuroprotective agents, or gene therapy) by giving rise to measurable behavioural deficits against which the functional efficacy of any particular strategy can be evaluated.

Striatal function was initially studied in experimental animals with the use of basal ganglia lesions.[2,26] However, there were major difficulties in interpreting the consequences of lesions made by electrolysis, radiofrequency, or direct surgical excision because of the inevitable damage of the immediately adjacent afferent and efferent fibres of the internal capsule connecting the cortex to subcortical structures including thalamus. However, this changed dramatically with the introduction of excitotoxic methods of lesions in the mid 1970s, opening the way for the modern era of basal ganglia research. The primary excitotoxins are amino acids, such as monosodium glutamate, N-methyl-D-aspartic acid (NMDA), and kainic acid that are glutamate agonists which are toxic when against glutamate receptor bearing neurones, a feature of most neurones of the nervous system. When administered directly into the striatum, excitotoxins specifically target and kill neurones within the striatum without damaging the axons of the corticofugal and corticopetal pathways passing through and adjacent to the striatum. Furthermore, injections of excitotoxins into the striatum produce neurochemical and pathological changes similar to those seen in HD. Initially, kainic acid was used for this purpose.[27-29] However, ibotenic acid and particularly quinolinic acid have since become the toxins of choice on account of numerous neurochemical studies that demonstrate a more selective neuronal loss within the striatum, with the medium spiny GABA neurones being particularly vulnerable and the large aspiny cholinergic, neuropeptide Y, and NADPH-diaphorase positive interneurones being relatively resisitant to these toxins, corresponding to the profile of degeneration observed in HD.[30-33]

While a single neurotoxic insult is able to mimic the neuropathology of HD, it cannot reproduce the slow and progressive degeneration that is a characteristic feature of the human disease. These features can be better mimicked by metabolic toxins, such as 3-nitropropionic acid, which target the striatal neurones selectively, even when administered peripherally.[34,35] Nevertheless, excitotoxins continue to be widely used, due to the fact that they typically produce more convenient, consistent, and reproducible lesions than appear achievable with the metabolic toxins.[36] The recent development of transgenic animals that mimic the genetic abnormalities of HD and reproduce specific aspects of the pathogenetic process of the human disease are likely to provide more powerful models for the future,[37-39] but the basic behavioural impairments observed in transgenic mice are only just beginning to be characterised.[40,41] Moreover, the precise mechanisms of cell death are still poorly understood and the relationship between the expression of expanded polyglutamine repeat, the formation of intranuclear inclusions, cell death, and behavioural symptoms remains to be clarified.

The destruction of striatal cells with excitotoxins not only produces some of the pathological hallmarks of HD, but also gives rise to behavioural sequelae which reflect many of the symptoms seen clinically. The first use of excitotoxins to model the pathology of HD showed unilateral striatal lesions to induce a marked rotation toward the ipsilateral side 48 hours after surgery.[27] This motor asymmetry reflected the inability of the lesioned striatum to mediate contralateral movement, and this biasing of motor output to the side ipsilateral to the lesioned striatum is evident in many indices of striatal dysfunction such as amphetamine-induced rotation[42-44] or the elevated body swing test.[45]

An additional advantage of unilateral striatal lesions is that they can allow a within-subject analysis of dysfunction and recovery. Paw-reaching,[44,46,47] as well as several of the operant tasks we will elaborate on below, typically take advantage of this asymmetry and allows performance mediated by the intact striata to be compared with performance that is under the control of the lesioned striata. However, this laterality of motor function is less applicable in a test of cognitive function or motivation, and paradigms that assess these aspects of behaviour typically employ bilateral lesions.

IV. Operant Conditioning and Operant Chambers

Although many standard tests such as rotation can provide a useful index of functional capacity, it is often desirable to examine a subject's ability to perform an action, or series of actions, that is more purposeful, or goal-directed. Thus, rather than assess a general and undefined motor capacity, one can examine the capability of an animal to produce specific motor responses which are required to achieve a particular outcome. Operant conditioning refers to this ability to train animals to perform purposeful actions, and the underlying learning processes involved.

In an operant task, a subject learns to respond in a particular way to gain reward (or to escape punishment) when a discrete cue, or stimulus, signals that this response will be effective.

A.   Operant Chambers

A standard operant chamber will typically contain:

- The *operandum*, such as a lever, on which an animal can operate, i.e., act on or respond to.
- *Reinforcers*, stimuli that increase the likelihood of responding. Animals may either act to obtain positive reinforcers (e.g., food or water) or alternatively they may act to terminate or postpone negative reinforcers (e.g., electric shock).
- *Discriminative stimuli*, typically lights, sounds, or olfactory cues, the properties of which signal the timing and location of reinforcers and thereby control the animal's response.

The operant chamber is an automated apparatus designed to present discriminative stimuli (e.g., lights that differ in colour, location, or intensity), to record and measure responses to the operanda (e.g., presses of the lever, presses of the

panel covering the food well, licks at a drinking tube), and to deliver reinforcements (e.g., by operating a dispenser to deliver food pellets). Although traditionally
run by electromagnetic relays, modern operant chambers are typically under microcomputer control.

The classic test apparatus for evaluating operant behaviours is the Skinner box,
an automated test apparatus first devised and developed by B.F. Skinner when
analysing the behaviour of rats responding to obtain food reward.[48] As illustrated in
Figure 13.1, a typical Skinner box provides two levers as operanda, to which the rat
may respond. The timing of responding into discrete trials can be achieved by making
the levers retractable and only available at discrete points within the trial. Discriminative stimuli are provided by a variety of different lights located above the levers,
above and within the food hopper, and in the roof of the test chamber. A loudspeaker
in the chamber can also present auditory stimuli either as white noise or discrete
tones of controlled frequency and intensity. A variety of different reinforcers can be
built in. This is most typically either a dispenser to deliver food pellets or a liquid
dipper to present water into the reward chamber. These will only be effective if the
animal is suitably motivated by hunger or thirst, achieved by some hours of food or
water deprivation, respectively, prior to the training session. However, other reinforcers are also possible, such as presentation of a receptive female rat to a male rat
in studies of hormonal control of sexual motivation.[49]

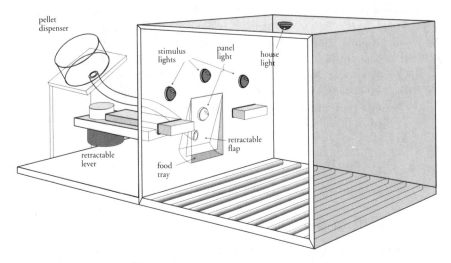

FIGURE 13.1
Schematic illustration of a two retractable lever operant chamber (Skinner box).

An alternative type of operant chamber that has proved highly effective is the
nine-hole box (Figure 13.2). The nine-hole box is conceptually similar to the standard
Skinner box except that instead of levers, the box is supplied with an arc of nine
holes. Discriminative stimuli are provided by lights at the rear of each hole, responses
(nose pokes) are monitored by infra-red beams at the entrance to each hole, and
food is delivered as the reinforcer to a well that is positioned at the rear of the

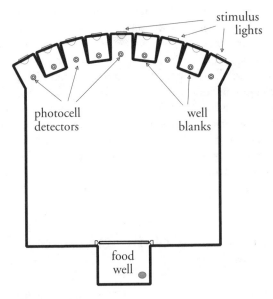

FIGURE 13.2
Schematic illustration of the nine-hole box operant chamber.

chamber. Rats typically investigate stimulus lights associated with food reward by producing a nose poke into individual holes, which is arguably a more ethologically natural action for a rat than pressing or releasing a lever. Although originally designed to assess attentional function, the physical configuration of the apparatus in the nine-hole box allows great specificity in defining both stimuli and responses and this has allowed for the laterality of motor function evident in structures such as the striatum to be analysed with great precision (see below).

The fact that behaviour can be controlled in an operant chamber enables a far more precise control of the factors determining behaviour than can be achieved by conventional observational and hand testing methods. Using different stimuli to signal the class of responses that will be reinforced, it is possible to determine the nature of the sensory discriminations that an animal can make, and subsequently its performance on cognitive tasks, as well as the effects of changes in reward value or magnitude.

B.   Operant Tasks to Assess Striatal Function in the Rat

Tasks that are administered in operant chambers usually require animals to perform a number of intermediate training sessions before they are able to perform the task in full. Since it is important to design and use tasks correctly to assess different specified functions, the rationale for a variety of tasks that assess striatal function is outlined below. Reference is made to those sources that contain more detailed descriptions of how to accomplish individual behavioural tasks.

Behavioural testing in operant chambers allows both a high degree of experimental control and a detailed and thorough behavioural analysis. Consequently,
operant conditioning has proved to be a valuable way in which to assess cognition,
motor function, and motivation in the rat. Simple operant responses, such as pressing
a lever to gain food reward, are unaffected by striatal lesions,[50,51] and more elaborate
schedules that require conditional responses which are under the control of spatial
and visual cues are also possible.[52,53] However, more complex operant tasks turn out
to reveal subtle but specific and robust behavioural impairments in many functional
domains after striatal lesions. Indeed, it is often not possible to define precisely the
nature of such impairments with alternative observational testing methods. This is
perhaps best illustrated with operant paradigms that were designed specifically to
ascertain the role of striatal dopamine in motor responding in the rat.

V.   Operant Analysis of Striatal Lesions: Deficits in Motor Responding

The unilateral depletion of striatal dopamine by central injection of 6-OHDA is
known to cause a general sensorimotor impairment on the side contralateral to the
lesion. This impairment takes the form of an ipsilateral rotation,[54] clumsy and
inefficient use of the contralateral paw[55-57] and also a failure to respond to stimuli
that are presented to the side of the rat contralateral to the lesion,[58] a phenomenon
that has been named striatal neglect. This syndrome of impairments has many
similarities to the sensorimotor produced by electrolytic lesions of the lateral hypothalamus.[59,60] Marshall and colleagues suggested in the 1970s that the neglect that
resulted from electrolytic lesions of the lateral hypothalamus might in fact be the
product of damage to the nigrostriatal projection.[55] Thus, tasks that used lateralised
stimuli and responses which were first introduced to define the precise nature of
deficits resulting from hypothalamic lesions,[61,60] came to influence subsequent behavioural paradigms which sought to analyse striatal function.

A.   Operant Analysis of the Sensory and Motor Aspects of Sensorimotor Striatal Neglect

Carli et al.[62] sought to distinguish the sensory and motor aspects that constituted the
sensorimotor components of striatal neglect. Inspired by the earlier design innova-
tions of Turner in a lateralised escape task,[61] Carli and colleagues designed a visual
choice reaction time task, an operant task in the nine-hole box apparatus that allowed
the spatial location of the discriminative stimuli and conditioned responses to be
separated.[62,63] In this task, only the central three holes were exposed, the remaining
holes being covered with a masking plate (Figure 13.3). Two groups of rats were
trained to poke their noses into the middle of the three holes for a variable period
until the presentation of a brief light cue which flashed unpredictably in one of the
holes, either to the left or the right of the centre hole. Rats were then required to

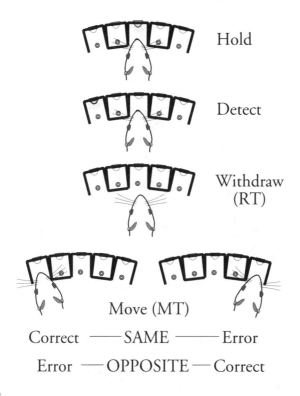

Hold

Detect

Withdraw
(RT)

Move (MT)

Correct ——— SAME ——— Error

Error —— OPPOSITE — Correct

FIGURE 13.3

Schematic illustration of the SAME and OPPOSITE versions of the Carli choice reaction time task. As well as measures of accuracy and response bias, the speed of initiating (reaction time) and executing (movement time) of correct responses to the two sides are also recorded. The test is based on Carli et al. 1985.[62]

produce a lateralised nose-poke response in order to gain food reward, but the rule defining a correct response was different for the two groups. The rats of one group were required to make a nose-poke into the same response hole that had been previously lit (the "SAME" condition), whereas the rats of the second group had to respond on the side which had *not* previously been lit (the "OPPOSITE" condition).

For animals trained in the SAME condition, because of the crossover of connections between the brain and the periphery, we would predict that unilateral lesions — whether of the dopamine system or the striatum itself — would produce deficits on the contralateral side of the body (Figure 13.4). This is equally true whether the deficit is sensory, motor, or associative in nature. However, the dissociation between the location of the stimuli and the response holes in the OPPOSITE condition allows differential predictions of the outcome depending on the nature of the underlying deficit. Thus, if the animals have a sensory impairment in the detection of the stimuli, then we would expect the rats with unilateral lesions to be impaired making an ipsilateral response to a contralateral stimulus, whereas a response to an ipsilateral stimulus would be unaffected (Figure 13.4). Conversely, if the deficit was primarily in the selection or initiation of motor response, then we would expect the animal to

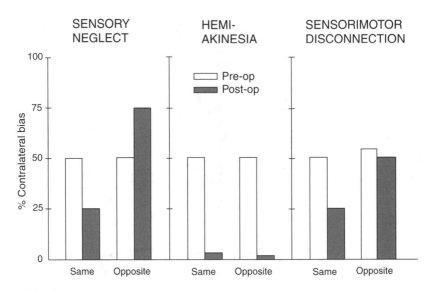

FIGURE 13.4

Graphic predictions of different outcomes on the Carli task in the nine hole box, dependent upon whether the functional impairment is one of sensory neglect (left), hemi-akinesia, or other impairment in response initiation (centre), or sensorimotor disconnection (right). (Based on Carli et al. 1989.[64])

be impaired making a contralateral response to an ipsilateral stimulus but be unimpaired in responding on the ipsilateral side even though the discriminative stimuli are presented in contralateral space.

Carli and colleagues[62] reported that in both the SAME and the OPPOSITE condition, rats were impaired in effecting responses to the side contralateral to the lesion, while neither group was impaired in producing ipsilateral responses (Figure 13.5). This pattern of impairments was not consistent with either a sensory impairment or a sensorimotor integration deficit; rather, the ipsilateral bias induced by unilateral striatal dopamine depletion was therefore interpreted as a bias in responding. The extent of this ipsilateral bias has subsequently been shown to correlate with the extent of dopamine depletion within the striatum.[64]

Further analysis of the nature of the deficit suggested that the general motoric deficits reported following striatal damage may reflect a specific deficit in the initiation of responses. Thus, Carli et al.[62] observed that unilateral striatal dopamine depletion resulted in a striking increase in the latency to execute contralateral responses to the contralateral side. However, a detailed analysis of this action showed that contralateral movement was not uniformly impaired. The time taken to effect each response was considered as two components. Reaction time or initiation time was defined as the latency to initiate a response by withdrawing the nose from the central hole (and therefore contains no lateralised component). Movement time was defined as the subsequent latency to complete the lateralised nose poke response into the response hole. The animals with unilateral nigrostriatal lesions were particularly impaired in the reaction time measure — the latency to withdraw the nose from the central hole — rather than in the movement time — the time taken to

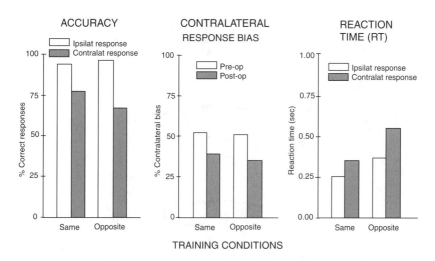

FIGURE 13.5

Unilateral 6-hydroxydopamine lesions of the nigrostriatal pathway produce marked deficits in initiating action (reaction time) specifically for contralateral responses in the Carli task. (Left) Post-op accuracy of responding to the ipsilateral and contralateral sides in the same and opposite task contingencies. (Centre) Degree of contralateral response bias pre- and post-op. (Right) Post-op reaction times to initiate ipsilateral and contralateral responses in the same and opposite task contingencies. (Data from Carli et al. 1985.[62])

execute the lateral movement to poke their noses in the contralateral hole.[62] This indicates that the deficit is attributable to an impairment in the planning, selection, or initiation of the lateralised response rather than in its execution, suggesting an executive impairment, similar to that seen after frontal lesions, rather than a pure motor deficit *per se*.

Unilateral striatal lesions have subsequently been shown to induce a similar pattern of deficits in biasing responding to the ipsilateral side and slowing of the reaction time on the contralateral side

B.   Operant Tasks to Delimit the Specificity of Striatal Neglect

The study of Carli et al.[62] demonstrated that unilateral striatal damage impaired animals in responding toward their contralateral side, while at the same time showing that — once initiated — movements made in contralateral space were not impaired *per se*. This suggested that if the motor actions that comprised the response itself were not themselves impaired, perhaps the striatum was more involved in some more abstract elements of responding that influence motor output at an early stage of response preparation. For example, it was possible that the co-ordinate frame in which responses were organised was disrupted by striatal damage.[66] We therefore sought to modify the experimental design to define the basis of this response space; what is the contralateral deficit seen in striatal neglect actually contralateral to? The

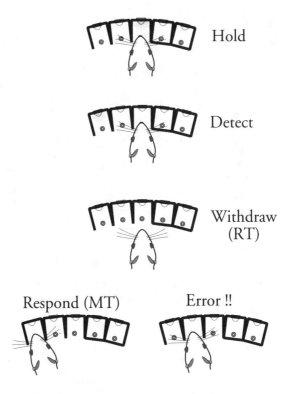

FIGURE 13.6
Schematic illustration of the Brasted lateralised choice reaction time task in the nine-hole box. The test is based on Brasted et al. 1997.[67]

nature of this neglect has been quantified using both unilateral lesions of the nigro-striatal dopamine neurones and excitotoxic lesions of intrinsic striatal neurones.

Animals were trained to perform two discriminations, independently, on alternate days. As in the Carli et al.[62] study, the task comprised a central hole and two response holes. However, unlike the Carli et al.[62] study, both response holes were on the same side. So, on one day, animals were required to respond to the holes on the left, and on the next day to the two holes on the right (see Figure 13.6).[67] All responses required rats to detect a stimulus light in one of the two response holes, and to make a nose poke response in the same hole. Once trained, animals received unilateral striatal lesions with central injections of quinolinic acid.

When testing resumed a week later, the lesion rats showed a severe impairment responding on the contralateral side. This impairment took the form of a marked bias toward the near hole, i.e., the hole closer to the centre, when the holes were on the side contralateral to the lesions. This response bias was so severe that lesion animals were rarely able to produce responses to the far hole on the contralateral side. In stark contrast to this impairment, lesion rats were able to respond efficiently and correctly when the holes were on the side ipsilateral to the lesion. It was this distinction in responding on the ipsilateral side and contralateral side that allowed the specific nature of the hypothesised response space to be revealed.[67]

The design of this operant task allowed a specific comparison to be made between two specific hypotheses concerning the nature of a striatally mediated response space. If responses were coded relative to an external referent within the animals' environment (allocentric coding) then animals would be expected to always neglect the relatively contralateral hole, regardless of in which side of space the holes were presented. (Allocentric coding is often seen in perceptual neglect, when patients with cortical lesions neglect the contralateral side of an object, regardless of where the object is located in space.[68,69]) In this task, an allocentric-based deficit would manifest itself as a bias toward the far hole when the task is performed to the ipsilateral side, and a bias toward the near hole when the task is performed to the contralateral side. Alternatively, if responses were coded with respect to the subject's body (egocentric coding), then one would predict responding to be disrupted only on the contralateral side.

The data clearly show that striatal neglect is not seen uniformly in all parts of space, but is restricted to the contralateral side and thus consistent with the latter, egocentric, hypothesis. When responding to the ipsilateral holes, animals showed no evidence of biasing their responding toward the far (i.e., relatively contralateral) hole (Figure 13.7). In contrast, animals were markedly impaired when performing on the contralateral side and were completely unable to select responses to the far (i.e., relatively contralateral) hole (Figure 13.7). A similar impairment was seen in studies which examined unilateral striatal dopamine lesions, using a between-subject design.[66]

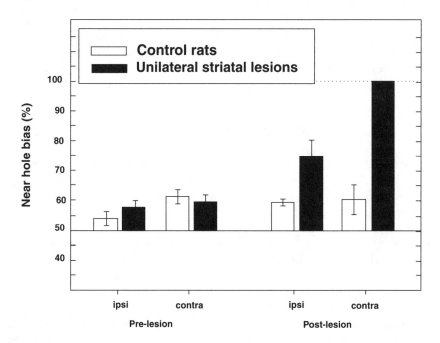

FIGURE 13.7
Unilateral striatal lesions produce a marked postoperative ipsilateral response bias which is more marked for discriminations on the contralateral than on the ipsilateral side. (Data from Brasted et al. 1997.[67])

These studies demonstrate the thorough and systematic analysis that is afforded by operant tasks, and illustrate how operant conditioning and appropriate experimental design can be combined to bring specificity to a general and undetermined behavioural deficit.

VI.   Operant Analysis of Striatal Lesions: Deficits in Cognitive Tasks

It was Rosvold's experimental work in the 1950s that seriously challenged the concept of the striatum as a structure involved purely in motor function. Based on the topographical nature of projections from prefrontal cortex to the caudate nucelus, it was shown that caudate lesions in monkeys produced impairments on the same tasks that were known to be sensitive to frontal lesions.[70] In particular, bilateral lesions of the caudate nucelus were seen to impair accuracy on Jacobsen's classic tasks of frontal function designed to assess spatial working memory, such as spatial alternation and delayed response.[71-73]

A corresponding functional organisation was demonstrated in the rat only when the prefrontal cortex came to be defined in terms of the projection areas of the mediodorsal nucleus of the thalamus — namely, to the medial and orbital walls of the frontal cortex rather than pole of the frontal lobe, as had previously been assumed.[74,75] Lesions in the rostral medial striatal areas that were innervated by medial wall prefrontal cortex again produced deficits in delayed alternation and spatial navigation tasks.[76-78] It is these advances in functional neuroanatomy, as well as the neuropsychological deficits reported in basal ganglia disorders such as HD and PD, that provide the theoretical foundation for operant tasks designed specifically to assess the impact of striatal lesions on cognitive function.

A.   Delayed Matching Tasks

The automation of delayed matching procedures represented one such attempt to assess the impact of striatal lesions on working memory. This requires an animal to be presented initially with a sample stimulus, and then after a variable delay, to be presented with a choice between two stimuli, one being the earlier sample and the other novel. Remembrance of the earlier stimulus presentation is then tested by requiring the animal to choose either the initial sample (delayed-matching-to-sample; DMTS) or the novel stimulus (Delayed-non-matching-to-sample; DNMTS). Correct responses result in the delivery of food rewards, whereas an incorrect response is signalled by "time out" or some other error signal. The introduction of a variable delay between sample presentation and choice response allows the rate of forgetting to be determined from the plot of decline in choice as the length of the delay interval increases. DMTS was first developed for assessing memory in primates, and required monkeys to discriminate between objects primarily on the basis of the visual properties (e.g., shape, colour) of objects.[70] Procedures for rats, however, can require animals to discriminate between objects in a maze[80] or between different odours.

Alternatively, choice discriminations can be made in the spatial modality, and this has led us to develop a delayed-matching-to-position (DMTP) task for use by rats.[81,82] Animals are trained in standard 2-retractable lever Skinner boxes (Figure 13.1). As shown in Figure 13.8, on each trial, one of the two retractable levers is inserted into the chamber as the sample, which the rat must press to register. The lever is then retracted and the rat must press the central panel until, after a random variable delay period, the next panel press causes both levers to be extended back into the chamber. A correct matching response on the same lever as that produced in the sample stage is rewarded with the delivery of a food pellet, whereas an incorrect response to the other lever results in "time out" (all lights are turned off for a short period to signal an error). After a further short interval, the next trial commences.

Delayed-matching-to-sample is sensitive to PFC damage in primates.[83-85] Consequently, we have developed DMTP to investigate the comparative contributions of the medial PFC and medial striatum in a working memory task in rats. Cortical lesions that are restricted to the anterior medial prefrontal cortex result in a progressive deficit such that accuracy is increasingly impaired at progressively longer delays.[86] This delay-dependent pattern of impairment suggests a rather specific disturbance in short-term memory. However, larger lesions that extend into the anterior cingulate cortex produce a broader impairment at all delays, reflecting a more generalised deficit in the animal's ability to perform the matching rule (Figure 13.9).[86] Similarly, when we looked at impairments following striatal lesions, lesions in the ventral striatum (which receives restricted prefrontal inputs) exhibited a clear delay-dependent deficit, whereas lesions in the dorsal striatum (which receives cingulate as well as frontal inputs) disrupted animals' performance at all delays (Figure 13.4).[86-88]

These studies illustrate, first, the way in which an operant task may be designed to tap discrete aspects of cognitive function — in this case, short-term working memory — and secondly that selective lesions within the fronto-striatal circuitry yield distinctive patterns of cognitive deficit associated with the particular cortical loop(s) that is/are disturbed.

B.   Delayed Alternation Tasks

The spatial delayed alternation task (DA) is another task that involves short-term working memory and requires animals to alternate their responding between two spatially distinct locations. Like the DMTP task, it is very sensitive to damage of the mPFC. Indeed, deficits in delayed alternation in Wisconsin boxes (for monkeys) and in T mazes (for rats) were among the defining features of the prefrontal deficit described by Jacobsen in his classic primate studies in the 1930s,[89,90] and replicated many times since.[21,91,92] Rats similarly exhibit clear deficits in delayed alternation after prefrontal lesions when tested in a T maze.[88,93-95]

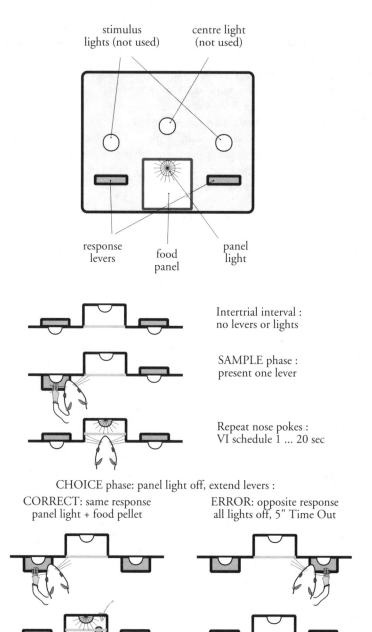

FIGURE 13.8

Schematic illustration of the delayed matching to position (DMTP) task in a two retractable lever Skinner box. (Left) The front panel of the Skinner box. (Right) the stages in sample and choice presentation and response in a single trial of the task. The test is based on Dunnett 1985[81] and 1993.[82]

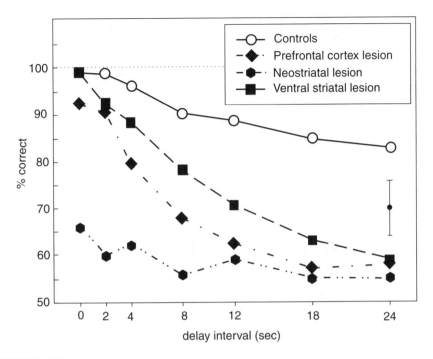

FIGURE 13.9

Prefrontal cortex and dorsal and ventral striatal lesions produce marked deficits on the DMTP task. Note that whereas prefrontal and ventral striatal lesions produce a delay dependent deficit, the neostriatal lesions induce a more global deficit at immediate as well as long delays. (Data from Dunnett 1990.[86])

There have been several attempts to adapt this classic task to the operant box,[96-99] although with varying degrees of success. In our adaptation,[100] rats are trained to press the centrally located food panel until the end of a variable delay period (5 to 20 s). A panel press subsequent to the end of the delay period results in the extension of both the left and the right levers. On the first trial of the day, pressing either lever produces a food pellet reward. On all subsequent trials, the rat is required to press the lever that was *not* pressed on the previous trial (Figure 13.10). A correct press is rewarded with a food pellet whereas an incorrect press (repetition of the same lesson as on the previous trial) has no consequence. In either case, after pressing one lever, both levers are withdrawn and the timer for the next variable delay interval is started. The distinctive feature of our variant of operant delayed alternation, as in the DMTP/DNMTP task described above, is that the animal is required to nose poke at the central panel during the delay in order to trigger presentation of the two choice levers. This serves to keep the rat centralised between the two response locations and reduces the opportunity for it to adopt a simple mediating response strategy during the delay (i.e., simply waiting at the location where the correct lever will next appear). We vary the delay interval on each trial and thereby accumulate information about the animal's level of accuracy over different lengths of time that the last trial response must be held in memory.

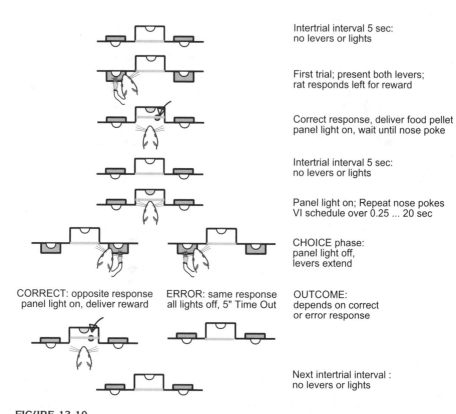

FIGURE 13.10
Schematic illustration of the delayed operant spatial alternation test in a two retractable lever Skinner box.

In recent studies we have begun to investigate differential roles of PFC and mSTR in this task. We find that lesions both of the mPFC and of the medial striatum disrupt performance on this task (Figure 13.11A). However, detailed analysis of the impairments suggests slightly different reasons for the deficit after each lesion.[100] In particular, the mPFC animals exhibit a relatively straightforward impairment in task accuracy, that may be related to an executive deficit in determining the correct response based on short-term memory for the last response. By contrast the rats with mSTR lesions exhibit a tendency to perseverate their responses. Thus, for example, if we analyse the errors from trial to trial, whereas the conditional probability of an error on a particular trial is, after prefrontal lesions, independent of how the animal performed on the previous trial, the chance of an error (involving repetition rather than alternation of the previous response), after a striatal lesion, increases as the animal makes a run of errors in a row (Figure 13.11B). It is worth noting that this pattern of errors is quite different from that which would be expected if the animals' deficits were due to a memory failure, for example by an increased susceptibility to proactive interference. Furthermore, analyses of other behavioural measures indicate that the time taken to collect a food reward is unaffected, suggesting that neither a general motor deficit nor a motivational deficit is the basis for these impairments.

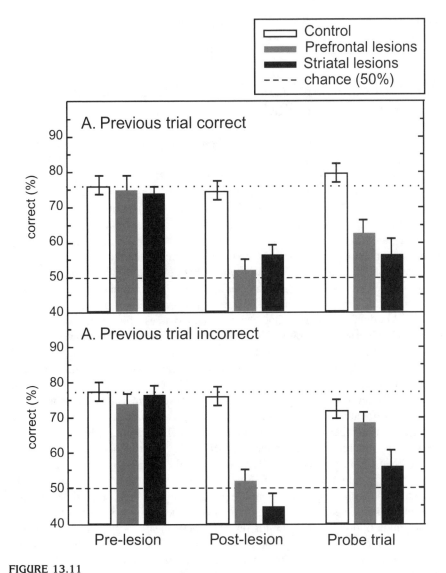

FIGURE 13.11

Prefrontal and striatal lesions both produce a marked decline in response accuracy in the operant delayed alternation test. The upper and lower panels illustrate choice accuracy depending on whether the previous trial was correct or incorrect. Whereas the prefrontal lesions induced a similar deficit on all trials irrespective of performance on previous trials, the striatal lesions induced a deficit whereby a deficit on the previous trial increased the chance of a deficit on the present trial. This is directly against an interference effect between trials and suggests a perseverative tendency of the rats with striatal lesions. (Data from Dunnett et al. 1999.[100])

Thus, the range of measures of different aspects of task performance allow not only a dissociation between different lesions — even though they all disrupt task performance — but also the beginnings of an analysis of the precise nature of the functional impairment that results following disturbance of each neuroanatomical component in a connected fronto-striatal circuit.

VII. Operant Analysis of Striatal Lesions: Deficits in Motivational State

One of the hallmarks of HD is that patients present with psychiatric as well as motor and cognitive disturbances. A true psychiatric impairment is of course difficult to evaluate in primates, let alone in the laboratory rat. However, as an alternative to trying to determine the animal's emotional state, it is certainly possible to evaluate motivational state, in particular the effects of striatal manipulations on changes in responding to stimuli with motivational significance. For example, the hyperactivity that is apparent following lesions of the medial striatum is not independent on the animals' motivational state, but is particularly associated with food deprivation.[101-103]

Traditionally, it is the ventral striatum that is believed to play a central role in reward-related processes.[104,105] More recently, research interest has turned to whether the ventral striatum not only mediates the evaluation of reward but also the 'effort' expended in obtaining it.[106,107] One way in which to assess this cost/benefit analysis is to employ a progressive ratio (PR) schedule of operant lever pressing.[108] Motivational deficits are a well-established symptom in HD, and often present at an early stage in the disease and as such these deficits are more associated with damage in the dorsal striatum. It is primarily this clinically based rationale which has led us to investigate reward-related processes in the dorsal striatum using a PR schedule.[109]

As illustrated in Figure 13.12, a PR schedule requires the rat to simply press a single lever a number of times to receive a food pellet reward. At first, each lever press results in a reward; then, after 5 rewards, 3 or 5 presses are required for each reward, then 10 presses per reward, and so on. As the session progresses, a progressively greater amount of work (e.g., number of lever presses) is required in order to gain an unchanging amount of reward (e.g., a single food pellet each time). Such a schedule makes it possible to examine a number of motivational measures that reflect the willingness of a rat to work for reward. These measures typically include the breakpoint of each animal (e.g., the greatest number of lever presses that an animal is prepared to make in order to obtain a single food pellet). A reduction in breakpoint is usually regarded as a general indication of lower levels of motivation.[108,110] A second useful measure is the post-reinforcement pause which describes the latency to resume lever pressing following the presentation of food. This is believed to reflect the reluctance to resume lever pressing as the cost of reward increases.

In contrast to lesions of the ventral striatum, we have found that focal lesions of the dorsal striatum do not induce deficits as revealed by these primary motivational measures.[109] Neither lesions of the dorsomedial striatum nor lesions of the dorsolateral striatum produced any changes in either breakpoint or the post-reinforcement pause. Nevertheless, these restricted striatal lesions did induce some specific performance deficits. Animals in both lesion groups took significantly longer to collect food rewards once a food pellet was delivered at the completion of a schedule.

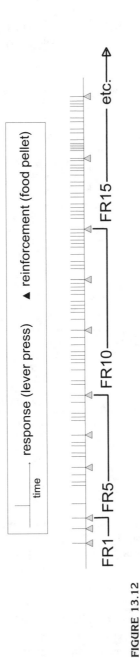

FIGURE 13.12

Schematic illustration of the contingencies in a typical progressive ratio (PR) schedule in a Skinner box. As the animal makes more responses, a greater number of responses is required (on an incrementing fixed ratio schedule) to achieve each additional food pellet.

Moreover, rats with dorsolateral striatal lesions also continued to press the lever once a schedule had been completed and food reward was available for consumption.[109] Thus, although striatal lesions did not cause any overall deficit in motivation, rats with striatal lesions again demonstrated a perseverative deficit. This suggested that while striatal lesions did not affect the ability of rats to regulate their rates of responding with changing reward cost, there was evidence that striatal lesions compromised the ability of rats to sequence and switch responding efficiently and appropriately.

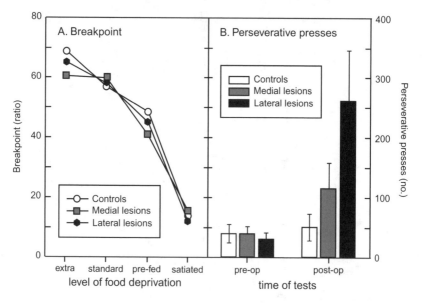

FIGURE 13.13

Data from medial and lateral lesions on the PR schedule. (Left) Neither lesion influences the level of basic motivation as measured e.g., by breakpoint. (Right) However, striatal lesions induce marked deficits in the animals' tendencies to perseverate responding once each reward has been earned. (Data from Eagle et al. 1999.[109])

VIII. Conclusion

The above examples have aimed to show the value of operant tasks in detecting specific deficits in well-defined functions. This chapter has examined only a few ways of examining striatal function in the rat. Tasks are being developed all the time to tap other aspects of function, particularly to try and align rat studies with ongoing primates and clinical research.

There are many advantages for using operant tasks to assess function. The impressive degree of experimental control afforded by such paradigms provides not only an extensive stimulus control over an animal's behaviour but also permits the quantification of the varied response options that an animal can choose to make. In addition, the automation of such paradigms overcomes the inherent biases in the observational recording of behaviour, and also allows for a greater efficiency in collecting and processing data.

In addition to allowing detailed functional studies, operant tests also allow a functional assessment of a wide variety of those surgical, cellular, or pharmacological interventions with potential clinical relevance. While hand testing and observational techniques may be more appropriate if the evaluation of a novel therapy is at an early stage, operant tasks allow for a fuller evaluation of the functional efficacy of treatments. Moreover, an understanding of how particular neural structures mediate function is crucial to the design of such interventions.[111]

References

1. Wilson, S. A. K., An experimental research into the anatomy and physiology of the corpus striatum. *Brain*, 36, 492, 1914.
2. Jung, R. and Hassler, R., The extrapyramidal motor system, in *The Handbook of Physiology. I. Neurophysiology, Volume 2*. The American Physiological Society, Washington, D.C., 1960, 863.
3. Huntington, G., On chorea. *Adv. Neurol.*, 1, 33, 1872.
4. Huntington's Disease Collaborative Research Group, A novel gene containing a trinucleotide repeat that is expanded and unstable on Huntington's disease chromosomes. *Cell*, 72, 971, 1993.
5. Vonsattel, J.-P., Myers, R. H., and Stevens, T. J., Neuropathologic classification of Huntington's disease. *J. Neuropathol. Exp. Neurol.*, 44, 559, 1985.
6. Myers, R. H., Vonsattel, J.-P., Paskevich, P. A., Kiely, D. K., Stevens, T. J., Cupples, L. A., Richardson, E. P., and Bird, E. D., Decreased neuronal and increased oligodendroglial densities in Huntington's disease caudate nucleus. *J. Neuropathol. Exp. Neurol.*, 50, 742, 1991.
7. Hedreen, J. C. and Folstein, S. E., Early loss of neostriatal striosome neurons in Huntington's disease. *J. Neuropathol. Exp. Neurol.*, 54, 105, 1995.
8. Fusco, F. R., Chen, Q., Lamoreaux, W. J., Figueredo-Cardenas, G., Jiao, Y., Coffman, J. A., Surmeier, D. J., Honig, M. G., Carlock, L. R., and Reiner, A., Cellular localization of huntingtin in striatal and cortical neurons in rats: Lack of correlation with neuronal vulnerability in Huntington's disease. *J. Neurosci.*, 19, 1189, 1999.
9. Davies, S. W., Turmaine, M., Cozens, B. A., DiFiglia, M., Sharp, A. H., Ross, C. A., Scherzinger, E., Wanker, E. E., Mangiarini, L., and Bates, G. P., Formation of neuronal intranuclear inclusions (NII) underlies the neurological dysfunction in mice transgenic for the HD mutation. *Cell*, 90, 537, 1997.
10. Ordway, J. M., Tallaksen-Greene, S., Gutekunst, C. A., Bernstein, E. M., Cearley, J. A., Wiener, H. W., Dure, L. S., Lindsey, R., Hersch, S. M., Jope, R. S., Albin, R. L., and Detloff, P. J., Ectopically expressed CAG repeats cause intranuclear inclusions and a progressive late onset neurological phenotype in the mouse. *Cell*, 91, 753, 1997.
11. Reddy, P. H., Charles, V., Williams, M., Miller, G., Whetsell, W. O., and Tagle, D. A., Transgenic mice expressing mutated full-length HD cDNA: a paradigm for locomotor changes and selective neuronal loss in Huntington's disease. *Phil. Trans. Roy. Soc. Lond. B*, 354, 1035, 1999.

12. Hackam, A. S., Singaraja, R., Zhang, T. Q., Gan, L., and Hayden, M. R., In vitro evidence for both the nucleus and cytoplasm as subcellular sites of pathogenesis in Huntington's disease. *Hum. Mol. Genet.*, 8, 25, 1999.

13. DiFiglia, M., Sapp, E., Chase, K. O., Davies, S. W., Bates, G. P., Vonsattel, J.-P., and Aronin, N., Aggregation of huntingtin in neuronal intranuclear inclusions and dystrophic neurites in brain. *Science*, 277, 1990, 1997.

14. Houk, J. C., Davis, J. L., and Beiser, D. G., *Models of Information Processing in the Basal Ganglia*, MIT Press, Cambridge MA, 1995.

15. Jahanshahi, M., Brown, R. G., and Marsden, C. D., A comparative study of simple and choice reaction time in Parkinson's, Huntington's and cerebellar disease. *J. Neurol. Neurosurg. Psychiat.*, 56, 1169, 1993.

16. Girotti, F., Marano, R., Soliveri, P., Geminiani, G., and Scigliano, G., Relationship between motor and cognitive disorders in Huntington's disease. *J. Neurol.*, 235, 454, 1988.

17. Penney, J. B., Young, A. B., Shoulson, I., Starosta-Rubenstein, S., Snodgrass, S. R., Sanchez-Ramos, J., Ramos-Arrovo, M., Gomez, F., Penchas-Zadeh, G., Alvir, J., Esteves, J., Dequiroz, I., Marsol, N., Moreno, H., Conneally, P. M., Bonilla, E., and Wexler, N. S., Huntington's disease in Venezuela — 7 years of follow-up on symptomatic and asymptomatic individuals. *Mov. Dis.*, 5, 93, 1990.

18. Thompson, P. D., Berardelli, A., Rothwell, J. C., Day, B. L., Dick, J. P. R., Benecke, R., and Marsden, C. D., The coexistence of bradykinesia and chorea in Huntington's disease and its implications for theories of basal ganglia control of movement. *Brain*, 111, 223, 1988.

19. Bradshaw, J. L., Phillips, J. G., Dennis, C., Mattingley, J. B., Andrewes, D., Chiu, E., Pierson, J. M., and Bradshaw, J. A., Initiation and execution of movement sequences in those suffering from and at risk of developing Huntington's disease. *J. Clin. Exp. Neuropsychol.*, 14, 179, 1992.

20. Brown, R. G. and Marsden, C. D., Subcortical dementia: the neuropsychological evidence. *Neuroscience*, 25, 363, 1988.

21. Butters, N., Sax, D., Montgomery, K., and Tarlow, S., Comparison of the neuropsychological deficits associated with early and advanced Huntington's disease. *Arch. Neurol.*, 35, 585, 1978.

22. Knopman, D. and Nissen, M. J., Procedural learning is impaired in Huntington's disease — evidence from the serial reaction time task. *Neuropsychologia*, 29, 245, 1991.

23. Lange, K. W., Sahakian, B. J., Quinn, N. P., Marsden, C. D., and Robbins, T. W., Comparison of executive and visuospatial memory function in Huntington's disease and dementia of Alzheimer-type matched for degree of dementia. *J. Neurol. Neurosurg. Psychiat.*, 58, 598, 1995.

24. Lawrence, A. D., Sahakian, B. J., Hodges, J. R., Rosser, A. E., Lange, J. W., and Robbins, T. W., Executive and mnemonic functions in early Huntington's disease. *Brain*, 119, 1633, 1996.

25. Harper, P. S., *Huntington's Disease*, W.B. Saunders, London, 1996.

26. Laursen, A. M., Corpus striatum. *Acta Physiol. Scand. suppl.*, 211, 1, 1963.

27. Coyle, J. T. and Schwarcz, R., Lesions of striatal neurones with kainic acid provides a model for Huntington's chorea. *Nature*, 263, 244, 1976.

28. Schwarcz, R. and Coyle, J. T., Striatal lesions with kainic acid: neurochemical characteristics. *Brain Res.*, 127, 235, 1977.

29. McGeer, E. G. and McGeer, P. L., Duplication of the biochemical changes of Huntington's choreas by intrastriatal injection of glutamic and kainic acids. *Nature*, 263, 517, 1976.

30. Beal, M. F., Kowall, N. W., Ellison, D. W., Mazurek, M. F., Swartz, K. J., and Martin, J. B., Replication of the neurochemical characteristics of Huntington's disease by quinolinic acid. *Nature*, 321, 168, 1986.

31. Beal, M. F., Ferrante, R. J., Swartz, K. J., and Kowall, N. W., Chronic quinolinic acid lesions in rats closely resemble Huntington's disease. *J. Neurosci.*, 11, 1649, 1991.

32. Schwarcz, R., Hökfelt, T., Fuxe, K., Jonsson, G., Goldstein, M., and Terenius, L., Ibotenic acid-induced neuronal degeneration: a morphological and neurochemical study. *Exp. Brain Res.*, 37, 199, 1979.

33. Schwarcz, R., Whetsell, W. O., and Mangano, R. M., Quinolinic acid: an endogenous metabolite that produces axon-sparing lesions in rat brain. *Science*, 219, 316, 1983.

34. Beal, M. F., Brouillet, E. P., Jenkins, B. G., Ferrante, R. J., Kowall, N. W., Miller, J. M., Storey, E., Srivastava, R., Rosen, B. R., and Hyman, B. T., Neurochemical and histologic characterization of striatal excitotoxic lesions produced by the mitochondrial toxin 3-nitropropionic acid. *J. Neurosci.*, 13, 4181, 1993.

35. Borlongan, C. V., Koutouzis, T. K., and Sanberg, P. R., 3-Nitropropionic acid animal model and Huntington's disease. *Neurosci. Biobehav. Rev.*, 21, 289, 1997.

36. Meldrum, A., Page, K. J., Everitt, B. J., and Dunnett, S. B., Malonate: profile and mechanisms of striatal toxicity, in *Mitochondrial Inhibitors as Tools for Neurobiology*, P.R. Sanberg, H. Nishino, C.V. Borlongan, Eds., Humana, Totowa, NJ, 1999.

37. Mangiarini, L., Sathasivam, K., Seller, M., Cozens, B., Harper, A., Hetherington, C., Lawton, M., Trottier, Y., Lehrach, H., Davies, S. W., and Bates, G. P., Exon 1 of the *HD* gene with an expanded CAG repeat is sufficient to cause a progressive neurological phenotype in transgenic mice. *Cell*, 87, 493, 1996.

38. Hodgson, J. G., Agopyan, N., Gutekunst, C. A., Leavitt, B. R., LePiane, F., Singaraja, R., Smith, D. J., Bissada, N., McCutcheon, K., Nasir, J., Jamot, L., Li, X. J., Stevens, M. E., Rosemond, E., Roder, J. C., Phillips, A. G., Rubin, E. M., Hersch, S. M., and Hayden, M. R., A YAC mouse model for Huntington's disease with full-length mutant huntingtin, cytoplasmic toxicity, and selective striatal neurodegeneration. *Neuron*, 23, 181, 1999.

39. Reddy, P. H., Williams, M., Charles, V., Garrett, L., Pike-Buchanan, L., Whetsell, W. O., Miller, G., and Tagle, D. A., Behavioural abnormalities and selective neuronal loss in HD transgenic mice expressing mutated full-length HD cDNA. *Nature Genet.*, 20, 198, 1998.

40. Carter, R. J., Lione, L. A., Humby, T., Mangiarini, L., Mahal, A., Bates, G. P., Morton, A. J., and Dunnett, S. B., Characterisation of progressive motor deficits in mice transgenic for the human Huntington's disease mutation. *J. Neurosci.*, 19, 3248, 1999.

41. Lione, L. A., Carter, R. J., Bates, G. P., Morton, A. J., and Dunnett, S. B., Selective discrimination learning impairments in mice expressing the human Huntington's disease mutation. *J. Neurosci.*, in press, 1999.

42. Dunnett, S. B., Isacson, O., Sirinathsinghji, D. J. S., Clarke, D. J., and Björklund, A., Striatal grafts in rats with unilateral neostriatal lesions. III. Recovery from dopamine-dependent motor asymmetry and deficits in skilled paw reaching. *Neuroscience*, 24 , 813, 1988.

43. Schwarcz, R., Fuxe, K., Agnati, L. F., Hökfelt, T., and Coyle, J. T., Rotational behavior in rats with unilateral striatal kainic acid lesions: a behavioural model for studies on intact dopamine receptors. *Brain Res.*, 170, 485, 1979.

44. Fricker, R. A., Annett, L. E., Torres, E. M., and Dunnett, S. B., The locus of a striatal ibotenic acid lesion affects the direction of drug-induced rotation and skilled forelimb use. *Brain Res. Bull.*, 41 , 409, 1996.

45. Borlongan, C. V., Randall, T. S., Cahill, D. W., and Sanberg, P. R., Asymmetrical motor behavior in rats with unilateral striatal excitotoxic lesions as revealed by the elevated body swing test. *Brain Res.*, 676, 231, 1995.

46. Montoya, C. P., Astell, S., and Dunnett, S. B., Effects of nigral and striatal grafts on skilled forelimb use in the rat. *Prog. Brain Res.*, 82, 459, 1990.

47. Kendall, A. L., Rayment, F. D., Torres, E. M., Baker, H. F., Ridley, R. M., and Dunnett, S. B., Functional integration of striatal allografts in a primate model of Huntington's disease. *Nature Med.*, 4, 727, 1998.

48. Skinner, B. F., *The Behavior of Organisms*, Appleton-Century-Crofts, New York, 1938.

49. Everitt, B. J., Fray, P., Kostarczyk, E., Taylor, S., and Stacey, P., Studies of instrumental behavior with sexual reinforcement in male-rats (*Rattus norvegicus*). 1. Control by brief visual stimuli paired with a receptive female. *J. Comp. Psychol.*, 101, 395, 1987.

50. Sanberg, P. R., Pisa, M., and Fibiger, H. C., Avoidance, operant and locomotor behavior in rats with neostriatal injections of kainic acid. *Pharmacol. Biochem. Behav.*, 10, 137, 1979.

51. Dunnett, S. B. and Iversen, S. D., Neurotoxic lesions of ventrolateral but not anteromedial neostriatum in rats impair differential reinforcement of low rates (DRL) performance. *Behav. Brain Res.*, 6, 213, 1982.

52. Döbrössy, M. D., Svendsen, C. N., and Dunnett, S. B., The effects of bilateral striatal lesions on the acquisition of an operant test of short-term memory. *NeuroReport*, 6, 2059, 1995.

53. Reading, P. J., Dunnett, S. B., and Robbins, T. W., Dissociable roles of the ventral, medial and lateral striatum on the acquisition and performance of a complex visual stimulus-response habit. *Behav. Brain Res.*, 45, 147, 1991.

54. Ungerstedt, U. and Arbuthnott, G. W., Quantitative recording of rotational behaviour in rats after 6-hydroxydopamine lesions of the nigrostriatal dopamine system. *Brain Res.*, 24, 485, 1970.

55. Marshall, J. F., Richardson, J. S., and Teitelbaum, P., Nigrostriatal bundle damage and the lateral hypothalamic syndrome. *J. Comp. Physiol. Psychol.*, 87, 808, 1974.

56. Evenden, J. L. and Robbins, T. W., Effects of unilateral 6-hydroxydopamine lesions of the caudate-putamen on skilled forepaw use in the rat. *Behav. Brain Res.*, 14, 61, 1984.

57. Whishaw, I. Q., O'Connor, W. T., and Dunnett, S. B., The contributions of motor cortex, nigrostriatal dopamine and caudate-putamen to skilled forelimb use in the rat. *Brain*, 109, 805, 1986.

58. Ljungberg, T. and Ungerstedt, U., Sensory inattention produced by 6-hydroxydopamine-induced degeneration of ascending dopamine neurons in the brain. *Exp. Neurol.*, 53, 585, 1976.

59. Marshall, J. F. and Teitelbaum, P., Further analysis of sensory inattention following lateral hypothalamic damage in the rat. *J. Comp. Physiol. Psychol.*, 86, 375, 1974.

60. Marshall, J. F., Turner, B. H., and Teitelbaum, P., Sensory neglect produced by lateral hypothalamic damage. *Science*, 174, 423, 1971.

61. Turner, B. H., Sensorimotor syndrome produced by lesions of the amygdala and lateral hypothalamus. *J. Comp. Physiol. Psychol.*, 82, 37, 1973.

62. Carli, M., Evenden, J. L., and Robbins, T. W., Depletion of unilateral striatal dopamine impairs initiation of contralateral actions and not sensory attention. *Nature*, 313, 679, 1985.

63. Robbins, T. W., Muir, J. L., Killcross, A. S., and Pretsell, D., Methods of assessing attention and stimulus control in the rat, in *Behavioural Neuroscience,* Volume I, A. Sahgal, Ed., IRL Press, Oxford, 1993, 13.

64. Carli, M., Jones, G. H., and Robbins, T. W., Effects of unilateral dorsal and ventral striatal dopamine depletion on visual neglect in the rat: a neural and behavioural analysis. *Neuroscience*, 29, 309, 1989.

65. Mittleman, G., Brown, V. J., and Robbins, T. W., Intentional neglect following unilateral ibotenic acid lesions of the striatum. *Neurosci. Res. Comm.*, 2, 1, 1988.

66. Brown, V. J. and Robbins, T. W., Deficits in response space following unilateral striatal dopamine depletion in the rat. *J. Neurosci.*, 9, 983, 1989.

67. Brasted, P., Humby, T., Dunnett, S. B., and Robbins, T. W., Response space deficits following unilateral excitotoxic lesions of the dorsal striatum in the rat. *J. Neurosci.*, 17, 8919, 1997.

68. Bisiarch, E. and Luzzatti, C., Unilateral neglect of representational space. *Cortex*, 14, 129, 1978.

69. Heilman, K. M., Neglect and related disorders, in *Clinical Neuropsychology,* K.M. Heilman, E.S. Valenstein, Eds., Oxford University Press, Oxford, 1979.

70. Divac, I., Rosvold, H. E., and Szwarcbart, M. K., Behavioral effects of selective ablation of the caudate nucleus. *J. Comp. Physiol. Psychol.*, 63, 184, 1967.

71. Rosvold, H. E. and Delgado, J. M. R., The effect on delayed alternation test performance of stimulating or destroying electrically structures within the frontal lobes of the monkey's brain. *J. Comp. Physiol. Psychol.*, 49, 365, 1956.

72. Rosvold, H. E., The frontal lobe system: cortical-subcortical interrelationships. *Acta Neurobiol. Exp.*, 32, 439, 1972.

73. Rosvold, H. E. and Szwarcbart, M. K., Neural structures involved in delayed response performance, in *The Frontal Granular Cortex and Behavior,* J.M. Warren, K. Akert, Eds., McGraw-Hill, New York, 1964, 1.

74. Leonard, C. M., The prefrontal cortex of the rat. I cortical projection of the mediodorsal neucleus. II Efferent connections. *Brain Res.*, 12, 321, 1969.

75. Krettek, J. E. and Price, J. L., The cortical projections of the mediodorsal nucleus and adjacent thalamic nuclei in the rat. *J. Comp. Neurol.*, 171, 157, 1977.

76. Divac, I., Markowitsch, H. J., and Pritzel, M., Behavioural and anatomical consequences of small intrastriatal injections of kainic acid in the rat. *Brain Res.*, 151, 523, 1978.

77. Dunnett, S. B. and Iversen, S. D., Learning impairments following selective kainic acid-induced lesions within the neostriatum of rats. *Behav. Brain Res.*, 2, 189, 1981.

78. Sanberg, P. R., Lehmann, J., and Fibiger, H. C., Impaired learning and memory after kainic acid lesions of the striatum: a behavioral model of Huntington's disease. *Brain Res.*, 149, 1204, 1978.

79. D'Amato, M. R., Delayed matching and short-term memory in monkeys. *Psychol. Learn. Motiv.*, 7, 227, 1973.

80. Aggleton, J. P., One-trial object recognition by rats. *Quart. J. Exp. Psychol. B.*, 37B, 279, 1985.

81. Dunnett, S. B., Comparative effects of cholinergic drugs and lesions of nucleus basalis or fimbria-fornix on delayed matching in rats. *Psychopharmacology*, 87, 357, 1985.

82. Dunnett, S. B., Operant delayed matching and non-matching to position in rats, in *Behavioural Neuroscience: A Technical Approach,* A. Sahgal, Ed., IRL Press, Oxford, 1993, 123.

83. Mishkin, M. and Manning, F. J., Non-spatial memory after selective prefrontal lesions in monkeys. *Brain Res.*, 143, 313, 1978.

84. Kowalska, D. M., Bachevalier, J., and Mishkin, M., The role of the inferior prefrontal convexity in performance of delayed nonmatching-to-sample. *Neuropsychologia*, 29, 583, 1991.

85. Goldman-Rakic, P. S., Circuitry of primate prefrontal cortex and regulation of behavior by representational memory, in *Handbook of Physiology — The Nervous System V,* American Physiological Association, Baltimore, 1989, 373.

86. Dunnett, S. B., Role of prefrontal cortex and striatal output systems in short-term memory deficits associated with ageing, basal forebrain lesions, and cholinergic-rich grafts. *Can. J. Psychol.*, 44, 210, 1990.

87. Döbrössy, M. D., Svendsen, C. N., and Dunnett, S. B., Bilateral striatal lesions impair retention of an operant test of short-term memory. *Brain Res. Bull.*, 41, 159, 1996.

88. Dunnett, S. B., Is it possible to repair the damaged prefrontal cortex by neural tissue transplantation? *Prog. Brain Res.*, 85, 285, 1990.

89. Jacobsen, C. F. and Nissen, H. W., Studies of cerebral function in primates. IV. The effect of frontal lobe lesions on the delayed alternation habit in monkeys. *J. Comp. Psychol.*, 23, 101, 1937.

90. Jacobsen, C. F., Studies of cerebral function in primates. I. The functions of the frontal association areas in monkeys. *Comp. Psychol. Monogr.*, 13, 3, 1936.

91. Campbell, P., Therapeutic horizons. *Nature suppl.*, 392, 1, 1998.

92. Brown, R. G. and Marsden, C. D., Subcortical dementia — the neuropsychological evidence. *Neuroscience*, 25, 363, 1998.

93. Larsen, J. K. and Divac, I., Selective ablations within the prefrontal cortex of the rat and performance of delayed alternation. *Physiol. Psychol.*, 6, 15, 1978.

94. Georgiou, N., Bradshaw, J. L., Phillips, J. G., and Chiu, E., Effect of directed attention in Huntington's disease. *J. Clin. Exp. Neuropsychol.*, 19, 367, 1997.

95. Brandt, J., Strauss, M. E., Larus, J., Jensen, B., Folstein, S. E., and Folstein, M. F., Clinical correlates of dementia and disability in Huntington's disease. *J. Cogn. Neuropsychol.*, 6, 401, 1984.

96. Mogensen, J., Iversen, I. H., and Divac, I., Neostriatal lesions impaired rats delayed alternation performance in a T-maze but not in a two-key operant chamber. *Acta Neurobiol. Exp.*, 47, 45, 1987.

97. Heise, G. A., Conner, R., and Martin, R. A., Effects of scopolamine on variable intertrial interval spatial alternation and memory in the rat. *Psychopharmacology*, 49, 131, 1978.

98. Numan, R. and Quaranta, J. R., Effects of medial septal lesions on operant delayed alternation in rats. *Brain Res.*, 531, 232, 1990.

99. van Haaren, F., de Bruin, J. P. C., Heinsbroek, R. P. W., and van de Poll, N. E., Delayed spatial response alternation: effects of delay-interval duration and lesions of the medial prefrontal cortex on response accuracy of male and female Wistar rats. *Behav. Brain Res.*, 18, 41, 1985.

100. Dunnett, S. B., Nathwani, F., and Brasted, P. J., Medial prefrontal and neostriatal lesions disrupt performance in an operant delayed alternation task in rats. *Behav. Brain Res.*, in press, 1999.

101. Jacobsen, C. F. and Haslerud, G. M., Studies of cerebral function in primates. III. A note on the effect of motor and premotor lesions on delayed response in monkeys. *Comp. Psychol. Monogr.*, 13, 66, 1936.

102. Isacson, O., Dunnett, S. B., and Björklund, A., Graft-induced behavioral recovery in an animal model of Huntington disease. *Proc. Natl. Acad. Sci. U.S.A.*, 83, 2728, 1986.

103. Sanberg, P. R. and Coyle, J. T., Scientific approaches to Huntington's disease. *CRC Crit. Rev. Clin. Neurobiol.*, 1, 1, 1984.

104. Mogenson, G. J., Jones, D. L., and Yim, C. Y., From motivation to action: functional interface between the limbic system and the motor system. *Prog. Neurobiol.*, 14, 69, 1980.

105. Berridge, K. C., Food reward: Brain substrates of wanting and liking. *Neurosci. Biobehav. Rev.*, 20, 1, 1996.

106. Salamone, J. D., Kurth, P., McCullough, L. D., and Sokolowski, J. D., The effects of nucleus accumbens dopamine depletions on continuously reinforced operant responding: Contrasts with the effects of extinction. *Pharmacol. Biochem. Behav.*, 50, 437, 1995.

107. Salamone, J. D., Cousins, M. S., and Snyder, B. J., Behavioral functions of nucleus accumbens dopamine: Empirical and conceptual problems with the anhedonia hypothesis. *Neurosci. Biobehav. Rev.*, 21, 341, 1997.

108. Hodos, W. and Kalman, G., Effects of increment size and reinforcer volume on progressive ratio performance. *J. Exp. Anal. Behav.*, 6, 387, 1963.

109. Eagle, D. M., Humby, T., Dunnett, S. B., and Robbins, T. W., Effects of regional striatal lesions on motor, motivational and executive aspects of progressive ratio performance in rats. *Behav. Neurosci.*, in press, 1999.

110. Skjoldager, P., Pierre, P. J., and Mittleman, G., Reinforcer magnitude and progressive ratio responding in the rat: effects of increased effort, prefeeding and extinction. *Learn. Motiv.*, 24, 303, 1993.

111. Dunnett, S. B. and Everitt, B. J., Topographic factors affecting the functional viability of dopamine-rich grafts in the neostriatum, in *Cell Transplantation for Neurological Disorders*, T.B. Freeman, J.H. Kordower, Eds., Humana Press, Totowa, NJ, 1998, 135.

Chapter

Use of Autoshaping with Non-Delayed and Delayed Reinforcement for Studying Effects upon Acquisition and Consolidation of Information

Sheldon B. Sparber

Contents

0-8493-0704-X/01/$0.00+$.50
© 2001 by CRC Press LLC

I.   Introduction and Overview

When Brown and Jenkins[1] first described auto-shaping of pigeons' key-pecks, they introduced the experimental protocol as one in which delivery of reinforcement was related to or "conditional on stimulus values but not on responses." Thus, the close temporal relationship between back-lighting a standard pigeon key and presentation of food, which engendered various behaviors directed toward the key, including eventually pecking at it, became the basis for autoshaping.

After magazine training, pigeons were exposed to 8 s of illumination of the key-light and 4 s access to a tray of pigeon grain upon the offset of the light. Intertrial intervals (ITI) generally varied randomly between 30 to 90 s, with a mean ITI of 60 s. One additional condition (protocol) used included the termination of the key-light and the immediate presentation of 4 s access to reinforcement. The third condition included a sort of punishing operation whereby a key-peck during the ITI delayed the presentation of the lighted key for 60 s. Brown and Jenkins[1] also explored the consequences of backwards conditioning (reverse pairing); if the pigeon pecked the lighted key, which was illuminated for 8 s after presenting the food tray for 4 s, the key-light was turned off and the pigeon again had access to 4 s of grain reinforcement. The forward pairings resulted in the emergence of behaviors near and, within two sessions of 80 presentations each, all 36 subjects eventually pecked the key at least once. Reverse pairing resulted in only 2 of 12 subjects pecking at the key within 160 trials.

In the discussion of this seminal paper, the authors argue in favor of a classically conditioned (first) key-peck response through either some sort of stimulus substitution, since pecking is presumably innately directed at the unconditioned stimulus (grain) used as the reinforcer or because intermittent presentation of a stimulus (key-light) followed shortly thereafter by non-contingent presentation of food (until the very first key-peck) leads to a multitude of (classically conditioned or superstitious?) motor behaviors, near and at the key, while it is lighted during an 8 s or 3 s trial. Brown and Jenkins[1] finally offer the procedure as a time and labor-saving, standardized method for obviating influences brought about by "individual differences among experimenters in the art of hand-shaping." This is one of the salient characteristics of the procedure that initially drew our attention to its potential utility.

Thus, "... autoshaping paradigms provide an unusually good opportunity to obtain parametric data on variables that influence the acquisition of such behavior, an issue that Williams refers to as the *psychophysics of association*."[2] Attempts to attain such a goal and theoretical issues and arguments related to the concept that autoshaped behavior is primarily a classically conditioned behavior, an instrumentally conditioned behavior, a bridge between the two (S-R vs. R-S contingencies) or a new class of behavior with components of each, can be gleaned from the many chapters in the book edited by Locurto, Terrace and Gibbon[3] entitled *Autoshaping and Conditioning Theory.*

Although the majority of studies dealing with the phenomenon of autoshaping have used a procedure whereby delivery of reinforcers was **not** contingent upon a specified response, my own laboratory has taken advantage of the automaticity and lack of experimenter-induced bias conveyed by autoshaping and we too have utilized a response-contingent component whereby the subject (normally a rat but we, as others[4] have, used autoshaping with chickens as well) is given the opportunity to manipulate its environment (e.g., touching the extended lever, initiating its retraction earlier than programmed to do so noncontingently) leading to the delivery of a reinforcing stimulus (e.g., food) slightly sooner than its delivery in the absence of an operant (lever touch) response. For protocols we generally use, the subject cannot earn more reinforcers during a daily session in which 12 to 30 or more trials are presented and their session lengths cannot be shortened by more than a fraction, thereby most likely obviating a shortened session length as a reinforcing outcome (i.e., trials without an autoshaped response typically are comprised of the presentation of a retractable lever for 15 s, followed by retraction and either immediate delivery of a reinforcing stimulus or delaying the reinforcer for up to 8 or 9 s) followed by a random or fixed time ITI, typically averaging 45 s. By introducing a delay of reinforcement we believe we have introduced a so-called working memory component as a variation of the task and this has enabled us to demonstrate selective and specific effects of neurotoxic insults to the hippocampus and associated structures (*vide infra*).

As is often the case, behavioral toxicologists, behavioral pharmacologists and psychopharmacologists borrow procedures and protocols from the experimental and cognitive psychologists and because of their training and reliance upon concepts like dose-response relationships, often discover a need to systematically manipulate antecedent and consequential variables which had not previously been carried out adequately by others. Likewise, it may be necessary to modify the typical physical environment used by others so that drug-behavior studies can be carried out in such a manner that ancillary measures of unconditioned behaviors can also be recorded concurrently, either automatically or via closed circuit television monitors and video recorders for later off-line analyses.

Because this contribution is for a book recommending and describing methods, I have chosen to discuss the advantages and disadvantages of the variations of autoshaping procedures we have used over the past 25 years. I have also included comments on theoretical and philosophical issues, where apparently appropriate, and discussions or comments about the technology used and choice of data collection, reduction, and statistical analyses.

The last points are not especially restricted to autoshaping as a behavioral paradigm, or to behavioral neuroscience in particular, but are important enough for a few comments at the outset because it has been my experience, in common with some colleagues, that as technological advances have enabled us to collect more and more data, at faster rates, not enough attention has been paid to the importance of experimental design, how the voluminous data will be handled once they are collected and the validity or appropriateness of whichever statistical analytical procedures are used. In my laboratory this is a continuing, evolutionary process and we continue attempting to devise/design experiments which enable us to rely upon pre-planned statistical contrasts. To the credit of most psychologists and behavioral neuroscientists, at least reasonable attempts at dealing with these issues are more commonplace than has been my experience when attempting to interact or collaborate with colleagues of a more reductionist persuasion. I have, on more than a few occasions, asked a colleague who points to the **obviously** more intense band or spot on a gel or chromatogram, derived from a single experiment (single subject's or pooled tissue) that it is also obvious (at least living in the midwestern plains of the U.S.) that the world is flat. All one has to do is to look out of the window to confirm it. Of equally great concern (frustration?) is trying to convince a colleague that 5 to 10 experiment(s) carried out with xenopus oocytes expressing one or another receptor, but upon oocytes derived from a single female frog, is nothing more than one experiment with 5 to 10 replications and that standard parametric or nonparametric statistical analytical procedures designed for independent observations from individuals (or parts thereof), randomly assigned to one or another treatment condition, are not appropriate, even if he/she is willing to consider using such statistical contrasts.

Over the years we have used variations of autoshaped behavior acquisition and/or maintenance to study the actions of drugs or toxins upon this class of behavior(s) in mature experimental subjects, either exposed to a drug or toxin insult during development (e.g., *in ovo* [chick] or *in utero* [rat]) or after exposure to drugs, withdrawal from them or exposure to toxins postnatally, at a more mature age. Other laboratories have followed our lead and have used identical or similar protocols to study the autoshaped acquisition of intravenous cocaine self-administration,[5] to study the effects of experimenter-administered cocaine upon autoshaped behavior acquisition,[6] and to study effects of pyrogens/cytokines, which reportedly specifically disrupt or interfere with acquisition of food-reinforced autoshaped behavior based upon the fact that the same behavior, once established, was not affected by the same treatment.[7]

It has been the policy of my laboratory to attempt to give something back to the discipline from which we have borrowed theoretical and/or methodological underpinnings. I believe our contributions in this regard can be gleaned to some extent from the publication list appended to this chapter. Thus, I have chosen to emphasize some of the ways we have devised to maximize the use of autoshaped behavior in order to control for as many potential confounding variables which can interfere with design, implementation, and, most importantly, the interpretation of data derived from such studies. Almost of necessity, it is a quasi-chronology of the evolution of protocols currently in use in my laboratory (and those of colleagues who have asked for advice or patterned their work from one or more of our publications in which autoshaped behavior was one of the protocols we utilized).

A. Choice of Autoshaping as a Behavioral Assay

Why use autoshaping, with a multitude of variations, for pharmacological, toxicological, or behavioral neuroscience studies, generally?

It was not long after the original publication by Brown and Jenkins[1] describing
auto-shaping of pigeons' key-pecks that we had a need for a purely objective means
to test learning and/or performance of chickens hatched from eggs injected with
drugs known to affect the synthesis, storage, or release of the catecholamine neu-
rotransmitters to determine if such *in ovo* neurochemical manipulations[8] produced
functional changes, manifest behaviorally, long after the perinatal period. To this
end we incorporated a variation of autoshaping by 55 to 60 day-old chickens hatched
from injected eggs and, in addition to observing facilitated detour learning at a young
age caused by drug treatment *in ovo*, we also observed facilitated key-peck
autoshaped behavior at a later age (in the absence of altered unconditioned open-
field behaviors at 3 weeks of age), attesting to a probable permanent functional
outcome.[9] Additionally, we were again[10] cautious in our interpretation of facilitated
learning/performance, as we have been to this day,[11] believing that performance
enhancement, under laboratory conditions, may not be compatible with survival or
normal development in the real world and any significant deviation from what may
be normal behaviors by otherwise healthy untreated or unlesioned subjects, including
enhanced or facilitated acquisition or performance, should be interpreted as potentially deleterious. For example, there are many reports of enhanced learning or
performance by brain-lesioned subjects, including enhanced autoshaped behaviors
of one or another sort,[12-14] often interpreted as secondary to increased locomotor
activity or to greater reactivity to the experimental environment, thereby presumably
enhancing the saliency of the conditioning stimulus (e.g., a retractable lever) and/or
interfering or slowing the rate of habituation to the conditioning stimulus when it
is not paired with a reinforcing stimulus. This would be the case with a latent
inhibition (or learned irrelevance) variation of an autoshaping task, as used by us to
help characterize the effects of different doses of the neurotoxin trimethyltin (TMT),
and discussed below.

Our interest in behavioral or functional teratology[10,15] predated our choice to
use autoshaping as a means to systematize and objectively measure learning and/or
performance capabilities of subjects exposed to one or another developmental insult
which did not cause obvious structural deformities. The disastrous human experiment, caused by exposure to methylmercury (CH_3Hg) in Minamata and Niigata,
Japan, then later in Iraq (and to at least one farming family in the Southwestern
U.S.) led investigators to the realization that age and degree of exposure determined
the extent of pathology, ranging from tunnel vision and blindness to ataxia, delirium,
and death of older children or adults poisoned by food-contaminated with CH_3Hg.
There were many cases of so-called fetal Minamata disease in children born to
mothers who were seemingly unaffected (presumably because fetuses acted as sinks
or repositories of the CH_3Hg). Symptoms associated with so-called fetal Minamata
disease were diverse and ranged from extreme mental retardation and cerebral palsy
to less severe but nevertheless obvious consequences of *in utero* exposure. While

the Chisso Chemical Company (responsible for dumping tailings into Minamata Bay and its surrounds) and the Japanese government accepted and confirmed the cause of this strange malady decades afterward, controversy existed (and still exists) over the likelihood that more subtle (CNS) dysfunction was much more widespread and insidious than acknowledged. Thus, in the late 1960s and early 1970s we initiated studies upon the developmental effects of exposure to CH_3Hg during embryogenesis of chickens,[16] mice,[17] and rats.[18]

There were three main reasons for using autoshaping as a behavioral assay for purposes of studying autoshaped behavior in exposed rats. Firstly, we wanted to determine if we could demonstrate subtle cognitive or other dysfunctional behavioral effects of prenatal exposure to otherwise apparently nontoxic doses of CH_3Hg without the potential for introducing an experimenter bias by hand-shaping operant behavior during the learning phase. Secondly, we were interested in determining if there were critical or sensitive periods during development when such exposure caused behavioral dysfunctional effects. Lastly, we wanted to determine if autoshaping acquisition or performance, alone or upon a psychopharmacological challenge, could have predicted a human functional teratogenic effect of *in utero* exposure to CH_3Hg.[18]

At this juncture I'd like to discuss some of the advantages and disadvantages of the choice of "technology," experimental design, and the use of a pharmacological probe for studying mature offspring exposed to doses of CH_3Hg, given to their pregnant mothers at different stages of pregnancy, which resulted in fetal exposure to the same concentrations of CH_3Hg during the perinatal period. Only those (male and female) offspring whose mothers were given a single dose of CH_3Hg (by oral gavage) on the first day of gestation or on the seventh day proved to have been affected in some way. While they autoshaped at rates comparable to control offspring, they were significantly less sensitive to the disrupting action of d-amphetamine upon established performance of food-reinforced autoshaped behaviors, compared with controls. In these early experiments the food pellet reinforcers were delivered immediately upon retraction of the lever, whether or not the rat touched the extended lever. Offspring whose mothers were dosed on the 14th day of gestation and whose perinatal body and/or brain concentrations of CH_3Hg were not different from offspring of the other treatment groups, were not noticeably different from controls, including their sensitivity to the autoshaped behavioral-disrupting action of d-amphetamine. Therefore, we concluded that doses of CH_3Hg that had absolutely no effect upon standard parameters used in such studies (e.g., body weight gain by dams, litter size, litter weight, sex ratios, etc.) but which led to equivalent perinatal concentrations of CH_3Hg in offspring caused effects only in mature offspring whose mothers were dosed on the first or seventh, but not on the 14th day of gestation. These results indicated that a critical or sensitive period existed, manifested as reduced sensitivity to the disrupting action of d-amphetamine upon performance of already acquired autoshaped behavior. The outcome of the experiment with d-amphetamine also confirmed our hypothesis that such pharmacological probes might unmask a silent functional teratogenic effect and at the same time point to possible neurochemical sources of the lesion. The use of a range of doses of d-amphetamine enabled us to observe prenatal effects in both sexes, since it turned out that males were generally less sensitive, on a mg/kg basis, than females. Had we used only up to 2.5 mg d-amphetamine/kg instead of up to 3.75 mg/kg, we might have erroneously concluded that a sexual dimorphic effect of *in utero* exposure existed, females having been affected and males not. The experimental design was to incorporate doses of d-amphetamine, which we established in pilot and previous experiments would have a significant effect upon control rats of the same strain, age, sex, etc.

As it turned out, in this series of experiments we also exposed the rats to two sessions in which the levers were presented for 15 s and then retracted, with an average ITI of 90 s, ranging from 45 to 135 s; parameters subsequently to be used for autoshaping. These two sessions, comprising 32 lever presentations and retractions after 15 s, with no programmed consequences if the rat contacted/touched the lever, were used to determine the operant or baseline level of exploratory lever-directed behaviors but, in retrospect, undoubtedly initiated the process of latent inhibition or learned irrelevance, which we have utilized in subsequent experiments (*vide infra*) and which may have influenced, to some extent, the negative outcome of the acquisition component of the study. We also concluded that while there were no noteworthy effects of *in utero* CH₃Hg exposure upon operant lever-touch measures and subsequent acquisition/performance of autoshaped behavior, using food pellet reinforcers immediately upon retraction of the lever (after a lever touch response or after being extended into the operant chamber for 15 s), the procedure offered sufficient advantages for continued use, planning to systematically alter some of the contingencies, levels of reinforcer deprivation, and physical aspects of the operant chambers. In addition, we reasoned that using higher but otherwise non-toxic doses of CH₃Hg with one or more of the variations alluded to above would probably have allowed us to determine if neurotoxic lesions upon areas important for cognitive or mnemonic processes altered acquisition and/or performance measures of autoshaped behavior that did not require pharmacological probes to unmask otherwise subtle or silent damage.[19]

B. Wrongful Expectations of Simple, Linear Dose-Response Relationships

Unlike lower doses, we observed that a larger dose of TMT caused a greater reduction in hippocampal weight but a lesser reduction in ^{3}H-corticosterone binding to hippocampal cytosolic glucocorticoid receptors. There was also considerably less evidence of autoshaped learning or performance retardation or blockade, as was observed after lower doses of TMT.[20] Nevertheless, by judicious use of a latent inhibition manipulation [21,22] we were able to demonstrate a different kind of impaired conditioning, autoshaping when it should **not** have occurred. The obviously different neuropathological consequences of the larger dose of TMT interfered with the ability of the rats to learn to ignore the insertion and/or retraction of the lever during the sessions which were devoid of the delivery of food pellets after the lever was retracted (i.e., during latent inhibition training sessions). Non-lesioned rats were latently inhibited, not forming the association between the lever retraction (CS) during subsequent sessions when lever retractions were followed by delivery of food pellets.

This apparent facilitatory effect of the large dose of TMT is reminiscent of the reports of facilitated autoshaping (or other measures of learning or performance) by rats with CNS lesions induced by other means and referred to above. It may also be related to possible roles of hippocampal granule cells and pyramidal cells in the function of the hippocampus (and related structures), if a low dose of TMT destroys relatively more pyramidal cells, while the high dose destroys relatively more of the granule cells, sparing some of the pyramidal cells (which contain more of the glucocorticoid receptors) and their function(s) for mnemonic processes.[23]

We have also used autoshaped behavior as a convenient way of training and stabilizing performance of moderate to large cohorts of rats for various behavioral pharmacological studies and we have concluded the following generalizations. Firstly, stabilized (i.e., already learned) autoshaped behavior, which requires a single response on the manipulandum, is generally less sensitive to pharmacological alterations compared with stabilized operant behaviors maintained by schedules of reinforcement which engender higher rates of responding (e.g., FR 20). Secondly, the introduction into the operant chamber of a stainless steel strip located on one or more walls of the chamber, but high enough above the stainless steel bars comprising the floor and connected to the floor with a high resistance drinkometer circuit, like that between the retractable lever and the floor (via the rats' contacts with both), enables us to measure unconditioned exploratory or adjunctive rearing behaviors prior to, during acquisition, and/or maintenance of autoshaped behavior (or any conditioned behavior being studied in the same operant chamber). Thirdly, by extending the retractable lever and leaving it in the extended position for one or more sessions before autoshaping sessions begin, we can measure unconditioned, horizontal exploratory, as well as exploratory rearing behaviors without, or greatly diminishing, the risk of introducing a latent inhibition component. This optional capability enables us to interpret more assuredly the nature of the outcome of acquisition or performance experiments. For example, treatments or manipulations that result in apparently enhanced acquisition may be secondary to increased exploratory/unconditioned behaviors which then increase the likelihood of the (accidental?) occurrence of the required response and/or increase the subjects' level of arousal and reactivity to the extension and/or retraction of the lever. This can easily increase the rate of acquisition through a process other than a specific association-enhancing effect. We have interpreted this to possibly be the case after injection of the high dose of TMT (above) and after injection of yohimbine, which caused enhanced acquisition after a low dose and enhanced acquisition after discontinuing a higher dose of yohimbine. The higher dose of yohimbine initially suppressed unconditioned behaviors and autoshaping. Enhanced autoshaping was associated, in both cases, with increased concurrent exploratory (rearing) activities.[24]

C.   Vasopressin and Enhanced Cognition or Arousal?

In the 1980s we also embarked upon a series of studies addressing the possibility that vasopressinergic peptide analogs could facilitate learning and/or memorial processes. The historical controversies surrounding the issues related to whether or not vasopressin had a specific and positive modulatory effect, as advocated by the Dutch group of de Wied and colleagues. This was challenged by other groups, who contended that too great a reliance was placed upon behaviors elicited or maintained by aversive stimuli; the aversiveness being enhanced by the pressor and/or endocrine properties of the peptide and/or a generalized arousal effect of vasopressin, somewhat akin to what we interpreted to be the case with yohimbine described above. Individuals interested in these controversies and their heuristic impact are directed to the literature, including our report[25] on the facilitatory effect of the vasopressin analog, desglycinamide-arginine vasopressin, which describes the status of the controversies and introduces, along with a companion paper,[26] the use of delaying the delivery of the reinforcer for several seconds after non-contingent or touch-contingent retraction of the extended lever.

From a physiological and pharmacological viewpoint, it is interesting to note that the des-gly analog, which is virtually devoid of both pressor and endocrine properties, nevertheless enhanced autoshaping in a less ambiguous fashion when used with a 6 s delayed reinforcement variation, compared with a 0 s delay variation,[27] which allowed acquisition to proceed at a considerably faster rate (and without a working memory component, which we believe is introduced by the delay of reinforcement). By introducing the delay of reinforcement, we were not only able to more assuredly detect the facilitatory effect of the des-gly vasopressin upon learning and/or memorial processes,[25] we were starting to realize that superstitious or adventitiously reinforced behaviors directed toward the retracting and retracted lever (mostly as nose-pokes, although observation of rats' behaviors during the reinforcement delays and inter-trial-intervals via closed circuit TV also revealed substantial numbers of lever touches with their forepaws) might be better or more convincing measures of the degree or strength of conditioning than discrete-trial extended lever touches during the 15 s the lever was extended into the operant chamber. Because of the reciprocal relationship between increased lever-directed behaviors and exploratory rearing activities, as conditioning progressed toward asymptotic levels of extended lever-touching, we have concluded that such lever-directed behaviors during the reinforcement delays and ITIs are probably a more open-ended measure of the strength of conditioning and much less constrained by the ceiling effects of the discrete trial (extended) lever touches we also use, at least during the early acquisition phase, as a measure of autoshaping.

D.   Trimethyltin Neurobehavioral Toxicity and the Learning-Performance Distinction

At the same time we were using variations of autoshaping tasks to study vasopressinergic influences upon learning and/or performance variables, as described above, we were using TMT as an experimental tool because it had been reported to interfere with learning and/or memory as a result of occupational exposure.[28] Space and purpose do not allow a review of the extensive neurotoxicological literature related to this and other organometals and the reader is directed to a treatise edited by Tilson and Sparber[29] which reviewed the status of most of the relevant literature up to that date. Considerably less attention has been paid to consequences of occupational or accidental exposure to TMT, but its use as an experimental tool continues to this day. A National Library of Medicine-Medline search, using trimethlytin and toxicity as index terms, brought up 229 citations, with about half appearing between 1990–1999.

The learning-performance distinction or conundrum is all too often casually dispensed with, being considered less and less these days as investigators unschooled in behavioral neuroscience attempt to parcel out phenotypic manifestations of transgenic or knock-out manipulations. Additionally, learning and memory are grouped together in unitary fashion when data are presented or discussed, even though the concept of multiple stage, separate processes has been acknowledged by cognitive psychologists and neuroscientists for decades. In this section I'd like to describe some experiments we carried out which enabled us to demonstrate what can best be described as evidence for an acquisition (aka learning) deficit caused by treatment with TMT in contradistinction to a memory storage, retrieval, sensorimotor or other performance-interfering effect caused by this neurotoxin, even though it has recently been concluded that TMT produces non-specific effects on learning.[30] However, it should be pointed out that Cohn and MacPhail treated rats that had been over-trained with a multiple repeated acquisition and performance paradigm; rats that had already essentially learned the task, most likely had the basic information processing, storage, and retrieval requirements distributed to areas other than those limbic structures damaged mostly by TMT. Moreover, they used a large dose of TMT (8mg/kg, i.v.), which may have caused a very different pattern of cell loss than a lower dose would have caused and paradoxically, a less selective and/or less interpretable behavioral outcome. As alluded to above and discussed in detail elsewhere,[20,31,32] the dose-effect relationship between TMT and neurochemical or learning-performance variables is anything but linear or quasi-linear.

What we had concluded by the end of the 1980s was that doses of TMT that had the greatest effect upon binding of ^{3}H-corticosterone to hippocampal cytosolic receptors interfered most with acquisition of autoshaping with a working memory component (i.e., a reinforcement delay after lever retraction), without **apparently** affecting acquisition or performance of the same task, in the same chambers, using the same reinforcers, and under the same level of food deprivation but without a substantial working memory component. In other words, such outcomes led us to conclude that moderate doses of TMT (e.g., 3 to 6mg/kg) most likely did not interfere with sensorimotor integration, reinforcement efficacy, and basic association processes, but rendered the rats less capable or incapable of learning or performing as control subjects when there was temporal discontiguity between the CS (lever retraction) and the UCS (food pellet delivery). Higher doses of TMT cause a larger lesion (measured by hippocampal weight alone) but, if anything, seem to enhance initial rates of delayed reinforcement autoshaping (although the quasi-learning is not sustained). As described above, high doses of TMT cause greater reactivity to the environment, including the retractable lever-CS, thereby causing a deficit in their ability to learn to disregard the lever after several sessions of exposure to its extension into the operant chamber and retraction without programmed consequences (i.e., retraction upon a lever touch or delivery of reinforcer after lever retraction). When, after several such sessions the autoshaping protocol is initiated (i.e., food pellets are delivered 6 s after lever retraction), control subjects are latently inhibited[21,33] and demonstrate this by refraining from engaging in lever-directed behaviors, having learned about the irrelevance of the lever.[34] The rats treated with the high doses of TMT continue to engage in lever-directed exploratory behaviors, indicative of a reduced ability to learn to suppress behaviors toward stimuli which ordinarily should have lost their saliency or novelty.[20]

The divergent autoshaped lever touch behavioral effects of low vs. high doses of TMT were also manifested as divergent progressive fixed ratio (PFR) lever press responding when TMT treated rats were switched to a lever press, fixed ratio schedule of reinforcement, and probed with daily-changing demands under the PFR schedule. Rats given 6mg TMT/kg months earlier, which showed a deficit in autoshaping, pressed the lever significantly less than controls as the ratio requirements incremented upward, while rats given 7.5mg TMT/kg responded upon the lever (now continuously extended into the chamber) significantly more than controls as the ratio requirements incremented upward. Lever press behavior was not different until ratio requirements increased to and beyond FR16-FR32. Interestingly, a significant correlation between response output and corticosterone receptor density in hippocampal cytosol emerged, again suggesting a relationship between behavioral performance and this neurochemical variable[31] and, while the basis for this relationship is currently without explanation, it appears as though the high dose of TMT produced an effect typical of the so-called hippocampal animal, which for decades has been produced by large aspiration or electrolytic lesions of the dorsal hippocampus, resulting in behavioral perseveration and hyper-reactivity.[35]

The reduced PFR performance by the rats treated months earlier with 6 mgTMT/kg upon FR16-FR32 demands could again have been due to the temporal gap between their initiation of lever pressing and the eventual delivery of the reinforcer which, at typical response rates of 2 to 3 per s, would have required 6 to 12 s to complete the run. Perhaps it is necessary for rats to use working memory to maintain responding for the reinforcer to be delivered in the future and their working memory capacity was overwhelmed by the lower dose of TMT but the lesion was not large enough, or was different enough so that perseveration was not maintaining higher rates of PFR responding, for even greater than control rats' PFR rates, as was demonstrated by rats in the 7.5 mg TMT/kg group.

Every student taking an introductory psychology course learns about the learning-performance distinction. The effects of TMT (6 mg/kg) upon autoshaping with at least a 6 s delay of reinforcement was robust and reproducible, but we could not determine from the aforementioned experiments if the deficit in performance was a learning vs. a storage, retrieval, and performance effect(s) of TMT. This is an important distinction because the possible use of rats treated with moderate doses of TMT as a model of the early stages of dementia hinged upon their ability to recall and use previously learned information, but they would be unable to learn/perform the exact same tasks after lesioning their limbic and forebrain structures with TMT. This possibility was tested in an experiment that utilized rats which had previously acquired the association, reaching asymptotic levels of performance of the working memory variation (i.e., 6 s delay of reinforcement) shortly **before** a 6mg dose of

TMT/kg. One month after being treated with TMT they were experimentally "asked" to perform a task that they would have been incapable of had they been treated with the same dose of TMT one month prior to acquisition (i.e., learning) sessions. Their performance was asymptotic before and also one month after TMT and virtually identical to an equally autoshaped cohort given water by gavage instead of TMT. This straightforward experiment, with its unambiguous outcome, enabled us to conclude that TMT (6mg/kg) did not affect storage, retrieval, and utilization of the information necessary to perform the task, once it had been consolidated and probably distributed to and retrievable from sites other than those necessary for initial processing and consolidation of information which required normal working memory capabilities, namely the hippocampus and associated limbic forebrain structures, demonstrating a clear cut learning-performance distinction.

E.   Effects of Pharmacological Manipulations Upon Performance: Problems of Interpretation

Choosing a dose of a drug to test for positive or negative effects upon mnemonic processes, instead of upon performance variables, should be easy. Simply use subjects that have already acquired the task (as above) and determine dose-response relationships upon performance. Thus, a dose of a drug that is devoid of effects upon established (i.e., already learned) behaviors theoretically does not affect sensorimotor integration, reinforcement efficacy, storage, retrieval, or utilization (i.e., performance) of the learned association. One might logically expect/predict interference with learning an autoshaped association by the psychotogenic NMDA antagonist phencyclidine, at a dose that does not affect already learned autoshaped behavior.[36] However, a similar experimental outcome with a drug that should theoretically have enhanced learning and/or memory speaks to the importance of rethinking and re-evaluating many of the assumptions upon which experiments designed to discover pharmacological interventions for retarding or reversing neurodegenerative-associated cognitive impairments are based. The following descriptions are of two series, one which led to the conclusion that an agent already in use for treating patients might be toxic itself, based upon its use against TMT-induced behavioral, neurochemical, and neuropathological effects. These have been published and will be tersely reviewed. The other series will be discussed in greater detail and has examples of a multitude of important concepts, ideas, and interpretations of behavioral pharmacological experimental outcomes.

   The first series of experiments led us to conclude that gangliosides are of questionable efficacy against TMT-induced acquisition and/or performance deficits and its associated neurochemical and neuropathological markers. We also concluded that these glycosphingolipids may cause neurobehavioral toxicity by themselves, when administered to rats which had not been treated with TMT, because rats treated only with gangliosides showed some effects reminiscent of TMT's effects and/or effects which suggested toxicity, because behavioral performance deviated significantly from control rats' performance. Moreover, a critical review of the experimental literature, including (mis)interpretations within a large basic research literature, and a review of the sparse clinical literature, was equally discouraging.[37,38] Use of gangliosides for treatment of human neurodegenerative disorders and CNS trauma has essentially been discontinued or terminated by government decrees because of potential toxicity (i.e., because of its association with neuropathological symptoms appearing after administration of gangliosides to patients), except for perhaps a few clinical trials in the U.S. and elsewhere for Parkinson's disease.

Thus, our experiments with CH_3Hg, TMT, and gangliosides, among other agents, had convinced us of the utility of variations of this class of behavior(s) and associated manipulations for psychopharmacological, behavioral toxicological, and behavioral neuroscience experimentation.

The last example of the use of autoshaping demonstrates the importance of considering several of the ideas and conclusions alluded to above and in our various publications, as well as those from other groups, so as not to possibly be misled into "throwing the baby out with the bath water" or perhaps prematurely concluding that a drug is toxic and/or devoid of potential efficacy for treatment of one or another type of neurodegenerative disorder, including dementia. Alternatively, it may be a confirmation that physostigmine, and possibly other drugs in its class, lack sufficient efficacy and convey too much potential toxicity to seriously consider them for treatment of, for example, Alzheimer's disease or other forms of dementia. I will leave it to the reader to interpret the outcome in this regard. However, the following experiment and its results incorporate examples of the need to systematically manipulate pertinent variables when doing learning and memory studies, no matter which behavioral neuroscience protocols one chooses to use. Since we used autoshaping as the paradigm for studying physostigmine's behavioral effects, I present it as one example of various methods we have used to ask a series of behavioral neuroscience questions and to test several provocative hypotheses about the nature of drug-behavior interactions.

F.   Physostigmine and Autoshaping: Examples of Hypotheses Testing, Complex Experimental Design, and Data Analyses

The importance of acetylcholine as a neurotransmitter in the central nervous system no doubt was responsible for its introduction into theories related to the biological bases of cognitive function. Deutsch[39] suggested a cholinergic basis for learning and memory and many studies in humans and animals have focused on the role of cholinergic processes in memory formation or retrieval. Investigations with both anticholinergic and cholinomimetic drugs indicate that fluctuations in cholinergic activity can profoundly affect memory formation, storage, and/or retrieval of information in animals[40] or normal human subjects.[41,42]

That acetylcholine plays a pivotal role in cognitive function is supported by studies that have found a deficit in choline acetyltransferase in brains of people who had been diagnosed as having senile dementia of the Alzheimer type (SDAT[43]), although several other monoaminergic and peptidergic transmitter systems are also altered in disease states associated with cognitive impairment.[44,45]

In humans, tests of the ability of the cholinesterase inhibitor physostigmine to enhance memory have produced equivocal results. Some authors claim significant and replicable memory enhancement in patients with SDAT after administration of the drug.[46,47] Other studies, however, report no improvement in memory with this drug.[48-50] However, any enhancement of human memory is limited in extent and to a narrow dose range.[41,46,51] It has also been suggested that a dose of physostigmine which engenders the maximum improvement in memory may be quite different among subjects. Individual differences of this type may be due to inherent genetic determinants, to a steep dose-effect curve, and to as yet undefined variables, conveying a small therapeutic index. Differences in the severity of SDAT or other dementias may also be a factor in accounting for the variable results. For example, it has also been reported that there is a correlation between the severity of the dementia (as measured by the Mattis Dementia Rating Scale) and physostigmine's effectiveness in improving recent memory.[52]

In general, older animal studies with both rats and mice failed to provide a consistent or easily understood analysis of the cholinergic involvement in learning/performance.[53-55] The use of many different experimental protocols may have contributed to this, although the equivocal nature of the animal data may be related to some of the factors described above, which appear to have influenced the outcome of human studies of learning and memory with physostigmine (e.g., individual differences, a narrow effective dose range, limited improvement, short duration of action, etc.). Moreover, it is possible that little or no cognitive-enhancing effects would be evident if the non-human subjects were healthy and ordinarily able to learn an association optimally (i.e., a ceiling effect) under the conditions established by the experimenter. This was the basis for using poor-performing rats by Rech.[56] A general lack of a cognitive-enhancing effect would not be readily apparent unless it could be demonstrated, via an acceptable drug standard or other manipulations, that the experimental protocol is sensitive and, therefore, is capable of demonstrating enhancing effects of physostigmine (e.g., the use of TMT-treated or aged rats).

Another factor that must be taken into account is the inverted U-shaped function often associated with drug studies on learning or memory, and the possible reason for the descending arm of the dose-effect curves, even if there is some evidence of learning enhancement. If "high" doses of a drug like physostigmine are toxic (i.e., would fall on the descending arm of an inverted dose-effect curve) and, by definition, punishing in relation to behaviors which occur immediately prior to or in association with treatment, the drug might act both as an aversive unconditioned stimulus, causing sickness,[57,58] and as a conditioned stimulus, signalling a discriminable internal state change,[59,60] which becomes associated (paired) with subsequent unconditioned aversive effects, as the drug is absorbed and distributed to target organs, attaining higher (aversive) concentrations. Although pharmacodynamic tolerance to the unconditioned, punishing properties is possible, if it develops too slowly the contiguous nature of the CS-UCS relationship is optimal for development of conditioned aversion or conditioned behavioral suppression. This would especially be the case in experimental subjects that are not learning-impaired to begin with (as might be the case because of ageing, some surgical or neurotoxic insult) and therefore capable of forming such an association, leading to conditioned suppression of behavior. If this association is formed, the same event (i.e., drug) would act as both discriminative stimulus (CS) and punisher (UCS). And, if administered systemically while in a novel environment, the injection-context may also become entrained as a conditioned punisher, thereby suppressing fragile behaviors not yet well conditioned, as in the early stages of autoshaping.

These ideas are the result of data derived from pilot experiments (unpublished observations) with rats that had already learned to associate the presentation and retraction of a lever manipulandum with the subsequent (delayed) delivery of a food pellet reinforcer in a forward autoshaping procedure, as described above. By determining which doses of physostigmine did and did not interfere with performance, we were able to choose a dose range which theoretically should not be aversive, not suppress the required motoric behavior, or otherwise interfere with the memory or retrieval process, thereby controlling for these sources of variation in acquisition (learning) experiments in which the same paradigm was to be used to test physostigmine's potential enhancing properties and/or conditioned punishing properties. In spite of our preliminary results, subsequent use of those innocuous doses resulted in what appeared to be evidence of behavioral toxicity, which was greatest in task-naive rats, at doses equal to or less than those which did not in the least affect rats that had learned the association and were performing at asymptotic levels. Our global (intuitive) interpretation of these results prompted us to test the hypothesis that task-naive rats were more susceptible to the behavioral toxic actions of physostigmine than were rats which had already learned the association to some extent before their first injection of the drug, which in turn were more sensitive to the drug's suppressant (behavioral toxic) effects than were better trained (i.e., autoshaped) subjects.

Coveney and Sparber[36] first introduced the need for determining a dose of a drug which should not suppress lever-touch (motoric) behavior during autoshaping (i.e., it is non-toxic in this regard). Because of the usual signs occurring with administration of high doses of anticholinesterase drugs or very low doses of chemical warfare agents (i.e., salivation, lacrimation, urination, and defecation[61]), it was necessary to avoid an obviously toxic dose of physostigmine which might suppress the required motor behavior or cause other toxic CNS or peripheral (aversive) parasympathetic effects in order to be able to facilitate acquisition, if it was to occur.

Additionally, we have previously reported that some sort of situational or context-related resistance to the disruptive effects of psychotropic drugs upon learned behavior emerged when the drugs' effects were first experienced in, or somehow paired, with the rats' familiar home cages, rather than with the less familiar environment of an operant chamber,[62,63] even in overtrained rats. Such observations suggest an interaction between novel environments and psychotropic drugs, regardless of the extent of conditioning. Thus, a combination of a drug-induced internal novel environment, paired with a novel external environment, may lead to an exaggerated effect, converting an otherwise innocuous or therapeutic dose into a toxic or punishing dose. The experiment reported herein was designed to test the several possibilities discussed above. We believe we have succeeded in demonstrating situational sensitization, which depends upon the drug's pharmacological properties being amplified by the degree of novelty of the external environment in which the subject experiences the drug's action for the first and subsequent injections. The more novel the gestalt (novelty of the drug-induced internal state change and the novelty of the external environment), the greater the possibility of a behavioral suppressant outcome.

When this occurs, there is a greater chance for development and maintenance of conditioned behavioral suppression, manifest as continued suppression even after the dose of drug is dramatically reduced so as to maintain its CS properties in the absence of its UCS properties. After a sufficient number of such pairings, the injection procedure itself may cause conditioned behavioral suppression.

Besides comparing a control group's learning to associate the retracting lever with reinforcer delivery during autoshaping with that of a group injected daily with a behaviorally innocuous dose of physostigmine shortly before acquisition sessions, we included two additional groups of subjects to determine if the behavioral/learning suppressant action of physostigmine would be diminished if they first experienced its effect in an environment other than the novel test chambers (i.e., while in their home cages), or if they were injected with the same dose after the acquisition sessions (i.e., after removal from the test chambers). We thought this latter manipulation might control for non-specific behavior-suppressing actions, if present, and at the same time allow for the pharmacological effects to enhance the so-called process of consolidation. Additional groups were injected for the first time with physostigmine after they had experienced 3 or 7 acquisition sessions, having been injected with saline before the physostigmine sessions and, therefore, after differing degrees of autoshaping/levels of performance. By comparing the magnitude of behavioral suppression caused by the first and subsequent injections of physostigmine into rats having the same number of exposures to the operant chambers but injected after different numbers of sessions (i.e., after having partially or more completely autoshaped but below maximally asymptotic levels of ELT), we were also able to demonstrate that resistance to the behaviorally toxic action of physostigmine is conveyed by greater strength of the conditioning, and perhaps is separable from the novelty variable, although the (aversive?) properties probably operate in tandem or in some multiplicative manner, under the right conditions. Others have also been intrigued by the combination of exteroceptive context interacting with stimulus properties of drugs.[64]

II.   Materials and Methods

Subjects — Fifty-four young, mature male Sprague-Dawley rats (Holtzman/Harlan, Madison, WI) were housed singly and maintained in a temperature (22°C) and humidity (40 to 50%) controlled environment, on a 12:12 hr light:dark cycle, with lights on at 0700 h. Food and water were initially available *ad lib*. Rats were gradually deprived to and maintained at 80% of their free feeding weight; testing was done between 1100 and 1600 h and supplemental feeding occurred at 1830 h.

Physostigmine — Physostigmine salicylate (Sigma Chemical Corp., St. Louis, MO) was dissolved in isotonic saline on the day of injection and administered subcutaneously in a volume of 1ml/kg body weight.

Apparatus — Subjects were tested in modified, standard operant chambers (Model 143-22, BRS/LVE, Laurel, MD) placed in specially constructed sound attenuating isolation chambers containing closed-circuit television cameras. The operant chambers are 31 cm wide by 25 cm deep by 25 cm high, with a grid floor consisting of 0.5 cm diameter stainless steel bars spaced 2 cm apart. Each chamber contains standard house and cue lamps, as well as a speaker which delivers a white masking noise. They also contain a metal strip 7.5 cm wide on two walls, the bottom edge of which was 15 cm above the grid floor. Contact with the strips (i.e., exploratory rearing activity) can be monitored using a high resistance drinkometer circuit (ca. 2 M). Each chamber is equipped with a retractable lever (Model RRL-015, BRS/LVE, Laurel, MD) and contacts with the lever, both while extended and while retracted, can also be monitored with drinkometer circuits. At this point, it should be emphasized that the type of food delivery trough or configuration (e.g., a simple, spoon-shaped trough as opposed to a recessed compartment with a hinged door) and/or the type of retractable lever (e.g., a motor-driven, cam operated, relatively quiet one as opposed to a solenoid-driven, louder one) may influence the rate and degree of autoshaping. While one should be able to have rats autoshape with variations of the mechanical devices in and configuration of the operant chambers, it should be remembered that magazine training, exposure to the reinforcer prior to its use (especially if it is different from the subjects' regular food) and other procedures which are capable of inducing neophobia should be avoided. On the other hand, too much habituation to the important physical characteristics and schedule components should also be avoided because the saliency of relevant stimuli (e.g., a retractable lever, lighted pigeon key, etc.) and state of the organism (i.e., partial arousal) are important considerations for optimal experimental outcomes. The contingencies are normally controlled with and the data collected by microprocessor-based computers connected to the chambers by custom built interfaces. Although the operant chambers are essentially physically identical, rats from each of the experimental groups were randomly distributed among the operant chambers to control for unknown sources of variation. A stratified assignment procedure was used such that approximately equal numbers of rats from each group were tested in each operant chamber.

Procedure — Two days prior to exposure to the behavioral apparatus, rats were allowed access to ten 45 mg food pellets (Formula 21, Bioserv, Frenchtown, NJ), which later served as the reinforcing stimuli. Animals were allowed 20 min to eat the pellets in their home cages. This procedure ensured that subjects were familiar with the food pellets prior to conditioning.

During autoshaping sessions the retractable lever was presented on a random time 45 s schedule, with ITIs ranging from 22 to 68 s, starting after the food pellet reinforcer was delivered. The levers remained extended until the subjects touched it or until 15 s had elapsed. A food pellet was then delivered 9 s after the lever was retracted. In this experiment subjects were given 12 daily trials for 13 consecutive days. Measurements of the number of extended lever touches (ELT), latencies to touch the extended lever, as well as lever touches during the intertrial and delay of reinforcement intervals are typically recorded for each session. Lever touches, when the lever is retracted (intertrial and reinforcement delay intervals) consist mainly of nose-poking behavior, although as stated above, the rats were also observed to touch the retracted lever with their forepaws. Latency to respond can be analyzed for all 12 trials. A failure to respond can be treated as a 15 s (maximum) latency. Additionally, response latency for lever touches which occur within 15 s (ELT) can be analyzed to determine if drug treatment affected sensorimotor capacity and/or motivation. To control for a practice effect upon speed of responding (i.e., reciprocal of latency), which will occur as subjects autoshape, speed of responding can be analyzed for that session during which an animal was determined to have learned, using an arbitrary but conservative learning criterion, regardless of in which session the criterion was reached. This method of analysis has been shown to control for a practice effect and has been interpreted to indicate that enhancement of autoshaping rates brought about by greater levels of food deprivation is most likely not due to greater motivation.[33] For this study we used a criterion for learning based upon ELT performance by individual rats when their ELT exceeded the 99% confidence interval, derived from session 1 ELT, on two out of three contiguous sessions. Also, for purposes of this study, an animal that failed to reach the criterion was assigned a value of 15 sessions (minimum possible). The 99% confidence interval used (6 ELT/12 trials) was calculated from the performance of the non-drug treated animals (N = 36) on the first day of autoshaping, at which time it was assumed that learning had not yet taken place, although this assumption is a rather conservative one, since it is probable that some learning had occurred in some rats during the very first session (*vide infra*).

All subjects were injected at least once with physostigmine salicylate (0.20 mg/kg, s.c.) 20 min prior to the autoshaping sessions. One group was injected with the cholinesterase inhibitor just after removal from the operant chambers (post-session group, PSG). Because previous reports (e.g.[65]) and our own pilot studies indicated that some sort of tolerance to the (behaviorally toxic) effects expected with this dose might develop after several injections, another group was administered drug daily for 3 days prior to initiation of autoshaping. This group was injected while in their home cages (HCG). Thus, at least the CS, if not the UCS properties of physostigmine were not experienced while in the novel (operant chamber) external environment for the first 3 injections. Various other groups were injected with drug at different stages of training (i.e., acquisition), since we were also testing the hypothesis that rats which had acquired the association between lever presentation/retraction and food pellet delivery would be increasingly resistant to the behaviorally disruptive effects of physostigmine. As introduced above, we carried out this experiment because it was observed that rats that had already learned the association and were performing at asymptotic levels were unaffected by the same dose of physostigmine which seemed to block learning or performance in experimentally naive animals.

Therefore, another group was injected with drug for the first time prior to the fourth day of the acquisition sessions (D4G) and one group prior to the eighth day (D8G). The Control group (CG/D13G) received injections of isotonic saline. The CG/D13G group, which approached asymptotic levels of performance by the 10th to 12th day, was given 0.20 mg of physostigmine/kg 20 min prior to the 13th session, and thus is being designated as the CG/D13G. This was done to verify that this cohort of subjects was, as predicted, also totally resistant to the behaviorally toxic action of the drug, which is manifest as blockade of learning in naive rats and of suppression of performance in rats that were in the midst of acquiring the lever-food association.

By the seventh session it was apparent that the 0.2 mg/kg dose was producing the experience-related results we predicted. Subsequently, all subjects (except the D8G and CG/D13G groups) were administered one third of the previous physostigmine dose (i.e., 0.067 mg/kg) to determine if a differential drug-behavior history would cause a differential response to the new, lower dose of the drug. As previously stated, we predicted that tolerance would emerge first for the rats injected in a familiar, home cage environment, while conditioned suppression would cause the performance of the rats injected for the first time and thereafter shortly before being placed into the novel environment to remain suppressed, in spite of the reduction by two thirds of the dose injected prior to Session 8. Table 14.1 depicts the various groups and an outline of the experimental protocol. All subjects received 2 injections per day (1 or 2 saline, 0 or 1 drug) so that a single control group (CG/D13G) could be used for the pre-session or post-session drug injection groups. In other words, the group injected with drug after sessions (PSG) was given saline 20 min prior to sessions, and the groups injected with drug prior to later sessions (i.e., D4G, D8G, and CG/D13G) were injected with saline before their first drug session and after each of the sessions. The D1G received drug 20 min prior to their 1st–12th sessions; 0.2 mg/kg prior to Sessions 1–7; and 0.067 mg/kg prior to Sessions 8–12.

TABLE 14.1
Physostigmine Treatment Protocols During Autoshaping Sessions 1 to 13

Treatment	Session 1	Sessions 2–3	Sessions 4–7	Sessions 8–12	Session 13
Cg/D13g[a]	Saline	Saline	Saline	Saline	0.2 mg/kg
D1G	0.2 mg/kg	0.2 mg/kg	0.2 mg/kg	0.067 mg/kg	Saline
HCG	0.2 mg/kg	0.2 mg/kg	0.2 mg/kg	0.067 mg/kg	Saline
PSG[b]	0.2 mg/kg	0.2 mg/kg	0.2 mg/kg	0.067 mg/kg	Saline
D4G	Saline	Saline	0.2 mg/kg	0.067 mg/kg	Saline
D8G	Saline	Saline	Saline	0.2 mg/kg	Saline

[a] Control Group injected once, prior to the 13th session.

[b] Physostigmine injected post-sessions.

Data Analyses — Because these rats had not generally achieved asymptotic levels of autoshaped behavior to a degree that allowed the optimal use of RD and/or ITI lever-directed behaviors, ELT data were used for most analyses presented here. Three (optional) methods of analyses are presented and the outcome of the different data analyses support each other and are consistent with the hypotheses stated above. Data were analyzed using repeated measures analyses of variance (ANOVA) and factorial ANOVA, followed by Dunnett's (hypothesized, unidirectional) tests to compare treatment groups with the controls (i.e., the CG/D13G) or, if appropriate, Fisher's Protected Least Significant Test (PLSD) to compare various treatment groups, including controls, with each other. A repeated measures ANOVA was also performed on ELT for sessions immediately prior to and on the drug days, regardless of which session day that fell upon. This was followed by individual group paired contrasts to determine which groups (i.e., D4G, D8G, and/or CG/D13G) showed a significant reduction in ELT as a result of treatment with 0.2 mg physostigmine/kg at that particular stage of acquisition. Regression analyses for average ELT made by each group injected immediately prior to sessions vs. the day those injections (of 0.2 mg/kg) were first given, were carried out for the 9th, 11th, and 13th sessions as a means of further characterizing the relationship between the degree of autoshaping and sensitivity to disruption by the drug. The data for sessions to criterion were also analyzed by ANOVA (using the session number upon which the criterion was reached). Additionally, the number of rats in each group, out of 9, that achieved criterion performance during 12 sessions was analyzed by Chi Square contingency tests.

Unconditioned, exploratory rearing behaviors were analyzed by ANOVA, at various stages during acquisition sessions, to determine if any of the treatment protocols affected this variable as well, or affected this variable at the exclusion of the autoshaping behavior. Correlation analyses were performed to determine if the exploratory behavior was (hypothesized to possibly be) inversely related to lever-directed behaviors, especially as the rats learned to associate the retractable lever with the delivery of the reinforcer and directed their attention to the lever and food pellet delivery trough. As it turned out, none of the correlation analyses between exploratory rearing behaviors and ELT showed a statistically reliable relationship. In all cases, a p value 0.05 was considered significant.

III.   Results

A.   Unconditioned Exploratory Rearing Activity (Strip Touches)

Repeated measures ANOVA for strip touches by all 6 groups, across all 13 autoshaping sessions, revealed a significant Treatment (Group) effect ($F_{5,48} = 5.66$; p<0.001), a significant Repeated Measures effect ($F_{12,576} = 9.17$; p<0.001); and a significant Interaction ($F_{60,576}$; p<0.001). When placed into the operant chambers for their first autoshaping session, the novelty of the environment elicited almost identical levels of strip touches, ranging between 42.7 (CG/D13G) and 49.1 (D4G) for the rats which had not yet been injected with the higher dose of physostigmine salicylate (i.e., 0.2 mg/kg, s.c.) 20 min before the session. However, the rats injected 3 times while in their home cages (HCG), receiving their fourth injection of the cholinesterase inhibitor, and the rats that were injected for the first time with drug 20 min before their first session (D1G) displayed a drug effect upon this measure of unconditioned exploratory behavior as a significant decrease, by about 55 to 75%, relative to the

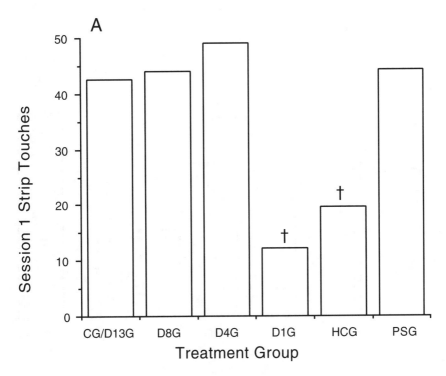

FIGURE 14.1

Exploratory rearing behaviors (strip touches) during autoshaping sessions 1(A), 2(B), 6(C), and 8(D) by the different groups administered saline and/or physostigmine (0.2 mg/kg, s.c.) prior to and/or after sessions. Prior to sessions 1 and 2, the control group (CG/D13G) and the D4G and D8G were injected with saline. The rats in the group injected with physostigmine while in their home cages (HCG) were injected with saline after their autoshaping sessions and the group injected with physostigmine after (post) autoshaping sessions (PSG) was injected with saline before the sessions. For a more complete description of the experimental protocol, see the text and Table 14.1. †$p<0.05$ or better vs. the CG/D13G; 2-tailed PLSD tests after significant omnibus ANOVA outcomes for individual sessions.

CG/D13G (Figure 14.1A). By the second autoshaping session the CG/D13G's rearing activity was reduced by almost 50%, relative to Session 1, as was this measure of behavior for the other groups also not yet injected with physostigmine (i.e., D4G and D8G), probably as a result of both habituation and attention to the retractable lever and pellet delivery, as it was apparent that these groups were already autoshaping during their first session (*vide infra*). Because of what may have been a partial floor effect, habituation to the drug-environment interaction, or some other form of tolerance for this variable for these 3 groups, rearing behaviors of the HCG and D1G were no longer significantly depressed during Session 2. However, the group injected after the first autoshaping session did not display evidence of such habituation (Figure 14.1B). Indeed, even during the sixth autoshaping session, the PSG still engaged in significantly more strip touching behavior (Figure 14.1C). It was during this session that ELT behavior of the PSG, which had not been different from the CG/D13G controls during the first 5 autoshaping sessions, dropped to its lowest

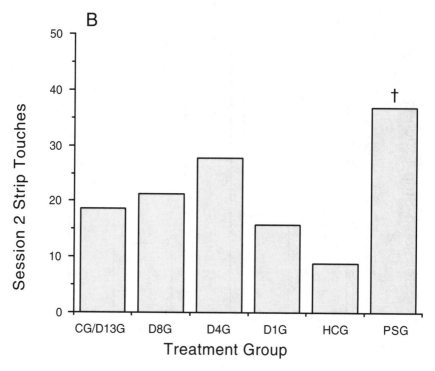

FIGURE 14.1 (Continued)

level, being significantly below that of the controls for this session and the seventh session. After the 7th session, ELT for the PSG was no longer significantly depressed, relative to the CG/D13G, perhaps because the dose of physostigmine was decreased to 0.067 mg/kg (Figure 14.2B).

The D8G was resistant to suppression of ELT by physostigmine during the eighth session (Figure 14.2A), when it received its first injection after having acquired the autoshaped task (*vide infra*). Figure 14.1D nevertheless demonstrates an effect of the drug upon the unconditioned behavior during this same session. While a floor effect may seem to have been responsible for a lack of a (reduced) difference in strip touching behavior for the D1G and HCG, relative to the CG/D13G during sessions subsequent to Session 1 (see, e.g., Figure 14.1B), it is unlikely. This is attested to by the fact that while the CG/D13G's strip touching behavior was down to an average of less than 20 per session during both sessions (2 and 8), D8G strip touching was depressed even further, to less than 4 per session, on average.

B.   Autoshaped Extended Lever Touches

The outcome of a repeated measures ANOVA, as part of the first of 3 types of data analyses for autoshaping being presented, resulted in a significant Treatment effect ($F_{5,48} = 5.26$; p<0.001) and a significant Interaction ($F_{60,576} = 2.06$; p<0.001). Planned

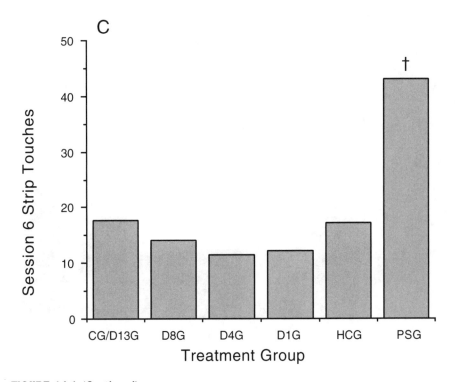

FIGURE 14.1 (Continued)

contrasts, using Dunnett's tests for individual sessions and based upon pilot experiments in which ELT was suppressed by the cholinesterase inhibitor, revealed significantly diminished ELT performance by the D1G during 12 out of 13 sessions; displaying no evidence of autoshaping at all, as a group, (Figure 14.2A,B). The fact that the D1G differed from the CG/D13G on all sessions except Session 12, but including Session 13, when the CG/D13G was injected with 0.2 mg physostigmine/kg and the D1G was injected with saline, suggests that the chronic injections of physostigmine prior to each autoshaping session, at 0.2 mg/kg before Sessions 1–7 and at 0.067 mg/kg for Sessions 8–12, followed by saline prior to Session 13, caused unconditioned behaviorally toxic effects initially, followed by conditioned suppression, as well as a conditioned effect of the injection procedure, possibly because of what may have been a "fading" process from high to low to no physostigmine prior to the 13th session.

When the D4G was injected with physostigmine instead of saline 20 min before Session 4, their ELT dropped from an average of 6.11 ELT during Session 3 to an average of 2.78 during Session 4 and were significantly less than those of the CG/D13G (2.78 vs. 7.11). Thereafter, their autoshaped ELT were less than those of the control group's during Sessions 5–7 but not during Sessions 8–10 and Session 12, after getting 0.067 mg/kg, or Session 13, after saline was injected.

Injection of 0.2 mg physostigmine/kg into the D8G caused their ELT to drop from an average of 8.44 during Session 7 to 4.89 ELT during Session 8. When

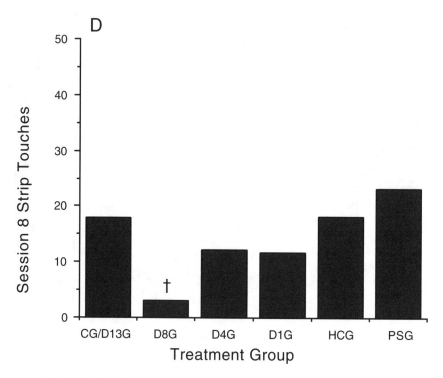

FIGURE 14.1 (Continued)

compared to the CG/D13G's 9.11 ELT during Session 8, the D8G's performance was significantly depressed only during Session 8. Thereafter, the D8G level of performance (ELTs) was not different from that of the CG/D13G.

As stated above, the group injected with the higher dose of physostigmine after each session, instead of before sessions, showed what appears to have been a sort of context-related, cumulative conditioning effect because its unconditioned exploratory rearing behavior remained elevated (*vide supra*) and its ELT performance was finally depressed during Sessions 6 and 7, returning toward that of the CG/D13G during the subsequent 6 sessions, after the dose was decreased to 0.067 mg/kg, or saline (i.e., Sessions 8–13).

The second and third methods of analyses of autoshaping behavior being presented, and how physostigmine affected it, utilized a learning criterion. While arbitrary in nature, it was felt to be conservative because it required that the number of ELT emitted by each rat exceeded the upper limit of the 99% confidence interval derived from the first session for the 36 rats which had not yet received drug prior to the session. Therefore, the criterion required that at least 6 ELT were performed during at least 2 of 3 contiguous autoshaping sessions. The conservative nature of this criterion (i.e., the demonstration that Session 1 performance was not devoid of some rats already learning the autoshaped relationships and not a pure measure of baseline, unconditioned ELT behavior) is attested to by the fact that 15 of 54 rats emitted 6 or more ELT during Session 1 and that 11 of 54 rats achieved criterion

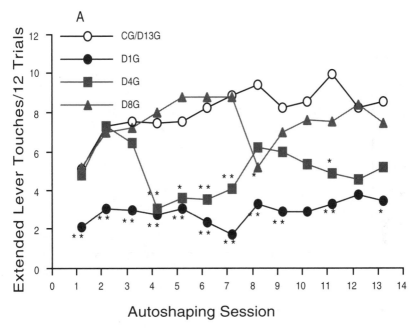

FIGURE 14.2

Extended lever touches (ELT) during autoshaping sessions. All groups were injected with 2 doses of saline or saline and 0.2 mg physostigmine/kg until Session 8, after which the dose of physostigmine was decreased to 0.067 mg/kg, except of the D8G and the CG/D13G. For a more complete description of the experimental protocol see the text and Table 14.1. *p< 0.05, **p<0.01 vs. the CG/D!#G; Dunnett's 1-tailed tests.

performance by Session 2, and 21 of 54 rats did so by Session 3. Interestingly, none of the 9 rats in the D1G achieved criterion ELT performance by the third session. The omnibus analysis for the number of rats in each group achieving criterion performance by Session 3 was significant ($X_5^2 = 12.86$; p<0.025), as was the contrast between the D1G and the controls (p<0.05). Omnibus X_5^2 analyses for the number of rats achieving criterion performance by the 6th, 9th, and 12th sessions were likewise significant (15.66, p<0.01; 13.82, p<0.02; 18.74, p = 0.01, respectively). Thus, while only 44% in the CG/D13G achieved criterion ELT performance by the third session, none in the D1G did so. The CG/D13G performance increased to 78%, achieving criterion by Session 6 and 89% for Sessions 9 and 12. Only 22% (2 of 9 rats) in the D1G demonstrated evidence of autoshaping by Session 12 (Figure 14.3A,B).

Figure 14.4 depicts the outcome of the ANOVA for the statistic defined as the session during which each rat first achieved criterion performance. It should be remembered that, while only 2 of 9 rats in the D1G did so by the 12th and 13th sessions, the other 7 rats in this group were assigned values of 15, the minimum possible if they emitted 6 or more ELT during the hypothetical Sessions 14 and 15. Thus, the average value for this group is (very conservatively) 13.1 and the omnibus ANOVA for this statistic was, as expected, significant ($F_{5,48} = 5.26$; p<0.001).

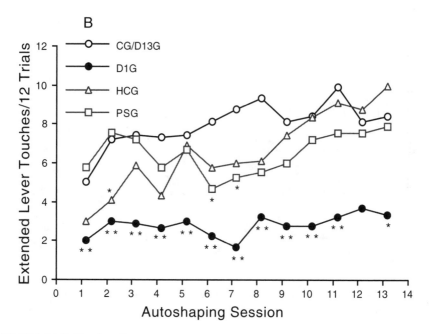

FIGURE 14.2 (Continued)

Additionally, Fisher's PLSD contrasts indicated that the ELT performance of the D1G, using this method of analysis, was significantly below that of all the other groups.

As a final way of characterizing the relationship between ELT performance and drug histories, a repeated measures ANOVA was carried out on ELT emitted by the D4G, D8G, and CG/D13G subjects during the autoshaping sessions immediately before and during the session immediately after the first physostigmine (0.2 mg/kg, s.c.) injection. The outcome was a significant Repeated Measures effect ($F_{1,24}$ = 15.45; p<0.001), accompanied by a significant Interaction ($F_{2,24}$ = 5.15; p = 0.014). Individual, paired t-tests indicated that only the D1G and D4G were significantly depressed after the first physostigmine injection, while the CG/D13G was unaffected (Figure 14.5).

When least squares linear regression analyses were carried out on the average number of ELT emitted during the 12 trials for Sessions 9, 11, and 13, additional support for one of the pivotal hypotheses being tested by this study was forthcoming. Figure 14.6 depicts the outcome. For Session 9 the result was not quite significant (Figure 14.6A), but the relationship between the day each of the 4 groups received their first (or only) physostigmine injection and the average number of ELT emitted during Sessions 11 and 13 was significant (Figure 14.6B,C). Such data support the supposition that the greater the degree of autoshaped learning to have taken place prior to physostigmine, the greater the resistance to interference with additional autoshaping acquisition and/or performance of measures of autoshaped behavior (i.e., ELT).

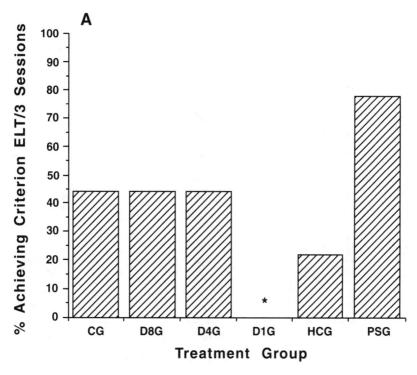

FIGURE 14.3
Percentage of rats in each group achieving the autoshaping criterion of emitting 6 or more ELT during
2 of 3 contiguous sessions. By Session 3 (Figure 14.3A) the percentage of rats in the D1G achieving the
criterion was significantly below the CG/D13G (0% vs. 44%, *p<0.05). These values increased to 11%
vs. 78% during Session 6 and 22% vs. 89% for Session 9 (**p<0.01 in both cases). By Session 12 the
percentage achieving criterion for these two groups was the same as for Session 9 and is depicted in
Figure 14.3B. None of the other groups were significantly different from the control group, using this
criterion analysis.

IV. Discussion

The results of this experiment indicate that, as animals learn to form the association
between lever extension/retraction and food pellet delivery during autoshaping they
become increasingly resistant to the behaviorally toxic effects of physostigmine. As
shown above, only the D1G and D4G groups were significantly suppressed, compared to the D8G on more than the first day that each of these groups received
physostigmine. The D8G group was significantly suppressed on only the first day
of its treatment with physostigmine and it took 4 sessions (and a reduction of the
dose of physostigmine to 0.067 mg/kg) before the D4G group returned to levels of
responding not different from the CG/D13Gs. The D1G group remained suppressed
virtually throughout the study, even after their dose was also reduced to one third
for Session 8 and beyond and even on Session 13, when they received a saline
injection. This decrease in the number of days of behavioral suppression with

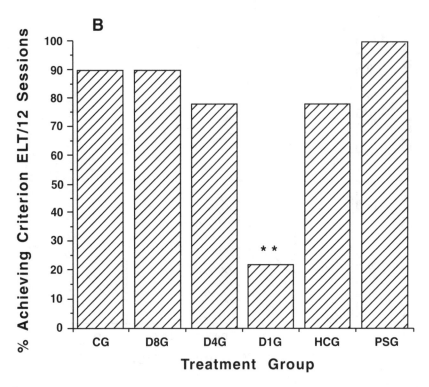

FIGURE 14.3 (Continued)

repeated training in the operant chamber, combined with the fact that the control group was not suppressed on their first drug treatment day (i.e., Session 13), compared to the session before, supports our hypothesis that as animals acquire the autoshaping task, the behavioral suppression due to physostigmine is lessened.

Furthermore, it was observed that treatment with physostigmine in the home cage for three days prior to autoshaping also reduced the behaviorally toxic effects of the drug. This was no surprise, given our previous observations related to drug treatment/experience while in the home cage vs. a relatively novel testing environment, but the observation that on equivalent drug days the HCG made significantly more ELT than did the D1G group (HCG Session 3 vs. D1G Session 6; 5.67 vs. 2.00 ELT, $p<0.01$; Fisher's PLSD test), in spite of the fact that the D1G group had three additional sessions in which to learn the association between the lever and food pellet delivery, showed that the development of tolerance to the behaviorally toxic effects of physostigmine was probably not dispositional and more than simply pharmacodynamic in nature.

One possible explanation for the differences observed between the HCG and the D1G group is the effect of the different environments experienced by the animals when they first received the drug. Previous reports have shown that external stimuli (e.g., home cage vs. operant chamber) at the time of, or shortly after, drug administration, can have as much of an effect on learning and performance as the direct, pharmacodynamic effects of the drug itself.[62-64]

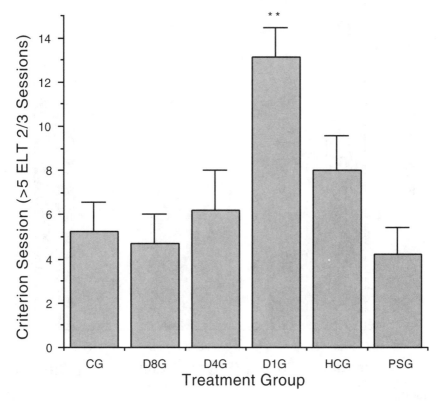

FIGURE 14.4

When an ANOVA was performed upon the session number in which the criterion was attained by each rat, assigning a value of 15 for those rats which had not achieved it by Session 13, the outcome is depicted in this figure. **p<0.025 or better vs. all other groups; PLSD following a significant omnibus outcome ($F_{5,48}$ = 5.26; p<0.001).

The suppression of the D1G group, at least after several sessions, is an indication that conditioned aversion/suppression was probably established and maintained. Previous studies[57,58] have shown that conditioned taste aversion takes place when physostigmine is presented immediately after the presentation of a novel stimulus (saccharine solution). Because this evidence suggests that pairings between a novel stimulus and physostigmine are aversive, it is quite possible that the pairing of the novel environment during the first days of autoshaping with the novel drug, physostigmine, produced conditioned aversion in the D1G group. In this case the aversive properties of physostigmine could be acting as a type of unconditioned stimulus which is affecting their behavior. This particular group showed the greatest amount of behavioral suppression, throughout the 13 sessions, and were the only group that experienced the effects of physostigmine and the novel environment of the operant chamber together from the very beginning. All of the other groups had previous experience with either the drug or the chamber before the two were paired.

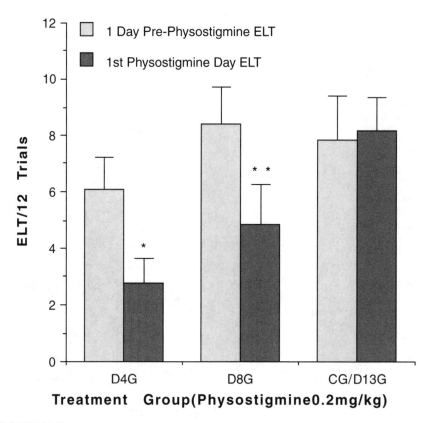

FIGURE 14.5
Autoshaped ELT were significantly depressed by physostigmine (0.2 mg/kg) during the session when first administered, relative to ELT performance during the previous session, for the D4G and D8G. The CG/D13G was unaffected by the injection of the cholinesterase inhibitor after 12 autoshaping sessions. *p<0.05; **<0.01.

Surprisingly, even after the dose of physostigmine was dropped on Session 8, the D1G group still showed no evidence of acquisition. As was suggested at the beginning of this section of the chapter, a high dose of physostigmine which is aversive, and thus acts as an unconditioned stimulus, may also signal a discriminable internal state change as the drug is absorbed, which can also act as a conditioned stimulus.[59,60] The animals that received a high dose of the drug for 7 days before the dose was dropped could thus discriminate an internal state change, even after the dose was dropped. Although they actually may not have been as adversely affected (i.e., by the high dose) as they had been previously, the expectation of the aversive effects to follow, and thus the discriminative drug stimulus, was sufficient to elicit the conditioned response (i.e., no performance). This pairing of CS and UCS may be responsible, at least in part, for the lack of autoshaping by the D1G group throughout the 13 sessions.

The PSG's autoshaping performance was almost identical to the CG/D13G's through the first 5 sessions. Thereafter, their performance dropped to a level

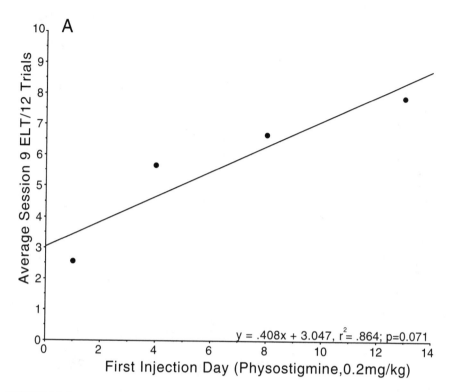

FIGURE 14.6

Regression analyses of average ELT for groups injected prior to sessions (D1G, D4G, D8G, CG/D13G) vs. the day the first injection of physostigmine was administered, for Session 9 (Figure 14.6A), Session 11 (Figure 14.6B), and Session 13 (Figure 14.6C).

significantly below the control group (i.e., during Sessions 6 and 7), coming back when the dose was decreased. It was during the deterioration of their autoshaped performance that their exploratory rearing behaviors were significantly elevated, so a different sort of interaction between their behavioral and drug histories was acting to prevent a similar habituation that was observed when the CG/D13G and D4G were injected with saline before and after the sessions. These two groups showed very similar exploratory rearing activity as did the PSG during the first autoshaping session but this behavior diminished thereafter, while the PSG, given physostigmine injection after the sessions, showed impaired habituation to the chamber.

Invoking the idea of aversive conditioning, it could be that the number of pairings of the chamber and the drug are determinants of the suppression in the D1G group, and that the D4G group was not as suppressed because they had not had enough pairings of the drug and the chamber for conditioned aversion to take place. Yet the HCG had more injections, in addition to the same number of pairings of drug and chamber as did the D1G group but nevertheless showed significant learning. Additional studies will have to be undertaken to more fully interpret the bases for the PSG's and the HCG's drug-behavior interactions, which deviate substantially from the other groups' performance.

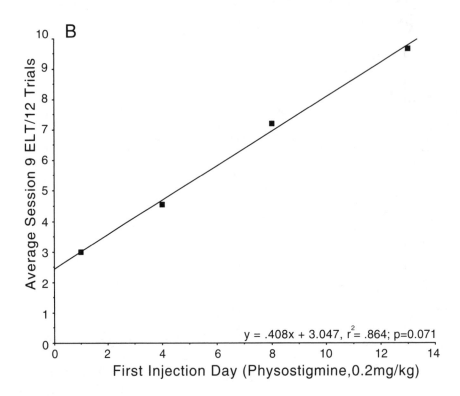

FIGURE 14.6 (Continued)

A.   Conclusions

In this portion of the chapter I have attempted to demonstrate by example the complex way in which a drug, one which theoretically should enhance cognitive function (based upon the neurochemical and neuroanatomical pathological literature on Alzheimer's disease), can affect a behavioral assay that has proven to be a reliable predictor of cognitive dysfunction, as described in earlier portions of the chapter. Perhaps more importantly, I have pointed out how to design experiments that consider the behavioral histories and drug histories and how these, among other variables, can influence the outcome of experiments in which learning a new association is being used as the dependent variable and how one might determine if a behavioral outcome is based upon acquisition or performance effects, and various stages in between.

Because this book is primarily methodological in nature, I have not spent too much effort on interpretation of data derived from such experiments because that depends heavily upon specific hypotheses for which behavior methods are being used to test and upon the state of knowledge/literature which is the basis for such hypothesis testing.

Finally, it should be pointed out that I think too much over-interpretation of data derived from studies with radial arm or water mazes has been reported. While it is

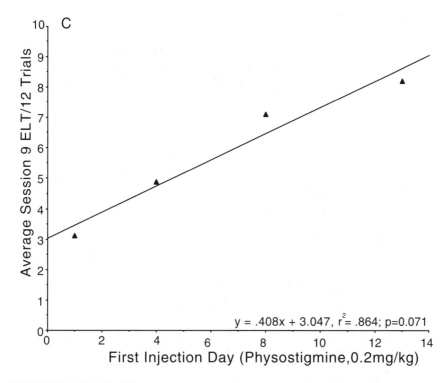

FIGURE 14.6 (Continued)

probable that limbic structures (e.g., the hippocampus, amygdaloid complex, fimbria-fornix) are important for the formation and/or use of spatial information or spatial mapping, it should be remembered that time is passing between the insertion of the experimental subject into the "start" area of the maze and its discovery of the reinforcer (food or platform). If there is a so-called working memory-consolidation dysfunction and the subject cannot bridge the temporal gap between the start area and goal because of a working memory defect, it most probably will be interpreted as a spatial mapping defect. Depending upon the specific experimental question being asked, this may or may not be an important distinction. It was because of this potential confound that we chose to use a procedure (autoshaping in an operant chamber) which did not require as much motoric behaviors as do mazes, and at the same time allowed us to control the magnitude of the temporal gap, and to measure motoric behaviors prior to and during learning or performance, with or without prior treatment with a drug or toxin.

It should also be obvious that no single experimental protocol will be useful for all experimental questions within the realm of behavioral neuroscience, but I recommend that few, rather than many, behavioral methods be mastered and used to their fullest, even if it/they require systematic manipulations of antecedent and consequential variables in order to make better use of the science upon which it/they are based and to more fully interpret the outcome of such studies when it/they are used to study higher nervous system function.

Acknowledgments

Some of the research upon which this chapter is based, and its preparation, was supported in part by the Office of Naval Research N00014-86-K-0407 and USPHS grants RO1 ES 00760; RO1 DA 01880; T37 DA 07097; R37 DA 04979. Technical assistance by D. Rosenfeld is likewise acknowledged and appreciated.

References

1. Brown, P. L. and Jenkins, H. M., Auto-shaping of the pigeon's key-peck. *Journal of Experimental Analysis of Behavior,* 11, 1, 1968.
2. Terrace, H. S., Introduction: Autoshaping and two-factor learning theory. In: *Autoshaping and Conditioning Theory,* Locurto, C. M., Terrace, H. S., and Gibbon, J., Eds., Academic Press, New York, 15, 1981.
3. Locurto, C. M., Terrace, H. S., and Gibbon, J., Eds., *Autoshaping and Conditioning Theory,* Academic Press, New York, 1981.
4. Cunningham, C. L., Ostroff, R. L., and Harris, D. F., Thermoregulation and performance of heat-reinforced autoshaped key pecking in chicks. *Behavior and Neural Biology,* 51, 54, 1989.
5. Carroll, M. E. and Lac, S. T., Autoshaping i.v. cocaine self-administration in rats: effects of nondrug alternative reinforcers on acquisition. *Psychopharmacology,* 110, 5, 1993.
6. Janak, P. H., Rodriguez, W. A., and Martinez, Jr., J. L., Cocaine impairs acquisition of an autoshaped lever-touch response. *Psychopharmacology,* 130, 213, 1997.
7. Aubert, A., Vega, C., Dantzer, R., and Goodall, G., Pyrogens specifically disrupt the acquisition of a task involving cognitive processing in the rat. *Brain Behavior and Immunity,* 9, 148, 1995.
8. Sparber, S. B., Is prenatal induction of tyrosine hydroxylase associated with postnatal behavioral changes? In: *Maturation of Neurotransmission.* Vernadakis, A., Giacobini E., and Filagamo, G., Eds., S. Karger, Basel, 200, 1978.
9. Lydiard, R. B. and Sparber, S. B., Postnatal behavioral alterations resulting from prenatal administration of dl-alpha-methylparatyrosine. *Developmental Psychobiology,* 10, 305, 1977.
10. Sparber, S. B., Postnatal behavioral effects of *in utero* exposure to drugs which modify catecholamines and/or serotonin. In: *Drugs and the Developing Brain,* Vernadakis, A. and Weiner, N., Eds., Plenum Press, New York, 81, 1974.
11. Bollweg, G. and Sparber, S., Ritanserin blocks DOI-altered embryonic motility and posthatch learning in the developing chicken. *Pharmacology Biochemistry and Behavior.* 55, 397, 1996.
12. Goldstein, L. H. and Oakley, D. A., Autoshaping in microencephalic rats. *Behavioral Neuroscience,* 103, 566, 1989.
13. Poplawsky, A. and Phillips, C. L., Autoshaping a leverpress in rats with lateral, medial, or complete septal lesions. *Behavioral and Neural Biology,* 45, 319, 1986.

14. van Haaren, F., van Zÿderveld, G., van Hest, A., de Bruin, J. P., van Eden, C. G., and van de Poll, N. E., Acquisition of conditional associations and operant delayed spatial response alternation: effects of lesions in the medial prefrontal cortex. *Behavioral Neuroscience,* 102, 481, 1988.
15. Sparber, S. B., Effects of drugs on the biochemical and behavioral responses of developing organisms. *Federation Proceedings,* 31, 74, 1972.
16. Rosenthal, E. and Sparber, S. B., Methylmercury dicyandiamide: Retardation of detour learning in chicks hatched from injected eggs. *Life Sciences,* 11, 883, 1972.
17. Spyker, J. M., Sparber, S. B., and Goldberg, A. M., Subtle consequences of methylmercury exposure: Behavioral deviations in offspring of treated mothers. *Science,* 177, 621, 1972.
18. Hughes, J. A. and Sparber, S. B., d-Amphetamine unmasks postnatal consequences of exposure to methylmercury in utero: Methods for studying behavioral teratogenesis. *Pharmacology Biochemistry and Behavior,* 8, 365, 1978.
19. Reuhl, K., Delayed expression of neurotoxicity: The problem of silent damage. *Neurotoxicology,* 12, 341, 1991.
20. Cohen, C. A., Messing, R. B., and Sparber, S. B., Selective learning impairment of delayed reinforcement autoshaped behaviors caused by low doses of trimethyltin. *Psychopharmacology,* 93, 301, 1987.
21. Lubow, R. E. and Moore, A. U., Latent inhibition: effects of non-reinforced preexposure to the conditioned stimulus. *Journal of Comparative and Physiological Psychology,* 52, 415, 1959.
22. Weiner, I. and Feldon, J., The switching model of latent inhibition: an update of neural substrates. *Behavioral Brain Research,* 88, 11, 1997.
23. Collier, T. J., Quirk, G. J., and Routtenberg, A., Separable roles of hippocampal granule cells in forgetting and pyramidal cells in remembering spatial information. *Brain Research,* 409, 316, 1987.
24. Huang, M., Messing, R. B., and Sparber, S. B., Learning enhancement and behavioral arousal induced by yohimbine. *Life Sciences,* 41, 1083, 1987.
25. Messing, R. B. and Sparber, S. B., Greater task difficulty amplifies the facilitatory effect of des-glycinamide arginine vasopressin on appetitively motivated learning. *Behavioral Neuroscience,* 99, 1114, 1985.
26. Messing, R. B., Kleven, M. S., and Sparber, S. B., Delaying reinforcement in an autoshaping task generates adjunctive and superstitious behaviors. *Behavioural Processes,* 13, 327, 1986.
27. Messing, R. B. and Sparber, S. B., Des-gly-vasopressin improves acquisition and slows extinction of autoshaped behavior. *European Journal of Pharmacology,* 89, 43, 1983.
28. Ross, W. D., Emmett, E. A., Steiner, J., and Tureen, T., Neurotoxic effects of occupational exposure to organotins. *American Journal of Psychology,* 138, 1092, 1981.
29. Tilson, H. A. and Sparber, S. B., Eds., *Neurotoxicants and Neurobiological Function: Effects of Organoheavy Metals.* John Wiley & Sons, New York, 1987.
30. Cohn, J. and MacPhail, R. C., Acute trimethyltin exposure produces nonspecific effects on learning in rats working under a multiple repeated acquisition and performance schedule. *Neurotoxicology and Teratology,* 18, 99, 1996.

31. Gerbec, E. N., Messing, R. B., and Sparber, S. B., Parallel changes in operant behavioral adaptation and hippocampal corticosterone binding in rats treated with trimethlytin. *Brain Research,* 460, 346, 1988.

32. Messing, R. B., Bollweg, G., Chen, Q., and Sparber, S. B., Dose-specific effects of trimethyltin poisoning on learning and hippocampal corticosterone binding. *Neurotoxicology,* 9, 491, 1988.

33. Sparber, S. B., Bollweg, G. L., and Messing, R. B., Food deprivation enhances both autoshaping and autoshaping impairment by a latent inhibition procedure. *Behavioral Processes,* 23, 59, 1991.

34. Solomon, P. R., Neural and behavioral mechanism involved in learning to ignore irrelevant stimuli. In: *Classical Conditioning,* Gormezano, I., Prokasy, W. F., and Thompson, R. F., Eds., Lawrence Erlbaum Assoc., Hillsdale, NJ, 117, 1987.

35. Swartzwelder, H. S., Hepler, H. S. J., Holahan,W., King, S. E., Leverenz, H. A., Miller, P. A., and Myers, R. D., Impaired maze performance in the rat caused by trimethyltin treatment: Problem solving deficits and perseveration. *Neurobehavioral Toxicology and Teratology,* 4, 169, 1982.

36. Coveney, J. R. and Sparber, S. B., Phencylidine retards autoshaping at a dose which does not suppress the required response. *Pharmacology Biochemistry and Behavior,* 16, 937, 1982.

37. Sparber, S. B., O'Callaghan, J. P., and Berra, B., Ganglioside treatment partially counteracts neurotoxic effects of trimethyltin but may itself cause neurotoxicity in rats: Experimental results and a critical review. *Neurotoxicology,* 13, 679, 1992.

38. Bollweg, G., Balaban, C. D., Berra, B., and Sparber, S. B., Potential efficacy and toxicity of GM1 ganglioside against trimethyltin-induced brain lesions in rats: Comparison with protracted food restriction. *Neurotoxicology,* 16, 239, 1995.

39. Deutsch, J., Higher nervous function: The psychological basis of memory. *Annual Review of Physiology,* 24, 259, 1962.

40. Deutsch, J., The cholinergic synapse and the site of memory. *Science,* 174, 788, 1971.

41. Davis, K. L., Hollister, L. E., Overall, J., Johnson, A., and Train, K., Physostigmine: Effects on cognition and effects in normal subjects. *Psychopharmacology,* 51, 23, 1976.

42. Davis, K. L., Mohs, R. C., Tinklenberg, J. R., Pfefferbaum, A., Hollister, L. E., and Kopell, B. S., Physostigmine: improvement of long-term memory processes in normal humans. *Science,* 201, 272, 1978.

43. Davies, P. and Maloney, A. J. F., Selective loss of central cholinergic neurons in Alzheimer's Disease. *Lancet,* 2, 1403, 1976.

44. Roberts, G. W., Crow, T. J., and Polak, J. M., Location of neuronal tangles in somatostatin neurones in Alzheimer's Disease. *Nature,* 314, 92, 1985.

45. Yamamoto T. and Hirano A., Nucleus Raphe Dorsalis in Alzheimer's Disease: Neurofibrillary tangles and loss of large neurons. *Annals of Neurology,* 17, 573, 1984.

46. Mohs, R. C., Davis, B. M., Johns, C. A., Mathé, A. A., Greenwald, B. S., Horvath, T. B., and Davis, K. L., Oral physostigmine treatment of patients with Alzheimer's Disease. *American Journal of Psychiatry,* 142, 28, 1985.

47. Muramoto, O., Sugishita, M., and Ando, K., Cholinergic system and constructural praxis: a further study of physostigmine in Alzheimer's Disease. *Journal of Neurology, Neurosurgery, and Psychiatry,* 47, 485, 1984.

48. Caltagirone, C., Gainotti, G., and Masullo, C., Oral administration of chronic physostigmine does not improve cognitive or mnesic performances in Alzheimer's Presenile Dementia. *International Journal of Neuroscience*, 16, 247, 1982.

49. Jenike, M. A., Albert, M. S., Heller, H., Gunther, J., and Goff, D., Oral physostigmine treatment for patients with presenile and senile dementia of the Alzheimer's type: a double-blind placebo-controlled trial. *Journal of Clinical Psychiatry*, 51, 3-, 1990.

50. Jotkowitz, S., Lack of clinical efficacy of chronic oral physostigmine in Alzheimer's Disease. *Annals of Neurology*, 14, 690, 1983.

51. Peters, B. H. and Levin, H. S., Effects of physostigmine and lecithin on memory in Alzheimer's Disease. *Annals of Neurology*, 6, 219, 1979.

52. Schwartz, A. S. and Kohlstaedt, E. V., Physostigmine effects in Alzheimer's Disease: relationship to dementia severity. *Life Sciences*, 38, 1021, 1986.

53. Cox, T. and Tye, N., Effects of physostigmine on the acquisition of a position discrimination in rats. *Neuropharmacology*, 12, 477, 1982.

54. Leaf, R. C. and Muller, S. A., Effects of scopolamine on operant avoidance acquisition and retention. *Psychopharmacologia*, 9, 101, 1966.

55. Stratton, L. D. and Petrinovich, L., Post trial injections of an anticholinesterase drug and maze learning in two strains of rats. *Psychopharmacologia*, 5, 47, 1963.

56. Rech, R. H., Effects of cholinergic drugs on poor performance of rats in a shutttle box. *Psychopharmacologia*, 12, 371, 1968.

57. Parker, L. A., Hutchinson, S., and Riley, A. L., Conditioned flavor aversions: A toxicity test of the anticholinesterase agent, physostigmine. *Neurobehavioral Toxicology and Teratology*, 4, 93, 1982.

58. Romano, J. A., King, J. M., and Penetar, D. M., A comparison of physostigmine and soman using taste aversion and nociception. *Neurobehavioral Toxicology and Teratology*, 7, 243, 1985.

59. Locke, K. W., Gorney, B., Cornfeldt, M., and Fielding, S., Characterization of the discriminative stimulus effects of physostigmine in the rat. *Journal of Pharmacolology and Experimental Therapeutics*, 250, 241, 1989.

60. Tang A. H. and Franklin S. R., Discriminative stimulus properties of physostigmine in rats. *European Journal of Pharmacology*, 153, 97, 1988.

61. Sivam, S. P., Hoskins, B., and Ho, I. K., An assessment of comparative acute toxicity of diisopropylflorophosphate, tabun, sarin, and soman in relation to cholinergic and GABAergic enzyme activities in rats. *Fundamental and Applied Toxicology*, 4, 531, 1984.

62. Sparber, S. B. and Tilson H. A., Environmental influences upon drug-induced suppression of operant behavior. *The Journal of Pharmacology and Experimental Therapeutics*, 179, 1, 1971.

63. Sparber, S. B., Tilson, H. A., and Peterson, D. W., Environmental influences upon morphine or d-amphetamine induced suppression of operant behavior. *Pharmacology, Biochemistry and Behavior*, 1, 133, 1973.

64. Jarbe, T. C. U., Laaksonen, T., and Svensson, R., Influence of exteroceptive contextual conditions upon internal drug stimulus control. *Psychopharmacology*, 80, 31, 1980.

65. Maayani, S., Egozi, Y., Pinchasi, I., and Sokolovsky, M., On the interaction of drugs with the cholinergic nervous system. VI. Tolerance to physostigmine in mice. *Psychopharmacology*, 55, 43, 1977.

Chapter 15

Assessing Frontal Lobe Functions in Non-Human Primates

Jay S. Schneider

Contents

I.   Introduction

The frontal lobes, which comprise the anterior (precentral) cortical regions of the mammalian brain are functionally diverse and heterogenous structures. The entirety of the frontal cortex, including the prefrontal region (defined as the part of the cortex that receives projections from the mediodorsal nucleus of the thalamus) is motor cortex in the broadest sense.[1] The frontal cortex is involved in the generation of skeletal and eye movements as well as in the mediation of a variety of cognitive functions, the temporal organization of behavior, and the expression of emotion. Cognitive functions represented in the prefrontal cortex include short-term (working) memory, preparatory set, inhibitory control, visual-spatial functions, and executive functions of planning and problem solving.

The study of the functions of the non-human primate frontal cortex has relied heavily on ablation studies,[1-6] and more recently on single unit electrophysiology studies.[1] However, in order to learn about the role of frontal cortical regions in behavior, regardless of the analytical technique, the research subject must be trained to perform specific behavioral tasks. In the case of ablation studies, an animal is trained to perform a behavior and then the effects of a specific lesion on the performance of that behavior are evaluated. In the case of electrophysiological studies, single or multiple unit recordings are obtained from awake animals performing a previously learned behavioral task and the neural processes associated with various aspects of task performance are studied. The common feature of ablation, electrophysiological, or even behavioral pharmacology studies (where effects of drugs on specific aspects of behavior are studied) is that the animal must first be trained to perform specific behavioral tasks. In this chapter the most commonly used behavioral tasks to assess prefrontal cortical cognitive functions in monkeys will be described. Tests of attention, which are included among tests that assess prefrontal function, will not be described here since these are covered in the chapter by Prendergast (Chapter 8).

Based primarily on the results of ablation studies, the prefrontal cortex has been implicated in the mediation of short-term (working) memory, that is, the short-term active retention of information to be used in the immediate future. Loss of recent memory was one of the first deficits to be described in monkeys with frontal lesions.[4,5] However, it is difficult, if not impossible to design a short-term memory task for use with monkeys that does not also contain elements that tap into other purported functions of the prefrontal cortex, such as attention and response inhibition. Despite this caveat, the classic delayed response, delayed alternation, and delayed matching-to-sample tests have been used repeatedly over the years to assess the role of the prefrontal cortex in working memory. The delayed response deficit in frontal lesioned animals remains one of the best-documented phenomena in physiological psychology.[1] Yet, even in this task, the memory component of the delayed response deficit has been questioned and the performance deficit may likely involve failure of other prefrontal functions including attention and inhibitory control functions.[1,7]

II. Assessing Short-Term (Working) Memory

A. Delayed Response Tasks

Delayed response tasks (as well as a number of other cognitive tests on monkeys) are typically performed in a Wisconsin General Test Apparatus (WGTA)[8,9] (Figure 15.1). In this situation, the animal sits in a sound-attenuated room with background masking noise (white noise) facing an opaque screen. The animal can sit unrestrained in a small testing cage attached to the apparatus or can sit in a primate restraint chair. If the animal is going to be seated in a chair, then it should be adapted to

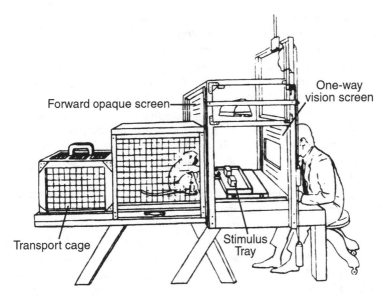

One-way vision screen

Forward opaque screen

Transport cage

Stimulus Tray

FIGURE 15.1

Standard version of a Wisconsin General Testing Apparatus (WGTA), showing the position of the monkey and the experimenter, the stimulus tray, and the one-way vision and opaque screens. The WGTA is typically used in testing delayed response, delayed alternation, delayed matching-to-sample, and discrimination tasks. In some versions of the WGTA, the monkey sits in a restraint chair located in a sound attenuated chamber rather than in a testing cage. (From Stuss and Benson: *The Frontal Lobes*. Raven Press, New York, 1986, used with permission. Originally published by Harlow: *Psychol. Rev.*, 56: 51–65, 1949.)

chair restraint prior to attempting any behavioral training or testing. The animal should become accustomed to the restraint and be relaxed in the testing situation prior to the initiation of training. The experimenter sits on the opposite side of the apparatus at a specified distance from the monkey and outside the view of the monkey. The experimenter is hidden behind a two-way mirror so that he/she can observe the monkey's behavior. There is a small opening at the bottom of the wall in front of the examiner so that the experimenter can place the food rewards into the wells on the sliding board. The distance between the monkey and the experimenter should be such that the experimenter can comfortably push the stimulus tray towards the monkey. To initiate a trial, the experimenter raises the opaque screen (attached by a pulley system) that, when raised, allows access to the sliding tray. The tray contains recessed food wells and is equipped with identical sliding covers over the wells. These can be made out of metal, Plexiglas, or any other sturdy material. The covers serve as stimulus plaques that can be displaced by the animal to obtain rewards (raisins, dried fruit, peanuts). The choice of reward should be arrived at empirically since different animals have different preferences.

Early in the training process, the animals will first need to be trained to displace the well covers to remove the food rewards from the wells. This behavior is first shaped by leaving food in open wells and then by progressively covering the wells until the monkeys learn to displace the covers to retrieve the food. Once this behavior is learned, task training can begin. Initially, the delay should be kept as short as possible. The experimenter raises the screen, shows the animal the food reward, places the food in one of the wells, covers the wells, and lowers the screen. The food wells at this point are out of the reach of the monkey. Care should be taken to ensure that the animal attends to the food placement. At the end of the delay period, the screen is raised and the stimulus tray is pushed toward the monkey. The monkey must now select the well that was baited and displace the cover to retrieve the reward. If the monkey makes an error, the opaque screen is immediately lowered and the next trial begins. Right and left wells are baited in a balanced order. A typical daily test session can consist of 30 trials, although this number can vary depending on the level of cooperation of the animals.

A typical session might have a two-second cue period and a five-second delay, although manipulating the cue duration and delay duration will influence the attentional and short-term memory components of the task, respectively. Arnsten[10] has described a modified version of the delayed response task in which five different delay lengths (arrived at empirically, depending on abilities of individual monkeys) are quasi-randomly distributed over the 30 trials that make up a daily test session. Arnsten has also described a variation of this task that included the introduction of distractors.[11] Other variations of the delayed response task have the monkey face a panel containing lighted stimulus/response buttons[12] (Figure 15.2) or a touch-sensitive computer monitor[13] instead of using the WGTA. Non-automated and automated tests each have their relative strengths and weaknesses. In general, monkeys seem to learn tasks quicker when there is interaction with an experimenter rather than interaction with an inanimate object such as a computer screen. Automated tests may be more objective, since there is no interaction or potential bias exerted by an examiner. Automated tests may also provide additional information, such as reaction time or response time data that cannot be obtained with manual testing. There are no rules for the use of automated or non-automated testing. The choice of testing methods depends on the type and extent of information you hope to gain from the testing. In our laboratory, monkeys have performed similarly in automated and non-automated delayed response tests and performances on both types of tasks have responded similarly to neurochemical lesions and to drug treatments (Figure 15.3).

B.   Delayed Matching-to-Sample

Another type of delay task used to assess working memory in monkeys is the delayed matching-to-sample task.[14] Whereas the delayed response task assesses memory for spatial location, the delayed matching-to-sample task assesses short-term visual memory without a positional component. This task can be performed in a WGTA or as an automated task[15] as described above for the delayed response task. In the delayed matching-to-sample task performed in the WGTA, the animal is presented with a stimulus (ex: a blue cover located in a central position between the two food wells) at the start of each trial. This serves as the cue for the subsequent match. A delay is imposed and then the animal is presented with the stimulus tray and allowed to displace the blue cover from the food well (match) or a cover of another color.

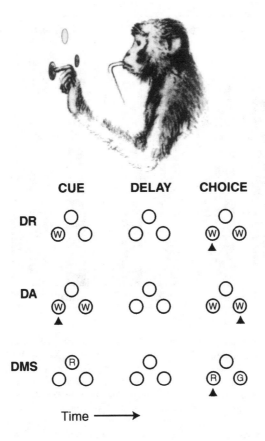

CUE   DELAY   CHOICE

FIGURE 15.2
Monkey facing a panel equipped with stimulus-response buttons for performance of delay tasks. The lower part of Figure 15.shows the sequence of events in three typical delay tasks: delayed response (DR), delayed alternation (DA), and delayed matching-to-sample (DMS). The black triangles mark the site of a correct response, which is rewarded with a squirt of juice automatically deivered through a tube at the monkey's mouth. Letters located inside the circles indicate the color of the light (W = white, R = red, G = green). (From Fuster: *The Prefrontal Cortex. The Anatomy, Physiology and Neuropsychology of the Frontal Lobe.* Lippincot-Raven, Philadelphia, 1997, used with permission.)

Variations of this task can use objects or patterns as stimuli instead of colors. The position of the correct choice is changed randomly between trials so that its position with respect to the comparison stimuli varies.

C.   Delayed Alternation

A third type of delay task is the delayed alternation task. This task can also be performed in a WGTA[16] or as an automated task.[17] In this task, the trials are temporally related to one another and a correct response is dependent upon the previous response. In the WGTA, both wells are baited out of view of the monkey, the screen is raised, and the stimulus tray (with identical covers over the wells) is

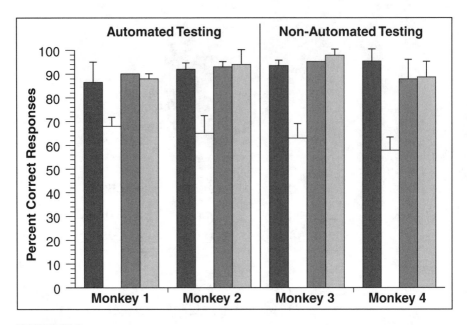

FIGURE 15.3

Data from performance of a computer automated delayed response (DR) task (left side of figure) and a nonautomated DR task performed in a Wisconsin General Test Apparatus (WGTA, right side of figure). In both situations, the monkey was seated in a restraint chair with the non-preferred hand restrained during testing. In the computerized task, the monkey faced a touch sensitive computer screen with a lever located beneath it. The animal initiated a trial by holding down the lever for one to three sec. This caused the cue, a filled white circle one inch in diameter, to appear on the right or left side of the screen for two sec. The cue was then extinguished for a delay period (five sec) and then identical left and right choice stimuli (filled red circles, one inch in diameter) were presented. The monkey was rewarded with a fruit flavored drink from an automatic dispenser if it touched the response light located on the same side as where the cue appeared. In the WGTA, the monkey sat behind an opaque screen that, when raised, allowed access to a sliding tray. The tray contained food wells and identical sliding red Plexiglas covers that served as stimulus plaques which could be displaced by the animal to obtain rewards (raisins, dried fruit). The monkey had to retrieve food from one of the wells after observing the experimenter bait it.

The graphs show that baseline performance of the tasks was similar in all monkeys (black bars = normal performance) and that exposure to the neurotoxin MPTP caused a significant performance deficit regardless of the type of apparatus used (unfilled bars = post-MPTP performance). These graphs also show that the response to a pharmaceutical agent (the neuronal nicotinic acetylcholine receptor agonist SIB-1508Y) also did not vary as a function of the type of apparatus used (lightly shaded bars = best dose response to SIB-1508Y at 30 min after drug administration; diagonal shaded bars = best dose response to SIB-1508Y at 24/48 hrs after drug administration).

presented to the monkey. The monkey then selects one of the food wells and if correct, obtains the reward. Whether the choice is correct or incorrect, the screen is lowered immediately after the response is made. If the response was correct, food is placed in the alternate well for the next trial and both wells are again covered. If the choice was incorrect, the food remains in the original well and the trials are repeated until a correct response is made. There is a definite relationship between performance and intratrial delay length, such that even frontal lesioned monkeys will perform this task well if the delay is short enough.[18,19] Thus, this task is typically used with a delay duration of at least 5 sec.

III.   Assessing Inhibitory Control Functions

A.   Go/No-Go Tasks

Numerous studies have suggested that lesions to certain areas of the prefrontal cortex (particularly the ventral portion of the prefrontal cortex) cause behavioral deficits that are accounted for at least in part by an absence of inhibitory control of internal or external interference.[1] Because of the reversal factor involved in successful performance of the delayed alternation task, this task may also be used as a test of inhibitory control function. In this task, a problem with inhibitory control is evidenced by perseverative errors or the tendency to return to the site of the previous response even though that response is no longer reinforced. Impairments in go/no-go task performance in frontal lesioned animals (and humans) is thought to occur because the dominant response (go response) cannot be inhibited during no-go trials.[20,21]

As with other tasks, go/no-go tasks can be performed using automated or non-automated systems, but are most commonly performed in some type of automated test system. While there are too many variations of the go/no-go paradigm to be enumerated here, the classic go/no-go task is a visual discrimination task that requires the monkey to discriminate between go and no-go signals and produce appropriate responses within a specified period of time. In a simple version of this task, the monkey may be required to depress a lever whenever it sees a green light signal (go) and to withold or suppress the lever press response when it sees a red light signal (no-go). In some studies the go response to a certain visual signal may be the release of a lever while the no-go response is a continued press of the lever.[22]

B.   Discrimination Reversal Tasks

To perform a discrimination reversal task, the animal must first be trained to perform a simple visual discrimination task.[16] Using either the WGTA or an automated procedure, the monkey is trained to discriminate between two patterns (ex: a plus sign vs. a square) on two identical backgrounds. If using a WGTA, the monkey is trained to respond to one of the stimuli (plus sign) and this choice is always rewarded. There is typically a brief intertrial interval (5 sec) imposed. The positive stimulus cover appears over the left or right food wells in random order. Once criterion is achieved (ex: 90% correct responses), the reward contingency is reversed and every response to the previously reinforced stimulus is now scored as an error. The number of trials (or test sessions) needed to reach criterion with the new discrimination are recorded. Deficits in the ability to inhibit the previously reinforced response (perseverative errors) or to make the requisite cognitive shift necessary for performance of the discrimination reversal is thought to reflect frontal (or frontostriatal) dysfunction.[23,24]

C.   Object Retrieval Task

Another task that has been used to study potential deficits in inhibitory control functions is a reaching task such as the object retrieval task[25] (Figure 15.4). In this task, monkeys are required to reach outside of a testing cage (or WGTA) to retrieve a food reward from a clear Plexiglas box (8 by 8 by 4 cm inner dimensions, 0.64 cm thick) with one open side and mounted on a tray. A groove runs down the center of the tray so that the test box can be secured onto the tray in various positions using a plastic thumb screw. In this situation, it is obviously critical that if the animal is being assessed in a testing cage, the bar separation is great enough so that the monkey can reach out of the cage and perform the reach without hindrance from the cage. The position of the box on the tray (center, left, or right relative to the monkey), the direction of the open side (front, left, or right relative to the monkey) and the placement of the food reward within the box (front edge, center, rear edge) varies over the course of the trials comprising the test session. A test session can be designed to have trials with different degrees of cognitive and motor difficulty, depending upon the goals of the study. The object retrieval task can be presented directly to the monkey[26] or a modified WGTA box can be used.[27] If a modified WGTA is used, the tray can be secured to a table top on the WGTA so that all positions are within reach of the monkey. The WGTA can then be wheeled in front of the testing cage and secured in position. During testing, a transparent screen can be placed between the task and the monkey. This screen can be raised on each trial so that the monkey can perform the reaching task.

Response initiation time (time between presentation of the box and and the monkey's contact with the box), trial completion time (time between the monkey's first contact with the box and correct retrieval response), the number of successfully completed trials (those on which reward was retrieved on the first attempt), and the number of correctly performed trials (those on which reward was eventually retrieved but not on the first attempt) are some of the data that can be recorded. If a reward is not retrieved on the first attempt, the number of attempts made to retrieve the reward are then recorded.

Cognitive (inhibitory control) errors are recorded when an animal makes a barrier reach (i.e., the animal hits a closed side of the box instead of making a detour movement to reach into the open side of the box) or a perseverative error (a trial on which the first reach is made into the barrier and is a repeat of the last reach on the previous trial).[27] Trials can also be classified as easy or hard depending upon several conditions. Taylor[27] have defined easy trials as those in which the open face of the box is directed toward the monkey and which succeeds a trial on which the box faced toward the monkey or to the side. Also, if the reward is at the edge of the box, the trial is considered easy. In contrast, a hard trial might be one on which the open face of the box faces the side and follows a trial on which the open face of the box was directed toward the opposite side (ex: a right to left switch of the open face of the box on successive trials). Cognitive errors can also be dissociated from motor errors using this task. [27,28]

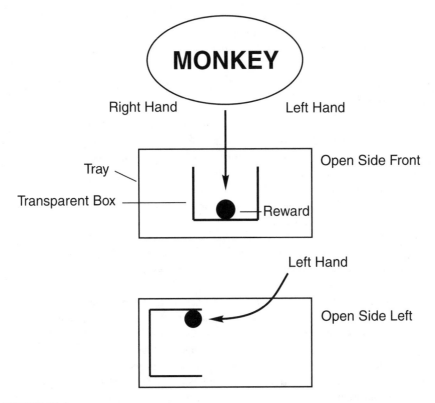

FIGURE 15.4

Diagram of the standard object retrieval reaching task, showing the postion of the transparent box on the test tray and the possible locations of the reward in the box. In condition A, the monkey can see the reward by looking directly into the open side of the box and can reach in its line of sight to retrieve the reward. In condition B, the open side of the box faces to the left side of the monkey. The monkey can see the reward through the transparent side of the box but must make a detour reach to retrieve the reward. If the monkey makes a forward reach and hits the closed side of the box facing it, that is scored as a barrier reach. The position of the reward in the box and the position of the box on the tray can be altered to produce trials that have different degrees of motor difficulty (see Taylor et al., 1990a for complete description of different cognitive and motor classifications of this task).

Normal monkeys can successfully perform this task and make the required detour reach to obtain food when the open face of the box is not directly in front of them. Monkeys with prefrontal cortical lesions have difficulty inhibiting the tendency to reach through the side they were looking at (direct line of sight reach) and make numerous barrier hits. [25] This deficit has also been observed in MPTP-treated monkeys. [26-29]

Although there are many variations of the tests described in this chapter, the ones outlined here are those that are most commonly used and which can be reproduced in the laboratory without great expense. This is particularly true for the WGTA which can be fabricated relatively inexpensively from standard materials. Automated and computer-controlled systems are more expensive and difficult to produce and, unfortunately, need to be custom programmed since there are no commercially available canned systems for running these kinds of tasks. However, testing cubicles, restraint chairs, manipulanda, and intelligence (test) panels can be purchased from commercial suppliers such as BRS/LVE, Inc. (Laurel, MD).

References

1. Fuster, J.M., *The Prefrontal Cortex. Anatomy, Physiology, and Neuropsychology of the Frontal Lobe*, Lippincott-Raven Publishers, Philadelphia, PA, 1997.
2. Bianchi, L., The functions of the frontal lobes, *Brain*, 18, 497–530, 1895.
3. Bianchi, L., *The Mechanism of the Brain and the Function of the Frontal Lobes*. Livingstone, Edinburgh, 1922.
4. Jacobsen, C.F., Functions of the frontal association area in primates, *Arch. Neurol. Psychiatry*, 33, 558–569, 1935.
5. Jacobsen, C.F., Studies of cerebral function in primates: I. The functions of the frontal association areas in monkeys, *Comp. Psychol. Monogr.*, 13, 3–60, 1936.
6. Konorski, J., Some hypotheses concerning the functional organization of the prefrontal cortex, *Acta. Neurobiol. Exp.*, 32, 595–613, 1972.
7. Stuss, D.T. and Benson, D.F., *The Frontal Lobes*, Raven Press, New York, 1986.
8. Harlow, H.F. and Settlage, P.H., Effect of extirpation of frontal areas upon learning performance of monkeys, *Res. Publ. Assoc. Nerv. Ment. Dis.*, 27, 446–459, 1948.
9. Schneider, J.S. and Kovelowski, C.J., Chronic exposure to low doses of MPTP. I. Cognitive deficits in motor asymptomatic monkeys, *Brain Res.*, 519, 122–128, 1990.
10. Arnsten, A.F.T., Cai, J.-X., and Goldman-Rakic, P.S., The alpha-2 adrenergic agonist guanfacine improves memory in aged monkeys without sedative or hypotensive side effects: evidence for alpha-2 receptor subtypes, *J. Neurosci.*, 8, 4287–4298, 1988.
11. Arnsten, A.F.T. and Contant, T.A., Alpha-2 adrenergic agonists decrease distractibility in aged monkeys performing the delayed response task, *Psychopharmocolgy*, 108, 159–169, 1992.
12. Stamm, J.S. and Rosen, S.C., Electrical stimulation and steady potential shifts in prefrontal cortex during delayed response performance by monkeys, *Acta. Biol. Exp. (Warsaw)*, 29, 385–399, 1969.
13. Schneider, J.S., Tinker, J.P., Van Velson, M., Menzaghi, F., and Lloyd, G.K., Nicotinic acetylcholine receptor agonist SIB-1508Y improves cognitive functioning in chronic low dose MPTP-treated monkeys, *J. Pharmacol. Exp. Ther.*, 290, 1999, in press.
14. Glick, S.D., Goldfarb, T.L., and Jarvik, M.E., Recovery of delayed matching performance following lateral frontal lesions in monkeys, *Commun. Behav. Biol.*, 3, 299–303, 1969.
15. Buccafusco, J.J. and Jackson, W.J., Beneficial effects of nicotine administered prior to a delayed matching-to-sample task in young and aged monkeys, *Neurobiol. Aging*, 12, 233–238, 1991.
16. Battig, K., Rosvold, H.E., and Mishkin, M., Comparisons of the effects of frontal and caudate lesions on delayed response and alternation in monkeys, *J. Comp. Physiol. Psychol.*, 53, 400–404, 1960.
17. Stamm, J.S. and Weber-Levine, M.L., Delayed alternation impairments following selective prefrontal cortical ablations in monkeys, *Exp. Neurol.*, 33, 263–278, 1971.

18. Meyer, D.R., Harlow, H.F., and Settlage, P.H., A survey of delayed response performance by normal and brain-damaged monkeys, *J. Comp. Physiol. Psychol.*, 44, 17–25, 1951.

19. Miller, M.H. and Orbach, J., Retention of spatial alternation following frontal lobe resections in stump-tailed macaques, *Neuropsychologia*, 10, 291–298, 1972.

20. Iversen, S.D. and Mishkin, M., Perseverative interference in monkeys following selective lesions of the inferior prefrontal convexity, *Exp. Brain Res.,* 11, 376–386, 1970.

21. Drewe, E.A., Go-no go learning after frontal lobe lesions in humans, *Cortex*, 11, 8–16, 1975.

22. Oishi, T., Mikami, A., and Kubota, K., Local injection of bicuculline into area 8 and area 6 of the rhesus monkey induces deficits in performance of a visual discrimination GO/NO-GO task, *Neurosci. Res.*, 22, 163–177, 1995.

23. Divac, I., Rosvold, H.E., and Szarcbart, M.K., Behavioral effects of selective ablation of the caudate nucleus, *J. Comp. Physiol. Psychol.*, 63, 184–190, 1967.

24. Mishkin, M., Perseveration of central sets after frontal lesions in monkeys. In: *The Frontal Granular Cortex and Behavior*, edited by J.M. Warren and K. Akert, McGraw-Hill, New York, pp. 219–241, 1964.

25. Diamond, A., Developmental time course in human infants and infant monkeys, and the neural bases of inhibitory control in reaching, *Ann. NY Acad. Sci.*, 608, 637–669, 1990.

26. Schneider, J.S. and Roeltgen, D.P., Delayed matching-to-sample, object retrieval, and discrimination reversal deficits in chronic low dose MPTP-treated monkeys, *Brain Res.*, 615, 351–354, 1993.

27. Taylor J.R., Roth R.H., Sladek, J.R. Jr., and Redmond D.E. Jr., Cognitive and motor deficits in the performance of an object retrieval task with a barrier-detour in monkeys (Cercopithecus aethiops sabaeus) treated with MPTP: Long-term performance and effect of transparency of the barrier, *Behav. Neurosci.*, 104, 564–76, 1990a.

28. Taylor, J.R., Elsworth, J.D., Roth, R.H., Sladek, J.R. Jr., and Redmond, D.E. Jr., Cognitive and motor deficits in the acquisition of and object retrieval/detour task in MPTP-treated monkeys, *Brain*, 113, 617–637, 1990b.

29. Saint-Cyr, J.A., Wan, R.O., Doudet, D., and Aigner, T.G., Impaired detour reaching in rhesus monkeys after MPTP lesions, *Soc. Neurosci. Abstr.,* 14, 389, 1988.

Chapter **16**

Validation of a Behavioral Test Battery for Monkeys

Merle G. Paule

Contents

I.   Introduction

While tremendous progress has been made with brain imaging techniques and we can now observe the apparent activation of certain brain areas during performance of specific tasks, it is still currently impossible to directly observe brain function. Additionally, the costs and technological requirements of such imaging procedures remain out of the reach of most investigators. Thus, surrogates of brain function are often used to infer their existence and to observe processes that influence them. The most directly observable surrogate of brain function is behavior: by defining specific parameters within which a particular behavior occurs, it can be readily observed and studied, often repeatedly over long periods of time in the same subject. In this way it is possible to study processes that affect the particular behavior (and presumably brain functions) associated with it, under a variety of experimental conditions.

When studying a phenomenon of interest it is most advantageous if that phenomenon can be generated or observed at will, thus expediting its examination. Toward this end, positive and negative reinforcement techniques can be used to shape, or train, very specific behaviors in most living organisms. These behaviors can then be elicited at will by provision of appropriate stimuli. For the purposes of this chapter, the discussion will focus on the use of nonhuman primates as experimental animals. However, all of the training principles discussed here are directly applicable to most living creatures. A positive reinforcer increases the likelihood that the event preceding it will occur again, and a negative reinforcer decreases the likelihood that it will occur again. For example, in training a subject to press a lever, positive reinforcers are provided immediately after a lever press. Merely placing a subject in proximity to a response lever is often sufficient to provide for the random or chance operation of the lever: most organisms (certainly monkeys, rodents, birds, and children) will explore a new environment into which they have been placed. During such exploration or during apparently random movement within the test environment, a response lever (or other type of manipulanda) will be actuated. If appropriately reinforced, that is, immediately upon the occurrence of the desired response (e.g., the lever press), with an appropriate reinforcer (food in a hungry subject, fluid for a thirsty one or a preferred treat), then the likelihood that the lever will be pressed again increases dramatically. In practice, robust lever-pressing behavior can usually be established within a single training session. Once lever-pressing behavior is established, behaviors associated with it can be dramatically shaped to fit the needs of the experimenter. Negative reinforcement can be used in conjunction with positive reinforcement to further shape specific behaviors. For example, after an incorrect or unwanted response is made, a simple time-out from access to positive reinforcement is sufficient to serve as a negative reinforcer. Likewise, mild aversive stimuli (electric shocks, air puffs, etc.) can be used.

In shaping or training subjects, the method of successive approximations is employed in conjunction with positive and negative reinforcement to guide the behavior of the subject to the desired product. In general, successive approximation is used to gradually change the rules under which a reinforcer can be obtained. Under such circumstances, subjects will alter their behavior in order to maximize reinforcement and, thus, express the desired behavior.

The number of functional domains attributable to the nervous system is, perhaps, unknowable. However, it is possible to define many that are thought to be of great importance such as learning, the ability to recognize specific stimuli, short- and long-term memory, etc. Additionally, it is reasonable to presume that many of these important functional domains will be shared across species. Since nonhuman primates are very close to humans on the phylogenetic tree, it may be that we share more common functional domains with them than with other species. Thus, they serve as the focus of the present chapter. While it may be impossible to ever study all of the functional domains of the nervous system/brain, it is possible to study several. Therefore, the use of a battery of behavioral tasks will be described here since the use of a battery increases the likelihood that multiple functional domains can be studied.

II. The NCTR Operant Test Battery (OTB)

Studies begun in the mid-1980s at the Food and Drug Administration's National Center for Toxicological Research (NCTR) incorporated a variety of behavioral tasks into an Operant Test Battery (OTB) for use with laboratory rhesus monkeys. (The term operant simply refers to the fact that subjects operate something in their environment, e.g., a response lever, in order to obtain reinforcers.) The initial studies were designed to assess the behavioral effects of delta-9-tetrahydrocannabinol (the primary psychoactive ingredient in marijuana smoke) and marijuana smoke. Initial OTB descriptions and drug effects can be found elsewhere.[1-3] Since those earlier studies, the OTB has been used to study the acute effects of a variety of prototypic psychoactive compounds (see Table 16.1) as well as the effects of chronic exposure to a variety of agents, both in a developmental context[4] and in adult subjects.[5] The specific brain functions thought to be modeled by performance of the NCTR OTB and the tasks utilized to assess them are described below.

A.   Motivation: Progressive Ratio Task

In the interpretation of the effects of an experimental manipulation on a given behavioral measure, it is important to have some indication of the level of motivation of the experimental subjects to perform the task of interest. If, for example, a drug treatment causes a decrease in overall response rate, an increase in reaction time, etc., it may be that the subject is simply less motivated to perform that task and not that the drug has otherwise affected specific cognitive processes associated with it. The NCTR OTB has incorporated a Progressive Ratio (PR) task specifically for the purpose of obtaining an assessment of behavior thought to be highly associated with a subject's motivation to work for the reinforcers being used (for the monkeys, the reinforcers are banana-flavored food pellets). Progressive Ratio tasks, initially described by others,[6,7] have been used in a variety of experimental settings to study processes associated with reinforcer efficacy and the interaction of deprivation states on PR performance. Obviously, if subjects are not motivated to perform for the reinforcer being used, then the behavior of interest will be hard to elicit: performance will not be maintained if the reinforcer becomes ineffective, as can be the case when subjects working for food are fully satiated prior to or during testing. In order to maintain a reproducible level of motivation to work for food, access to food is often restricted such that subjects earn a substantial portion of their daily rations during behavioral test sessions and are provided the rest via supplemental feedings immediately following the session.

For the Progressive Ratio task, responses are made on a single response lever. In the monkey OTB, it is the rightmost of four retractable levers aligned horizontally across a response panel (the other three levers are retracted during this task). Here, the number of lever presses required for reinforcer delivery (the response/reinforcer ratio) increases with each reinforcer. Initially, some small number of lever presses, e.g., two, is required for the first food pellet. The next reinforcer costs four lever presses, the next, six, and so on up to some predetermined ratio, some target number of reinforcers, or some maximum session time. The task can be defined in terms of both the initial ratio requirement (here two) and the ratio increment (here also two). The task just described is a PR2+2 task. As used in the NCTR OTB, the initial ratio and ratio increments are kept equal for a given subject (PR 3+3 or PR 5+5, for example) but the ratios can be adjusted for individual subjects such that they generally earn a specific number of reinforcers during any given session. Measures obtained include percent task completed (PTC, number of reinforcers obtained/maximum number obtainable × 100), response rate (RR), reinforcers earned, breakpoint (the size of the last ratio completed during the session), and post-reinforcement pause (time from reinforcer delivery to re-initiation of lever pressing). As part of the NCTR OTB, this task lasts 10 minutes.

1.   Relevance to Human Intelligence

The PR performance of 109 children at approximately 6.5 years of age has been studied using apparatus identical to that for monkeys. In these human studies, money (nickels) served as reinforcers. It was demonstrated that performance of the PR task was not correlated at all with either performance, verbal or full scale IQ scores.[8] Thus, performance of this task is not related to intelligence in humans. Clearly, however, it is thought to be providing information about how motivated subjects are to work under the testing circumstances.

B.   Discrimination Behavior: Color and Position Discrimination Task

It is often of great interest to know how well subjects can discriminate between different objects and make decisions based on those differences. Processes underlying these types of abilities are thought to reside in frontal-cortical brain areas.[9,10] In the NCTR OTB a Conditioned Position Responding (CPR) Task has been adapted to provide information as to how subjects discriminate colored stimuli. In this task, three press-plates that can be illuminated from behind are used as manipulanda. They are aligned horizontally on the response panel above the four retractable levers that are used in some of the other tasks. Initially, a red, yellow, blue, or green color is presented on the center press-plate. Subjects acknowledge the presence of this stimulus by pressing the plate (making an observing response), after which it is immediately extinguished and the two side keys are illuminated white. If the center press-plate had been red or yellow, then a left choice response is reinforced; if it had been blue or green, then a right choice response is reinforced. Incorrect choices result in a 10-second time out (negative reinforcement) and presentation of the next trial (colors presented randomly). Task solution is relatively simple for most subjects, as evidenced by rapid responding and average choice accuracies of greater than 90%. The PTC, observing response latency (ORL) and choice response latency (CRL) and choice accuracy (ACC) are the primary endpoints for this task. Choice response latencies can be further categorized as either correct or incorrect (CCRL and ICRL, respectively). As part of the NCTR OTB, this task is presented for no more than 5 minutes per test session.

1.   Relevance to Human Intelligence

The CPR performance of 107 children at approximately 6.5 years of age has also been studied using apparatus identical to that for monkeys. In these human studies, as previously mentioned, money (nickels) served as reinforcers. It was demonstrated that performance of the CPR task was highly significantly correlated with performance, verbal, and full scale IQ scores,[8] with correlation coefficients ranging from −0.14 for choice response latencies (shorter latencies, i.e., faster responses, were associated with higher IQ scores) to 0.58 for accuracy of responding (greater accuracies were associated with higher IQ scores). Thus, performance of this task is clearly associated with level of subject intelligence as measured by traditional IQ tests.

C.   Timing Behavior: Temporal Response Differentiation Task

The ability of organisms to time events in the seconds to hours range is advantageous because it provides for the anticipation and prediction of events. Many types of learning are sensitive to the timing of important events, suggesting that this capability provides organisms with the tools needed to make the correct response at the most optimal time (see[11] for a review of issues concerning experimental aspects of timing ability). The recent application of imaging techniques is beginning to provide information concerning the brain areas thought to be involved in the processing of temporal information.[11] The task used in the NCTR OTB for the assessment of timing behavior is referred to as the Temporal Response Differentiation (TRD) Task. This task requires subjects to press and hold a response lever in the depressed position for a minimum of 10 seconds but not more than 14 seconds. Thus, subjects must target a 4-second window of opportunity in order to obtain a reinforcer. Releasing the lever too early or too late has no programmed consequence and the subject may begin the next trial at any time.

The data obtained from the TRD task include PTC, RR, ACC, and a variety of measures associated with the distribution of lever-hold (time production) durations.

Typically, in well-trained subjects, the largest proportion of response durations occur within the 10 to 14-second reinforced window. These response durations are typically characterized by Gaussian distributions and it is thought that the distribution means represent the timing accuracy of the subject and that the spread or standard deviation of the distribution represents the precision of timing. Additionally, it has been suggested that the peak height is associated with aspects of motivation. Alterations in the characteristics of the response duration distribution are thought to provide insights into the mechanisms of timing. Shifts in the mean, for example, are thought to indicate changes in the speed of an internal clock: leftward shifts indicating a speeding up of the clock (8 seconds feels like 10 seconds) whereas rightward shifts are thought to indicate a slowing of the clock.[12] As part of the NCTR OTB, this task is presented for 20 minutes per session.

1. Relevance to Human Intelligence

The TRD performance of 86 children at about 6.5 years of age has also been studied using apparatus identical to that for monkeys. In these human studies, as previously mentioned, money (nickels) served as reinforcers. It was demonstrated that performance of the TRD task was highly significantly correlated with intelligence as measured using performance, verbal, and full-scale IQ tests.[8] Here, percent task completed, accuracy, response rate, and average lever-hold duration were all significantly correlated with IQ. Correlation coefficients ranged from –0.34 for average lever hold duration to 0.32 for response accuracy (number of correct lever holds/total lever holds $\times$ 100). Again, higher response accuracies were significantly correlated with higher IQ scores. Thus, several aspects of performance of this time estimation task are clearly associated with level of intelligence in humans.

D. Short-Term Memory: Delayed Matching-to-Sample Task

Clearly, memory is a very important brain function. Within the confines of even a relatively short test session, it is quite possible to assess processes associated with short-term memory using Delayed Matching-to-Sample (DMTS) procedures (see[13] for a mini-review of experimental approaches to the study of short-term memory). For this task, each trial begins with the illumination of the center of the three horizontally aligned press-plates described previously for the CPR task. Instead of using colors for this task however, the press-plates are illuminated with one of seven white-on-black geometric shapes. Subjects make observing responses to the shape (a sample stimulus) by pushing the illuminated plate, after which it was immediately extinguished. Following an interval (recall delay) that varies randomly (from 2 to 32 seconds, for example) all three plates are illuminated, each with a different stimulus shape, only one of which matches the sample stimulus. A choice response to the plate illuminated with the match results in reinforcer delivery. Incorrect choices are followed by a 10-second timeout, then initiation of a new trial.

Data from this task include PTC, ACC, RR, ORL, and CRL, as well as others associated with specific aspects of memory processes (e.g., encoding, forgetting). Accuracy of matching after no or very short delays, i.e., with no or little opportunity to forget, is thought to represent a measure of an organism's ability to encode the information to be remembered. Typically, response accuracy at very short delays is quite high. As the delay intervals increase, response accuracy decreases, with the slope of the decrease in accuracy over time representing a decay or forgetting function. Increases in the slope of the decay line without changes in the origin (position on the Y-axis) are thought to represent increases in the rate of forgetting in the absence of any difficulty in encoding. Changes in the origin with no change in slope would indicate changes in encoding, but no change in the rate of forgetting. Changes in both the slope and the origin indicate changes in both encoding and recall. As part of the NCTR OTB, this task is presented for 30 minutes per session.

1.   Relevance to Human Intelligence

The DMTS performance of 99 children at about 6.5 years of age was again studied using apparatus identical to that for monkeys. Money (nickels) served as reinforcers. In these studies it was found that performance of the DMTS task was also significantly correlated with intelligence.[8] Here, PTC, ACC, RR, ORL, and CRL were all significantly correlated with full scale, verbal, and performance IQ. Correlation coefficients ranged from −0.22 for observing response latency to 0.50 for percent task completed. Higher response accuracies and shorter response latencies were significantly correlated with higher IQ scores. Thus, again, several aspects of performance of this short-term memory task are significantly related to intelligence measures in humans.

E.   Learning Behavior: Repeated Acquisition Task

Obviously, the ability to learn is paramount to an organism's survival and thus an important aspect of brain function. The learning task used in the NCTR OTB is referred to as the Incremental Repeated Acquisition (IRA) Task and is a modification of more traditional repeated acquisition procedures.[14] As the name implies, subjects performing this task must repeatedly acquire knowledge in an incremental fashion. Four horizontally-aligned response levers serve as the manipulanda for this task and a row of six colored lights above the levers serve to indicate how many more correct lever presses are needed to obtain the next reinforcer. Correct and incorrect responses are also indicated by the illumination of white stimulus lights located slightly above and to the left and right of the response levers, respectively. Subjects are required to learn a new sequence of lever presses (generated randomly) every test session. The IRA session starts with presentation of a one-lever response sequence: a response to the correct one of the four levers results in reinforcer delivery. After this sequence has been mastered (for example, after 20 reinforcers have been earned), the sequence length (and presumably task difficulty), is incremented to a two-lever response sequence. Now the subject must learn which of the four levers is the new lever and remember to follow a response to it with a response to the correct lever from the previously learned one-lever sequence. Once this two-lever sequence has been performed correctly 20 times, the response requirement is again incremented to a three-lever sequence and so on up to a six-lever sequence or until the session times out.

Data for the IRA task include PTC, overall (collapsed across all lever sequence lengths) ACC, and RR, as well ACC and RR for each lever sequence length. Additionally, the number of acquisition (searching for the new lever) and recall (remembering the already learned lever) type errors are obtained for each lever sequence, as are learning curves (errors vs. number of correct sequences completed). As part of the NCTR OTB, this task is presented for 35 minutes per session.

1.   Relevance to Human Intelligence

The IRA performance of 92 children at about 6.5 years of age was again studied using nickels as reinforcers and the apparatus was identical to that for monkeys. Here, IRA performance was also significantly correlated with intelligence.[8] PTC, ACC, and RR were all significantly correlated with full scale, verbal, and performance IQ. Correlation coefficients ranged from 0.14 for response rate to 0.53 for accuracy. Higher response accuracies and response rates were significantly correlated with higher IQ scores. Thus, as demonstrated for the CPR, TRD, and DMTS tasks, several aspects of performance of this learning task are also significantly related to level of intelligence in humans.

III.   Correlations Between Tasks

The primary objective in utilizing a battery of tasks is to maximize the number of functional domains assessed during test sessions. Implicit in this approach is the assumption that each part, or task, of the battery is measuring something different than what is being measured with the other tasks. Each of the tasks just described has face validity; that is, taken at face value they appear to generate behaviors representative of the particular brain function of interest. In addition, each has content validity, having been accepted by a variety of experts as reasonable instruments for the assessment of the functional domains ascribed to them. Each task also has construct validity, in that performance is not highly correlated with performance of tasks that are supposed to measure other functions (discriminant validity).[15]

In determining the correlations of performance in each of the NCTR OTB tasks with all of the other tasks, a multitude of endpoints was examined for each task. In general it was found that, in most cases, performance in one task was not significantly correlated with performance in any of the other tasks.[15] Where significant correlations were noted, the correlation coefficients were not terribly striking. For example, performance (breakpoint) of the motivation task (PR) was significantly correlated with performance accuracy of both the short-term memory (DMTS) and timing

(TRD) tasks (p < 0.013 and 0.008, respectively). However, the respective Pearson correlation coefficients were only 0.310 and 0.396. Performance accuracy of the color and position discrimination task (CPR) correlated significantly (p<0.003) with performance accuracy of the short-term memory (DMTS) task, with a Pearson correlation coefficient of 0.365. Performance of the motivation task (breakpoint) was also significantly correlated (p<0.035), albeit negatively, with learning task (IRA) accuracy (correlation coefficient of –0.254). It is important to point out here that these were among the highest correlations observed among all comparisons made, and most were much less. Thus, even where significant associations between tasks were noted, the associations were not great. These data suggest, then, that performance of each task occurs relatively independently and, for the most part, behavior in each task represents functional domains that are different from each other.

IV.   Behavioral Profiles of Drug Effects

Another way of determining whether the behaviors contained in the NCTR OTB are independent of each other is to determine whether experimental manipulations can preferentially affect specific behaviors. Toward this end, a series of experiments have been conducted in monkeys in which the behavioral effects of psychoactive compounds have been determined. In these studies, a variety of doses of each compound were given and the relative sensitivities of each task to these treatments were determined. In short, the profiles of behavioral effects obtained were specific for the particular compound tested. For example, time estimation behavior was significantly affected (disrupted) by low doses of delta-9-tetrahydrocannabinol (THC, the primary psychoactive ingredient in marijuana, MJ, smoke) that had no effect on the other OTB behaviors. Likewise for other compounds, behavioral profiles suggest that certain drugs preferentially affect different behaviors, again suggesting that these behaviors are subserved by different neurological substrates, since they can be manipulated independently of each other. Table 16.1 summarizes the acute behavioral profiles for several of the drugs tested. Here, the order of task sensitivity indicates which behavioral task was affected at doses that did not affect the other tasks. Thus, for THC, timing (TRD) behavior was affected at the lowest doses tested, while learning (IRA), short-term memory (DMTS), and color and position discrimination (CPR) were all affected at the same but higher dose than that required to affect timing. Motivation (PR) behavior was unaffected by doses of THC that significantly affected all other tasks.

Thus, it can be seen that, depending upon the dose and drug administered, a particular OTB behavior may be significantly disrupted or totally unaffected, showing that it is possible to differentially affect specific behaviors. Interestingly, even compounds with similar mechanisms of action (e.g., d-amphetamine, cocaine, and methylphenidate) can show differential behavioral effects, suggesting that the behaviors monitored using the OTB are sensitive even to apparently subtle differences in drug action.

TABLE 16.1

OTB Task Sensitivity to the Acute Behavioral Effects of a Variety of Psychoactive Compounds in Monkeys

Compound	Order of OTB task sensitivity	Reference
THC	TRD > IRA = DMTS = CPR > PR	1
MJ smoke	DMTS > TRD > IRA = CPR > PR	2
diazepam	TRD > IRA = DMTS > CPR = PR	16
pentobarbital	TRD > IRA = DMTS = PR > CPR	17
caffeine	TRD > DMTS = IRA* = CPR* = PR*	18
cocaine	TRD = PR > IRA > DMTS > CPR	19
methylphenidate	TRD = PR > IRA = DMTS > CPR	20
d-amphetamine	TRD = IRA > PR = DMTS > CPR	21
chlorpromazine	TRD = IRA = DMTS = CPR = PR	22
d-fenfluramine	TRD = PR > DMTS = CPR > IRA	23
MDMA	TRD = PR = IRA > DMTS* = CPR*	24
LSD	TRD = PR > CPR = IRA > DMTS	25
atropine	IRA > CPR > DMTS = TRD = PR	26
physostigmine	TRD = IRA = PR = CPR = DMTS	27
morphine	TRD = IRA = PR > DMTS > CPR	28
naloxone	PR = CPR = DMTS > TRD* = IRA*	29
phencyclidine	TRD = IRA = DMTS = PR > CPR	30
MK-801	TRD = IRA > DMTS > CPR = PR	31

MJ = marijuana smoke; MDMA = methylenedioxymethamphetamine; LSD = lysergic acid diethylamide; MK-801 = dizocilpine.

* Denotes no significant drug effect noted over the dose range tested.

V. Drug Effects: Concordance with Human Data

A primary reason for using a battery of behavioral tasks in the animal laboratory is to determine the effects of experimental manipulations on several behaviors simultaneously. Ultimately, a goal is to determine whether such findings are generalizable to other species, especially humans. While comparable human data are not yet available for many of the drugs tested in the monkey model, there are important examples of cases where the same drugs have been studied in humans. In these studies, humans performed tasks designed to assess brain functions similar to those modeled in the OTB. A summary of these findings is presented in Table 16.2 where it can be seen that there is good concordance between findings in both monkeys and humans.

As the application of animal-appropriate tasks to human subjects becomes more widespread, it will be possible to collect data for a variety of experimental manipulations (i.e., drug administration) using exactly the same instruments in both humans and laboratory animals. Thus, direct interspecies comparisons will be more

TABLE 16.2
Cases of Comparable Behavioral Effects of Drugs in Both Humans and Monkeys

		Citations	
Drug	Primary acute effect	Monkey	Human
THC	overestimate time passage	1	32
MJ smoke	short-term memory impairment	2	33
cpz[a]	decrease response initiation	22	34
diazepam	learning and memory impairments	16	35
morphine	decrease response rates	28	36
atropine	learning disruption	26	37
pentobarbital	overestimate time passage	17	38
	Primary Chronic Effect		
MJ smoke	amotivational syndrome	39	40

[a] cpz = chlorpromazine

readily accomplished and the predictive validity of our animal model will be more directly assessable.

VI. Comparison of Monkey OTB Performance with that of Children

While a comprehensive comparison of the OTB performance of monkeys and children is currently underway (see, for example[41]), the data obtained to date indicate remarkable comparability of performance between the two species. In these comparative studies it is important to understand that the monkey data used are obtained only after extensive training (several months or more) when behavior has stabilized and is predictable (i.e., varies within known parameters). The data for children are obtained from naïve subjects during a single session after the presentation of audio and videotaped instructions have been provided. Thus, the assumption is made that monkeys have come to know the task rules after many practice sessions and that the children know the task rules after having been provided instructions immediately prior to task performance.

As might be expected, comparability in performance between the two species is dependent upon age, task, and endpoint. For example, as determined by response rate in the PR task, young adult monkeys work just as hard for banana-favored food pellets as seven-year-old children work for nickels.[42] Overall accuracy and rate of memory decay for these same monkeys performing the DMTS task is virtually indistinguishable from that of four-year old children, yet monkey response latencies for this same task are most similar to those for 13-year-old children. Thus, for the DMTS task, speed of monkey responding is comparable to that of older children, whereas aspects of encoding and retention in monkeys are more like those of younger children. In the color and position discrimination task (CPR), the accuracy of monkeys is comparable to that of children 6 and older, whereas monkey response rates in this task are similar to those of 5 to 6 year old children. Accuracy of learning task (IRA) performance for monkeys is comparable to that of 5-year-old children, while IRA response rates for monkeys are generally greater than those for even 12 to 13 year old children. There are, thus, clear differences in the patterns of OTB performance between monkeys and children, but in all cases examined, well-trained monkeys perform as well as or better than children aged 4 years and older.

The use of the NCTR OTB in the monkey laboratory produces information both relevant to and predictive of important aspects of brain function in humans. The degree to which monkey behavior can serve as a surrogate for the study of human brain function and dysfunction remains to be determined. Application of similar behavioral techniques in other animal models may identify additional surrogate species. Likewise, the application of different behavioral techniques should provide surrogates for additional brain functions.

References

1. Schulze, G. E., McMillan, D. E., Bailey, J. R., Scallet, A. C., Ali, S. F., Slikker, W., Jr., and Paule, M. G., Acute effects of delta-9-tetrahydrocannabinol (THC) in rhesus monkeys as measured by performance in a battery of cognitive function tests, *J. Pharmacol. Exp. Ther.*, 245(1), 178, 1988.

2. Schulze, G. E., McMillan, D. E., Bailey, J. R., Scallet, A. C., Ali, S. F., Slikker, W., Jr., and Paule, M. G., Acute effects of marijuana smoke on complex operant behavior in rhesus monkeys, *Life Sciences*, 45(6), 465, 1989.

3. Paule, M. G., Schulze, G. E., and Slikker, W., Jr., Complex brain function in monkeys as a baseline for studying the effects of exogenous compounds, *Neurotoxicology*, 9(3), 463, 1988.

4. Morris, P., Gillam, M. P., Allen, R. R., and Paule, M. G., The effects of chronic cocaine exposure during pregnancy on the acquisition of operant behaviors by rhesus monkey offspring, *Neurotox. Teratol.*, 18(2), 155, 1996.

5. Frederick, D. L., Ali, S. F., Slikker, W., Jr., Gillam, M. P., Allen, R. R., and Paule, M. G., Behavioral and neurochemical effects of chronic methylenedioxymethamphetamine (MDMA) administration in rhesus monkeys, *Neurotox. Teratol.*, 19(5), 531, 1995.

6. Hodos, W., Progressive ratio as a measure of reward strength, *Science,* 134, 943, 1961.

7. Hodos, W. and Kalman, G., Effects of increment size and reinforcer volume on progressive ratio performance, *J. Exp. Anal. Behav.*, 6, 387, 1963.

8. Paule, M. G., Chelonis, J. J., Buffalo, E. A., Blake, D. J., and Casey, P. H., Operant test battery performance in children: correlation with IQ, *Neurotox. Teratol.*, 21(3), 223, 1999.

9. Goldman, P. S., Rosvold, H. E., and Mishkin, M., Selective sparing of function following prefrontal lobectomy in infant monkeys, *Exp. Neurol.*, 29, 221, 1970.

10. Kojima, S., Kojima, M., and Goldman-Rakic, P. S., Operant behavioral analysis of memory loss in monkeys with prefrontal lesions, *Brain Res.*, 248, 51, 1982.

11. Paule, M. G., Meck, W. H., McMillan, D. E., McClure, G. Y. H., Bateson, M., Popke, E. J., Chelonis, J. J., and Hinton, S. C., Symposium overview: the use of timing behaviors in animals and humans to detect drug and/or toxicant effects, *Neurotox. Teratol.*, 1999, in press.

12. Meck, W. H., Neuropharmacology of timing and time perception, *Cognitive Brain Res.*, 3, 227, 1996.

13. Paule, M. G., Bushnell, P. J., Maurissen, J. P. J., Wenger, G. R., Buccafusco, J. J., Chelonis, J. J., and Elliott, R., Symposium overview: the use of delayed matching-to-sample procedures in studies of short-term memory in animals and humans, *Neurotox. Teratol.*, 20(5), 493, 1998.

14. Cohn, J. and Paule, M. G., Repeated acquisition: the analysis of behavior in transition, *Neurosci. Biobehav. Rev.*, 19(3), 397, 1995.

15. Paule, M. G., Use of the NCTR operant test battery in nonhuman primates, *Neurotoxicol. Teratol.*, 12(5), 413, 1990.

16. Schulze, G. E., Slikker, W., Jr., and Paule, M. G., Multiple behavioral effects of diazepam in rhesus monkeys, *Pharmacol. Biochem. Behav.*, 34, 29, 1989.

17. Ferguson, S. A. and Paule, M. G., Acute effects of pentobarbital in a monkey operant behavioral test battery, *Pharmacol. Biochem. Behav.*, 45, 107, 1993.

18. Buffalo, E. A., Gillam, M. P., Allen, R. R., and Paule, M. G., Acute effects of caffeine on several operant behaviors in rhesus monkeys, *Pharmacol. Biochem. Behav.*, 46(3), 733, 1993.

19. Paule, M. G., Gillam, M. P., and Allen, R. R., Cocaine (COC) effects on several 'cognitive' functions in monkeys, *Pharmacologist*, 34(3), 137, 1992.

20. Morris, P., Gillam, M. P., McCarty, C., Frederick, D. L., and Paule, M. G., Acute behavioral effects of methylphenidate on operant behavior in the rhesus monkey, *Soc. Neurosci. Abs.*, 21, 1465, 1995.

21. Schulze, G. E. and Paule, M. G., Acute effects of d-amphetamine in a monkey operant behavioral test battery, *Pharmacol. Biochem. Behav.*, 35, 759, 1990.

22. Ferguson, S. A. and Paule, M. G., Acute effects of chlorpromazine in a monkey operant behavioral test battery, *Pharmacol. Biochem. Behav.*, 42(1), 333, 1992.

23. Frederick, D. L., Ali, S. F., Gillam, M. P., Gossett, J., Slikker, W., Jr., and Paule, M. G., Acute effects of dexfenfluramine (D-FEN) and methylenedioxymethamphetamine (MDMA) before and after short-course, high-dose treatment, *Ann. N.Y. Acad. Sci.*, 844, 183, 1998.

24. Frederick, D. L., Gillam, M. P., Allen, R. R., and Paule, M. G., Acute effects of methylenedioxymethamphetamine (MDMA) on several complex brain functions in monkeys, *Pharmacol. Biochem. Behav.*, 51(2/3), 301, 1995.

25. Frederick, D. L., Gillam, M. P., Lensing, S., and Paule, M. G., Acute effects of LSD on rhesus monkey operant test battery performance, *Pharmacol. Biochem. Behav.*, 57(4), 633, 1997.

26. Schulze, G. E., Gillam, M. P., and Paule, M. G., Effects of atropine on operant test battery performance in rhesus monkeys, *Life Sci.*, 51(7), 487, 1992.

27. Frederick, D. L., Schulze, G. E., Gillam, M. P., and Paule, M. G., Acute effects of physostigmine on complex operant behavior in rhesus monkeys, *Pharmacol. Biochem. Behav.*, 50(4), 641, 1995.

28. Schulze, G. E. and Paule, M. G., Effects of morphine sulfate on operant behavior in rhesus monkeys, *Pharmacol. Biochem. Behav.*, 38, 77, 1991.

29. Morris, P., Gillam, M. P., Allen, R. R., and Paule, M. G., Acute effects of naloxone on operant behaviors in the rhesus monkey, *FASEB J.*, 9(3), A101, 1995.

30. Frederick, D. L., Gillam, M. P., Allen, R. R., and Paule, M. G., Acute behavioral effects of phencyclidine on rhesus monkey performance in an operant test battery, *Pharmacol. Biochem. Behav.*, 52(4), 789, 1995.

31. Buffalo, E. A., Gillam, M. P., Allen, R. R., and Paule, M. G., Acute behavioral effects of MK-801 in rhesus monkeys: assessment using an operant test battery, *Pharmacol. Biochem, Behav.*, 48(4), 935, 1994.

32. Hicks, R. E., Gualtieri, C. T., Mayo, J. P., and Perez-Reyes, M., Cannabis, atropine and temporal information processing, *Neuropsychobiology*, 12, 229, 1984.

33. Darley, C. F., Tinklenberg, J. R., Roth, W. T., and Atkinson, R. C., The nature of storage deficits and state-dependent retrieval under marijuana, *Psychopharmacologia*, 37, 139, 1974.

34. Tecce, J. J., Cole, J. O., and Savignano-Bowman, J., Chlorpromazine effects on brain activity (contingent negative variation) and reaction time in normal woman, *Psychopharmacologia*, 43, 293, 1975.

35. Gohneim, M. M., Hinrichs, J. V., and Mewaldt, S. P., Dose-response analysis of the behavioral effects of diazepam: I. Learning and memory, *Psychopharmacology (Berlin)*, 82, 291, 1984.

36. Golderg, S. R., Spealman, R. D., and Shannon, H. E., Psychotropic effects of opioids and opioid antagonists, in *Psychotropic agents. Part III: Alcohol and psychotomimetics, psychotropic effects of central acting drugs*, Hoffmeister, F. and Stille, G., Eds., Springer-Verlag, New York, 1982, 269.

37. Higgins, S. T., Woodward, B. M., and Henningfield, G., Effects of atropine on the repeated acquisition and performance of response sequences in humans, *J. Exp. Anal. Behav.*, 51, 5, 1989.

38. Goldstone, S., Boardman, W. K., and Lhamon, W. T., Effect of quinal barbitone, dextro-amphetamine, and placebo on apparent time, *B. J. Psychol.*, 49, 324, 1958.

39. Paule, M. G., Allen, R. R., Bailey, J. R., Scallet, A. C., Ali, S. F., Brown, R. M., and Slikker, W., Jr., Chronic marijuana smoke exposure in the rhesus monkey II: Effects on progressive ratio and conditioned position responding, *J. Pharmacol. Exp. Therap.*, 260(1), 210, 1992.

40. Lantner, I. L., Marijuana use by children and teenagers: A pediatrician's view, in *Marijuana and youth: Clinical observations on motivation and learning* (DHHS Publication No. ADM 82-1186), U.S. Government Printing Office, Washington, D.C., 1982, 84.

41. Chelonis, J. J., Daniels, J. L., Blake, D. J., and Paule, M. G., Developmental aspects of delayed matching-to-sample task performance in children, *Neurotox. Teratol.*, in press, 2000.

42. Paule, M. G., Forrester, T. M., Maher, M. A., Cranmer, J. M., and Allen, R. R., Monkey versus human performance in the NCTR operant test battery, *Neurotoxicol. Teratol.*, 12(5), 503, 1990.

Chapter 17

Theoretical and Practical Considerations for the Evaluation of Learning and Memory in Mice

*Robert Jaffard, Bruno Bontempi,
and Frédérique Menzaghi*

Contents

I.   Introduction

The purpose of this chapter is to introduce the reader to proper experimental procedures for the use of mice in learning and memory models. As the selection of appropriate learning and memory tasks requires knowledge of the cognitive processes modeled by these tasks, theoretical concepts will first be considered. The rest of the chapter describes tasks commonly utilized in our laboratories and includes general experimental issues related to the use of these models.

A.   Historical and Theoretical Issues: Implication for the Selection of Memory Tasks

At the end of the 19th century, the development of experimental methods for studying learning and memory in humans and animals led to the emergence of a rigorous empirical school of Psychology called Behaviorism. By concentrating on the relationships between specific stimuli (S) and responses (R), behaviorists (i.e., I. Pavlov, E. Thorndike, and F. Skinner) provided major insights into rules that govern simple forms of learning (i.e., habits). However, in the 1940s, the realization that all learning processes are not explained by simple S-R learning theories led to prevailing hypotheses that learning is not a unitary process. The best known of these hypotheses was proposed by Tolman[1,2] who distinguished between response learning based on simple S-R associations and place learning mediated by the construction and use of cognitive maps.

Nowadays, the idea that there exists more than one kind of memory has considerable support. Memory is not a unitary entity but is organized in multiple systems involving distinct brain areas or circuitries. Specific brain lesions in both animals and humans result in severe learning impairments in certain memory tasks but not in others. For instance, H.M., an amnesic patient who underwent a removal of most of his hippocampal formation and its associated medial temporal lobe structures, exhibited normal learning in acquisition of motor skills and Pavlovian conditioning despite a severe impairment in declarative memory tasks (i.e., representation of facts and events).

These observations have led to the emergence of several dual-memory theories, most of which separate memory into two components, one dependent on the activity of the hippocampus and the other independent of the activity of the hippocampus. These theories postulate that the hippocampus plays a specific role in complex forms of memory, such as spatial learning, declarative memory, or processes underlying the establishment of relational representations. The hippocampus allows for the organization of information (cognitive maps) in ways that are adaptable to novel circumstances. In contrast, hippocampal-independent forms of memory are simpler forms of learning (such as procedural memory) mediated by neuronal circuitry such as striatum and cerebellum and resulting in rather rigid and inflexible representations of knowledge and responses. Although there is still debate on how to characterize and contrast these hippocampal- and non-hippocampal-dependent systems, an appropriate understanding of the current dual-memory theories is absolutely required for the pertinent selection of learning and memory tests. The following section is a short review of some of the relevant theoretical notions in this field.

1.   Locale (Spatial) vs. Taxon Memory

The notion of locale memory is based on the concept that the hippocampus constitutes a *locale* system, which encodes and stores spatial information.[3] This *locale* system supports the acquisition of *knowledge* about an environment, defined as the relationships among various distal stimuli. In contrast, the *taxon* system refers to hippocampal-independent learning in which information relates to the positive and negative values of specific stimuli, thus enabling approach or avoidance behaviors. These two forms of memory can be modeled using variants of the Morris water maze (MWM) and radial maze (RM) tasks. Specifically, learning is assessed using either distal cues (place learning) or intra-maze cues.

2.   Working vs. Reference Memory

Olton and collaborators[4] assigned to the hippocampus a specific role in *working memory* as opposed to *reference memory*. Working memory refers to the temporary storage of information whereas reference memory refers to long-term storage of information through consolidation processes. Therefore, working memory is essential to achieve a certain level of performance on a specific trial (i.e., trial-dependent memory), whereas reference memory may be useful across all trials (i.e., trial-independent memory). Reference memory and working memory may be contrasted using simple two-choice paradigms termed discrimination learning and conditional discrimination learning, respectively. For example, reference memory is required if the animal must learn to always go to the left arm of a T-maze in order to collect a food pellet. On the other hand, working memory is involved if the animal must learn to alternate arms in a T-maze in order to receive a food pellet. In this case, the correct choice on a given trial is exclusively determined by the response made on the preceding trial, i.e., left, right, left, right, and so on.

3.   Relational vs. Procedural Memory

This theory, developed by Eichenbaum and collaborators,[5] postulates that the hippocampal system supports a *relational representation* of items in memory. This theory is conceptually similar to the cognitive mapping theory of O'Keefe and Nadel[3] except that the role of the hippocampus is no longer considered as strictly limited to spatial memory. Emphasis is made on the use of previously stored memories in novel situations (i.e., *behavioral flexibility*). In contrast, *procedural memory* (simple associations) involves the genesis of particular routines and is *inflexible*. These two forms of memory may be contrasted using two-stage paradigms. For example, hippocampal-lesioned animals can succeed in learning two-choice odor discriminations (i.e., A+ vs. B–, C+ vs. D–) but will fail to choose correctly when these familiar odors are paired in combinations not previously experienced (i.e., A+ vs. D– or B– vs. C+). According to Eichenbaum, this loss of behavioral flexibility is due to the lack of formation of encoding relationships between the two discriminations that were experienced separately during Stage 1.

B.   Selection of Pertinent Test Protocols

1.   Memory Systems

Consequently, *the nature of the to-be-processed information* constitutes the most important point in the design and selection of a learning and memory task. As pointed out by Eichenbaum,[6] "what matters is not *how* the experimenter formally characterizes a task (i.e., spatial or nonspatial, simultaneous or successive discrimination) but *which* representational strategies guide the animal's behavior." In other words, selection of a pertinent test protocol should be based primarily on the representational demands attached to a given task rather than the kind of apparatus (radial maze, water maze) in which this task is carried out. It follows that, in general, tasks that are thought to involve the same type of information processing and that therefore utilize the same memory system must be contrasted with tasks that require another (or other) memory system(s).

It is also necessary to keep in mind that the hippocampal-dependent and hippocampal independent systems are not exclusive.[7,8] For instance, damage to the hippocampal formation results in a facilitation of acquisition of S-R learning tasks indicating that the hippocampal-dependent memory system may inhibit brain memory systems subserving classes of procedural memory.[8] This has also been observed after intra-hippocampal administration of the somatostatin depletor cysteamine which markedly impairs spatial learning in a radial maze but strongly facilitates the acquisition of an appetitively motivated bar-pressing task.[9] Thus, conclusions with respect to improvement (promnesiant) or impairment (amnesiant) of learning in a given task are dependent on the specific memory systems required for performance in this particular task.

How the information is to be acquired should also be taken into consideration when an experimenter selects a protocol. In the past, reinforcement (either appetitive or aversive) was considered to be absolutely necessary for learning. This assumption was formalized at the beginning of the century by Thorndike's law of effect, according to which a S-R connection is significantly modified only if its activation is associated with outcomes important for the animal's behavior (i.e., reduction of need). Since that time it has become evident that motivations such as hunger, thirst, or safety are not necessary conditions for animals to acquire knowledge (i.e., cognitive maps) and that knowledge can be acquired in neutral situations. The use of conventional reinforcements in assessing such knowledge may thus be considered as providing a goal (i.e., finding food or avoiding a shock in a specific place) through which this knowledge is behaviorally expressed and measurable. This can be illustrated by the immediate shock deficit phenomenon. Mice do not exhibit classical Pavlovian conditioning to contextual cues when they are only exposed for a brief period to a novel environment (the context) in which they will subsequently receive a foot-shock (i.e., they do not exhibit conditioned freezing behavior when subsequently re-exposed to the context). However, contextual conditioning will occur if mice have the opportunity to explore this novel environment for a longer period of time before the conditioning session (and thus acquiring sufficient knowledge about it). In contrast, an extensive period of exploration of the neutral environment may result in latent inhibition, a diminished conditioning to this context. The diminished conditioning is thought to be due to inhibition of association of the context with an aversive experience, i.e., the knowledge that this context is safe becomes sufficiently strong to inhibit its association with the aversive event. Thus, the experimenter needs to be sensitive to the type of information to be acquired by the mice and to the potential impact of the acquired information on their behavior, as this could change the outcome of the study.

2.   Cognitive Processes

In designing a study to measure learning and memory in rodents, the experimenter needs to not only determine the memory system to be studied but also which cognitive processes are to be targeted. Acquisition and memory of novel information results from encoding, storing, and retrieval processes. As such, drugs can be administered before or during the acquisition phase, which may potentially affect all cognitive phases, from acquisition to retention. Drugs can also be administered immediately after training (acquisition) during the consolidation/storage phases or before a retention test, thus affecting the consolidation or retrieval of information, respectively. The term consolidation refers to the notion that memories become permanently fixed (consolidated) at a certain time after acquisition of a stimulus or event. As a general rule, the longer after acquisition a treatment occurs, the less retention performance is altered so that, beyond a certain time after acquisition, the treatment completely loses efficacy (i.e., temporal gradient of efficacy). Post-acquisition administration presents the advantage of eliminating any possible effects of treatments on non-mnemonic processes (i.e., perceptual, motivational, motor) that are critically involved during acquisition. On the other hand, it is clear that brain mechanisms underlying the consolidation processes are, at least in part, different from those that underlie the encoding of information during acquisition. This is a very important factor to consider when interpreting results. The hippocampus and its rhinal cortex-mediated backward projections to the neocortex are essential for the consolidation of hippocampal-dependent forms of memories, whereas the amygdaloid complex plays a critical role for the consolidation of emotionally based memories, whether these memories are hippocampal-dependent or not.[10]

II.   Practical Considerations Concerning Measurements and Interpretations

Before examining some of the tasks available for evaluating learning and memory in mice, it is important to revise the basic design and analysis of typical learning experiments.

A.   Subject Selection

The experimenter should be aware of several internal factors that may affect cognitive performances:

(a)   **Strain differences** — strain differences in sensory capabilities and performances in learning and memory tasks are well documented.[11,12] Before conducting any behavioral evaluation, the experimenter is strongly encouraged to consult the existing bibliography in order to select the mouse strain most appropriate for the proposed behavioral investigation. This is particularly important in the context of a drug-screening program aimed at identifying potential cognitive-enhancing molecules or a gene-targeting program aimed at profiling knockout or transgenic mice.[12] Failure to select a strain that is able to sufficiently process the sensorial stimuli necessary to solve the selected memory task may generate false negative results. For example, sufficient visual acuity is required for memory tasks involving the use of visuospatial cue, and intact auditory function is necessary to process tones. Motivation and anxiety levels and sensitivity to shock are other important strain-dependent factors that may affect the level of performance in a given task. Similarly, certain strains of mice may perform better than other strains. Good performers may not be appropriate in the context of characterization of cognitive enhancers as their performance may rapidly reach a maximum and cognitive improvement may be difficult to measure. Strain-dependent differences in the pharmacological effects of various drug treatments have also been reported. For example, the D1 agonist SKF 38393 and the D2 agonist LY 171555 were found to increase acetylcholine levels in the hippocampus of C57BL/6 mice, whereas no significant effect was observed in DBA/2 mice. These data support the notion of a complex genotype-related neuronal organization of dopamine-acetylcholine interactions in the mesolimbic system.[13] Thus, the large number of behavioral phenotypes among various inbred strains can be problematic if not properly controlled. On the other hand, it can also represent a significant advantage for the experimenter who knows how to benefit from naturally occurring behavioral differences.

In our laboratories, two strains are used routinely: the BALB/c and the C57/BL6 mice. Although BALB/c mice are albino and are likely to exhibit a rather poor visual acuity, we found that this strain presents satisfactory performance in the battery of behavioral tests routinely used in our laboratory, including visuospatial tasks such as the radial arm maze. The C57BL/6 strain was also selected for its average good performance in a variety of tasks as well as its particularly strong resistance to the aging process.

(b)   **Sex and age** — unless the experimental design requires it, young adult males should be utilized in order to avoid the effects of oestrous cycle and age on behavioral performance. Note that aged mice are less active than younger mice.

(c) **Social ranks** — another factor to be considered while designing an experiment using mice is the rank of the animals in the hierarchy of a social group (dominant vs. inferior animals) which could potentially confound the effect of a selected treatment.[14] One way to control for this effect is to house mice individually. Since this tends to increase the individual stress level, wait at least one week before starting behavioral experiments.

B. Importance of Non-Mnemonic Variables

A number of non-mnemonic variables may also confound the results. These include the following:

(a) **Environmental conditions** — behavioral experiments must be carried out in a controlled environment avoiding inconsistencies in temperature, background noise, odors, day of cleaning cages, and lighting conditions. The importance of odors is very often overlooked. Rats and mice should not be tested nor housed in the same room and experimenters with pets should shower and wear fresh clothes before contact with experimental animals.

(b) **Circadian rhythms** — mice are nocturnal animals and are considerably more active at night. Note also that performance level may vary along the day.

(c) **Food or water deprivation** — hungry animals are usually more active and investigative than satiated ones.

(d) **Handling** — mice should be handled prior to testing, in order to habituate the animals to experimental manipulation and reduce potential stress.

(e) **Motor/motivational effects** — mice exhibit both exploration and fear when exposed to a novel environment. This may affect their learning performance. Similarly, motor dysfunction (hypo- or hyper-activity) may affect the performance of test subjects. These non-mnemonic variables must be investigated and controlled in order to properly evaluate potential cognitive impairment or improvement. For example, the open field test is useful in exploring any motor or motivational aspects of the test conditions or treatments.

(f) **Previous exposure to drugs** — It is sometimes necessary to re-use animals in different experiments. The experimenter should keep in mind that previous treatments may affect the outcome of an experiment through long-term neurochemical changes.

(g) **Group size** — A suitable large group size (n) is needed for reliable statistical analysis. Although it is important to limit the numbers of animals used to the necessary minimum, as a rule, the larger the sample size is, the more robust the conclusions. We usually use an average of 8 to 10 mice per group. Caution should be used when interpreting results obtained with less than 6 animals per group, especially when characterizing transgenic or knockout mice. Such a small sample size might not truly represent the general behavior of the population studied.

III. Task Selection

Descriptions of some of the tasks that we currently utilize in our laboratories to target specific forms of memory follow.

A.   Selected Procedures Involving Positive Reinforcement

1.   Bar-Pressing Task

The bar-pressing task described below is a simple and easy way to evaluate the effects of treatments on procedural memory. The animal learns a simple association between a bar-press and the delivery of food reinforcement.

a) Apparatus   This task requires an operant test cage (12.5 by 13.5 by 18.5 cm) constructed of clear Plexiglas and equipped with a grid floor.[15] A metal bar and a food cup extend from one wall and are separated by a 5-cm long partition. After a bar-press, a mouse travels around the partition to reach the food cup where reinforcers (i.e., 20 mg food pellets, Dustless Precision, Bioserv, Frenchtown, NJ) are delivered. The cage is equipped with photoelectric cells, which detect the position of the animal in front of the bar or the food cup. This information is constantly transmitted to a computer, which automatically records each bar-press and consumption of food pellets.

b) Basic procedure

1.   Following a 2-week initial acclimation to collective conditions (20 subjects per cage), house mice individually in a temperature-controlled animal room (22 ± 1°C) on an automatic 12h:12h light/dark cycle (light period: 07.00–19.00) with *ad libitum* access to food and water. Mice are usually 8 weeks old at the start of the experiment.

2.   Handle and weigh mice daily for at least 1 week prior to the experiment. The same experimenter should perform handling and weighing at the same time each day.

3.   Gradually food deprive the animals to maintain body weight at 83–85% of their *ad libitum* weight throughout the entire duration of the experiment. During the deprivation period, animals should also be given pellets identical to those used as reinforcers for acclimation. Food deprivation should be well controlled as too much deprivation will disrupt their level of performance. Similarly, if the animals are not deprived enough, they will not be motivated to perform the task.

4.   Initiate the training paradigm, which consists of 15 reinforced training trials in the test cage. A reinforced trial is defined as a bar-press followed within 30 sec by the consumption of a food pellet. Restricted food ration is adjusted daily according to the animal's weight and given in the home cage 1 hour after the end of the training session.

5.   Twenty-four hours after the training session, expose the test subject to a 20-min retention test under a continuous reinforcement schedule (CRF1).

6.   Record the number of reinforced responses made during the 20-min period as a measure of retention of the bar-press training.

An example of the behavioral effects of a systemic injection of apamin, a neurotoxin extracted from bee venom, which specifically binds to Ca^{2+}-activated K^+ channels (K_{Ca} channel), is presented in Figures 17.1 and 17.2. Apamin binding sites are abundant in brain areas that have been implicated in learning and memory processes such as the septum, hippocampal formation, cingulate cortex, and thalamic nuclei.[16] As shown in Figure 17.1A, animals that received a pre-training injection of apamin (30 min prior to the acquisition session) completed the 15 reinforced

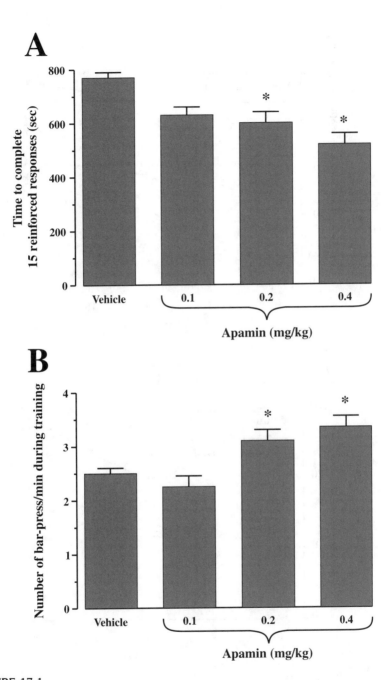

FIGURE 17.1

A. Average time (sec ± sem) taken to complete 15 reinforced responses. **B**. Mean number of bar presses per min (both reinforced and non-reinforced) made during the acquisition phase. Young adult BALB/c mice were injected with vehicle or apamin 30 min before training. *p<0.05, significantly different from vehicle (adapted from Messier et al.[16]).

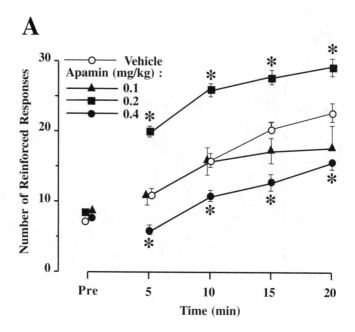

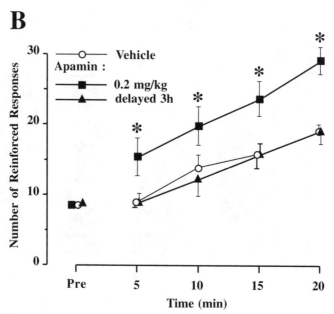

FIGURE 17.2

A. Mean number of reinforced bar-presses made during the retention test. Mice were administered with vehicle or apamin 30 min before training. **B.** Mean number of reinforced bar-presses made during the retention test. Young adult BALB/c mice were administered with vehicle or apamin (0.2 mg/kg) immediately after training or 3 h post-training. The number of reinforced responses made during the last 5 min of the training sesssion (Pre) is compared to the number made during the four 5-min periods of the retention test. *p<0.05, significantly different from vehicle (adapted from Messier et al.[16]).

training responses more rapidly than saline-injected animals. Retention performance measured 24 hours later was also significantly improved (Figure 17.2A). These data suggest that apamin facilitated the acquisition of the bar pressing task. However, this is a typical example in which caution should be exercised in interpretating results, as training in this task depends partly on general behavioral activity. Thus, drugs known to increase behavioral activity, and especially locomotor activity, will increase the bar-pressing rates of the animals although the acquisition rate will not be necessarily improved. This confounding effect can be evaluated by analyzing the number of bar presses (total reinforced or non-reinforced bar presses/min) recorded during the acquisition session (Figure 17.1B). Thus, in this example, the pre-training injection paradigm does not allow for true characterization of the effects of the drug on memory processes *per se*. As an alternative, the drug was injected immediately after the acquisition session and the animals were re-tested 24 hours later. It was verified that acquisition performance was similar across the different groups. As shown in Figure 17.2B, apamin facilitated consolidation as indicated by improved retention. Moreover, the absence of effect of delayed (3 hours) post-training apamin injections (Figure 17.2B) indicated that the facilitative effects of apamin injections depended on processes that took place shortly after training and eliminates the possibility of persisting effects of apamin on the retention session.

c) **Variants**   Many variants, using different schedules of reinforcement (fixed ratios, delayed responding) or more complex test cages equipped with several levers and other discriminative stimuli (light, tones) are available. Test procedures using several levers offer the advantage of exposing the animals to a choice situation, which reduces the bias inherent in simple motor responding.

d) **Scope and limitations**   The bar-pressing task described above is specifically tailored to evaluate procedural memory. This form of memory predominantly engages neuronal circuitry linking the neocortex and the basal ganglia and does not require the participation of the hippocampal formation, which has in fact been shown to interfere with acquisition of the task. Indeed, hippocampal-lesioned animals were shown to display better retention scores than controls when tested 24 hours following a partial acquisition session.[17] Similarly, post-training sub-seizure electrical stimulation of the hippocampus was demonstrated to produce a time-dependent enhancement of retention performance, thus suggesting that hippocampal dysfunction actually facilitates procedural memory consolidation processes. Consequently, this bar-pressing task is not pertinent for studies of the hippocampal's role in memory processes.

2.   Radial Arm Maze

The eight-arm radial maze was originally designed by Olton and collaborators[18] to measure spatial memory in rodents. Memory performance is assessed by evaluating the ability of the animal to find food pellets located at the ends of certain arms of the maze. This test is well designed for rodents, that possess excellent spatial abilities. This test is flexible and can be adapted to measure both spatial (i.e., place learning) and non-spatial (i.e., associative learning) memory. In addition, the test is easily automated, thus minimizing stress and disturbance of the animal due to experimental handling during the test.

a) Apparatus   Testing is conducted using an automated elevated eight-arm radial maze constructed of gray Plexiglas. The maze consists of a circular central platform (30 cm in diameter) from which eight arms radiate in a symmetrical fashion (50 cm long by 11 cm wide). A circular food pellet tray is situated at the end of each arm. Photoelectric cells are located along each arm to detect the position of each animal. This information is transmitted to a microcomputer, allowing for the automated recording of the sequence of arm choices, choice latencies, and running speeds. The maze is also equipped with vertical doors at the entrance of each arm, which are controlled by the computer program. The maze is located in a soundproof room (3 by 3 m). Various pictures and objects are placed around the room and serve as spatial cues. A closed-circuit video system placed above the maze allows the experimenter to observe the behavior of each animal from an adjacent room.

b) Basic procedures   We currently use two basic spatial discrimination procedures to measure spatial reference and working memory in a radial maze.

Reference memory — This widely used spatial reference memory procedure involves training the animal to discriminate a subset of constantly baited arms (spatial discrimination test). Our protocol is as follows:

1. Following a 2-week initial acclimation to collective conditions (20 subjects per cage), house mice individually in a temperature-controlled animal room (22 ± 1°C) on an automatic 12h:12h light/dark cycle (light period: 07.00–19.00) with *ad libitum* access to food and water. Mice are usually 8 to 10 weeks old at the start of the experiment. For our experiments involving aged animals, we usually use C57BL/6 mice, 22 to 24 months old at the start of the experiment.

2. After one week of handling and weighing the mice, gradually food deprive the animals to maintain body weight at 85% of their *ad libitum* weight. Be particularly careful when depriving aged mice, as they are fragile and sensitive to stress.

3. During the deprivation period, animals should be acclimated to the food pellets which will serve as reinforcements in the radial maze test. Place 3 or 4 of these food pellets into the home cages. Food pellets are available from many vendors but we recommend that experimenters try several brands, as taste for food varies across mouse strains. Another important factor is the size of the food pellets. As some protocols may require a large number of daily trials, it is important that the animal maintains its motivation for food throughout the entire training session. We therefore recommend the use of pellets no larger than 20 mg. It should be noted that during maze testing, weighing and feeding should always occur at least 1 hour after testing.

4. Once body weight is maintained at 85% of the *ad libitum* weight, allow free exploration of the radial maze on two successive days to familiarize the mice with the maze and then the environment. During this habituation phase, bait each arm of the maze with one food pellet. Place the animal on the central platform and after 1 min, open all 8 doors simultaneously so that the animal can freely enter the arms and find a food pellet reward at the end of each arm. Terminate each daily session when all eight arms have been visited and all eight food pellets have been consumed. After the second day of habituation, food rationing should be adjusted so that body weight is maintained at 90% of the *ad libitum* weight for the remainder of the study.

5. On the next day, initiate the spatial discrimination task. Prepare the radial maze by placing food pellets in only three arms (one pellet per arm) (for example, arm numbers 1, 4, and 6). Each experimental subject is assigned a different set of three baited arms and is submitted to daily sessions composed of six trials separated by 1 min intervals. Sets of 3 baited arms are chosen such that the 3 angles separating the 3 arms are always 90°, 135°, and 135°. For a given group of animals, we strongly recommend using a different set of baited arms for each animal to ensure that all arms of the maze are utilized and to minimize the possibility of confounding factors due to preference for a particular spatial location. In addition, the experimenter should scatter food pellets around the room to prevent animals from using food odor trails. Feces and urine should also be cleaned between animals.

6. Start each daily session by placing the subject on the central platform of the radial maze with all 8 doors closed. One minute later, open all doors simultaneously to allow the animal to freely locate the set of 3 baited arms.

7. After the third food pellet reward has been retrieved, close the other 7 doors. As soon as the animal returns to the central platform of the maze, close the final door to end the trial.

8. Re-bait the 3 arms while the animal remains in the center with all doors closed. One minute after the previous trial ended, conduct another identical trial. When the sixth daily trial is completed, return the animal to its home cage and bring it back to the animal room.

9. Repeat Steps 7 and 8 on subsequent days. We usually train animals for 9 consecutive days, including weekend days. An example of acquisition of the discrimination task is shown in Figure 17.4.

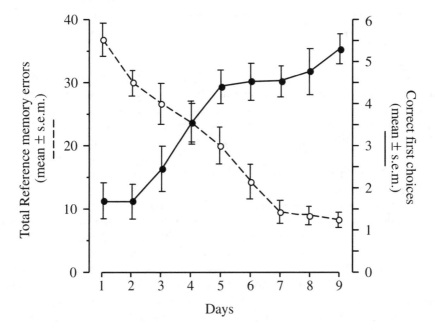

FIGURE 17.4
Number of reference memory errors and correct first choices over 9 days of training in the spatial discrimination task in young adult BALB/c mice.

10. Animals can be subsequently tested for retention at different time intervals after training, with retention sessions identical to acquisition sessions.

Sequences of arm choices as well as choice latencies and running speeds are recorded automatically by the computer. This information is subsequently utilized to generate measures of learning, namely: i) The number of total reference memory errors, defined as entries into non-baited arms, whether or not already visited during the trial; ii) The number of absolute reference memory errors defined as entries into non-baited arms during any one trial, with a maximum of 5 per trial; iii) the number of correct first choices, defined as the number of trials per session in which a baited arm is the first visit (maximum of 6).

Note: *Working memory performance can also be assessed by measuring the ability of the animal to avoid re-entries into arms within a given trial. Such repeated visits can be considered as working memory errors. Within these repetition errors, it is useful to distinguish between re-visits to baited and non-baited arms. Working memory is not absolutely needed to avoid repeated visits to non-baited arms since the knowledge that such arms are not rewarded (i.e., use of reference memory) is sufficient. In our opinion, the number of re-entries into baited arms appears to be a more precise index of working memory performance. However, if the goal of the experimenter is to evaluate working memory only, we recommend the use of the procedure that follows. A simple way to limit the working memory component in the spatial discrimination protocol is to close doors as arms are visited, thus preventing the animal from re-entering these arms (Figure 17.3).*

Working memory — A simple protocol for evaluating working memory performance in the radial arm maze involves measuring the ability of animals to not re-enter already visited arms during a given trial. A commonly used paradigm consists of baiting all arms of the maze and allowing the animals to freely explore the maze until all food rewards are collected from the arms. The apparatus, food deprivation schedule, and habituation phases are the same as those described above. Number of re-entries into already visited arms is considered as a measure of working memory performance.

Although this protocol is well established for rats, we have found that mice tend to develop a clockwise or counter-clockwise strategy, particularly in mazes without doors or with large central platforms. This strategy involves always entering adjacent arms (i.e., 45°-body-turn entries), thus minimizing any working memory use in the task. This strategy is highly efficient in a procedure in which all arms are baited but not in tasks in which some, but not all, arms are baited as in the spatial discrimination task described above. An alternative to the body-turn strategy involves confining the animal to the central platform for a limited period of time (i.e., 10 sec) by closing all doors after each arm visit.

c) Variants Other training paradigms that we frequently use to measure spatial reference and working memory are the concurrent spatial discrimination and the

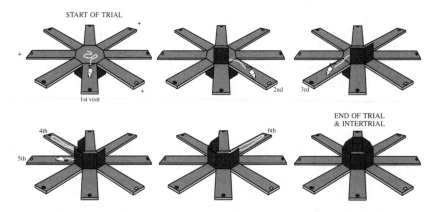

FIGURE 17.3

Schematic representation of a spatial discrimination task using an 8-arm radial arm maze. The figure illustrates a sequence of arm visits (from upper left to bottom right) in a single trial of this reference memory task in which the mouse is trained to discriminate between 3 baited arms (+) and 5 other never baited arms. In order to prevent a re-entry into an already visited arm, the door giving access to that arm is closed after the animal has returned to the central platform. Once the mouse has obtained the third food pellet (here, after the sixth visit) it is confined on the platform by closing all doors, until the next identical trial start after rebaiting the 3 arms.

delayed non-matching to place tasks, respectively. These paradigms are more complex than the tasks described above.

Concurrent spatial discrimination testing — This reference memory procedure measures the ability of the animals to learn and remember the position of specific baited arms presented in pairs. For each pair (2 adjacent arms separated by a 45° angle), the same arm is always baited. The apparatus, food deprivation schedule, and habituation phases are the same as those described above. A trial is initiated by placing the animal on the central platform of the maze with all arms closed. One minute later, the two arms of the first pair (i.e., 1^+2^-) are opened simultaneously. A choice is made when the animal reaches the food tray of either arm of the pair. This triggers the closure of the door of the other arm of the pair. When the animal returns to the central platform of the maze, the two arms of the second pair are opened simultaneously (i.e., 5^-6^+). As soon as the animal reaches the end of either arm, the door to the other arm is closed. Once the animal returns to the central platform, the door of the chosen arm is closed and the animal remains confined on the central platform for one minute, at which point the next trial begins. Each daily session consists of eight trials, with each trial consisting of the same two pairs of arms, presented in a pseudo-random order through the session. Care should be taken to balance the side of the baited arm within the two pairs, to prevent the animal from using an egocentric non-spatial strategy (i.e., always choosing the arm to its left or right).

Choice accuracy in a session is calculated by counting the number of correct responses which is expressed as percent correct. Animals are usually trained for nine consecutive days to ensure that they reach a performance level of at least 80% correct choices. Retention performance can be evaluated by testing the animals at different time points following acquisition of the concurrent discrimination.

Note: *In our fully automated maze, the computer opens a pair of arms only when the animal is situated in the opposite quadrant of the central platform to ensure that the animal will notice that it is confronted with a choice. In a partially automated maze, the experimenter must pay attention to the position of the animal on the central platform before opening a pair of arms. For this reason, we recommend selecting pairs of arms that are opposite to each other to ensure that the animal is aware of an additional arm choice upon exiting the arms of the first pair.*

Delayed non-matching to place (DNMTP) procedure — This procedure is similar to the delayed matching to sample procedure frequently used to evaluate working memory in monkeys. The procedure assesses the animal's ability to distinguish between a novel stimulus and a familiar stimulus on the basis of a single presentation. The apparatus, food deprivation, and habituation procedures are the same as described above. Each acquisition trial consists of a study phase (two forced runs) followed by a test phase (two choice runs). During the study phase, each mouse is given two consecutive forced runs in two different open arms. In each forced run, one arm is opened to allow the animal to collect the food pellet at the end of the arm. Once the animal returns to the central platform of the maze after the second forced run, two doors, one giving access to the first arm that was previously visited during the first forced run and one giving access to an adjacent non-visited arm, are opened simultaneously (first choice run). Once the animal has chosen one of these two arms and has returned to the central platform, the next pair of doors is opened, consisting of the second arm visited in the study phase and an adjacent novel arm. On both choice runs, the animal is rewarded when it enters the arm that was not visited during the study phase (non-matching to place). Incorrect choices are neither rewarded nor punished. Forced and choice runs are presented in a pseudo-random order. Forced and choice runs should be counterbalanced for left and right positions to prevent animals from using an egocentric strategy (i.e., always choosing left or always choosing right). If properly counterbalanced, the use of such a strategy would result in a choice accuracy of 50%. Daily sessions consist of eight trials (total of 16 choices), with each trail separated by a 1-min interval. It is important that the same sequence of door opening is not used twice to prevent the use of reference memory. As a general rule, we recommend rotating the sequence of choice arms by 45° on successive days of training.

Animals are usually trained until they reach a performance of at least 70% correct responses on two consecutive days. Adult C57BL/6 mice usually require no more than a week to reach this level of performance. This criterion is necessary to ensure that any decrease in performance during the DNMTP testing phase (see below) is the consequence of forgetting of information rather than due to a misunderstanding of the rule or an incapability to apply this rule. After mastering the DNMTP rule, the mnemonic demand of any one particular choice run can be manipulated by adding different delays between the relevant information (forced run) and the choice run. For each trial, upon returning to the central platform after the second forced run, the animal can be confined to the central platform of the maze for different delays ranging from 0 to 90 sec. We usually use three delays of either 0, 30, or 90 sec. Animals then complete the test phase as previously described. Following the imposed delay on the central platform, the animal is given one additional forced run before the two successive choice runs. This additional forced run is introduced to avoid response bias to the animal's position at the time of the opening of the doors for retention testing.

Daily sessions consists of nine trials (three trials per delay) separated by a 1-min interval. Within a test session, delays are presented in a mixed order. Animals are usually trained for at least four consecutive days. An alternative to the delay procedure that increases the difficulty of the task consists of interposing arm visits between a forced run and a choice run to serve as a potential source of retroactive interference. Two levels of difficulty, with eight problems for each, are assessed. In the low difficulty condition, the choice is separated from the relevant information run by one interposing visit (as during the acquisition of the DNMTP rule). In the high difficulty condition, five successive forced visits can be interposed between the choice run and its relevant information acquired during the corresponding forced run. An example of performance during the acquisition of the DNMTP rule and the DNMTP task involving interposed visits is shown in Figure 17.5.

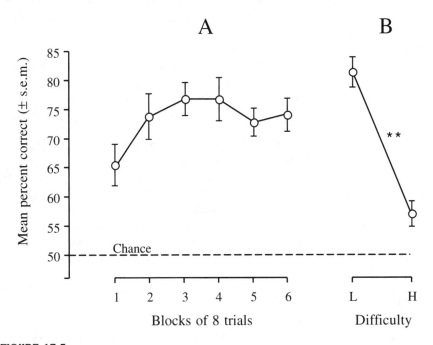

FIGURE 17.5

Percent correct choices obtained from C57BL/6 mice submitted to the delayed non-matching to place (DNMTP) task in the 8-arm radial maze. **A.** Non-matching rule acquisition over 6 days of training. **B.** Performance level in DNMTP task with either one (low difficulty) or five (high difficulty) forced visits interposed between the forced run and its subsequent recognition (choice run). **p<0.01, significantly different from performance on low difficulty problems (from Marighetto et al.[25]).

d) Scope and limitations   As described above, the radial arm maze allows for the design of a wide variety of testing procedures.[19] Animals are required to learn complex sequences of arm choices as opposed to the water maze (see next section), which is based on spontaneous exploration of a pool and its environment. Task difficulty in the radial maze can be easily manipulated by varying the amount of information (i.e., the number of arms) presented to the animal or by incorporating time delays between the presentation of information and its subsequent recall. This type of manipulation maximizes differences between groups of animals submitted to different experimental treatments. Increasing task difficulty to reduce the level of performance in control animals is particularly important when assessing drug treatments aimed at improving memory performance (suppression of a possible ceiling effect). Conversely, reducing task difficulty in order to increase the level of performance in controls is useful in identifying treatments aimed at reducing memory performance (suppression of a possible floor effect).

As an appetitive task, the radial maze requires the use of a food deprivation schedule to motivate animals for food reward. This could act as a confounding factor, especially in the case of drug treatments that affect satiety or taste for food. The experimenter should therefore always verify that food pellets are readily consumed during testing. Another disadvantage of the radial maze is the use of particular search strategies adopted by certain animals. Whereas these strategies can be easily suppressed in paradigms which used forced sequences of door opening, specific patterns of arm entry (i.e., 45°-body turns, sequential arm entries using specific angles) may appear in protocols which allow the animal to freely explore the maze. These strategies, which are not always easy to detect, are an efficient way for the animal to successfully locate reinforcement without the use of actual memory. From our experience, animals rarely abandon a successful motor strategy and, as a result, we recommend removing such animals from the study.

In some cases, the speed of learning may also be a disadvantage of the radial maze. Learning is rather slow as compared to learning in the water maze, for instance. This is likely because the radial maze is appetitively motivated, as opposed to the water maze in which failure to find the platform is potentially life-threatening. Slow learning may in some cases mask the existence of memory deficits in impaired animals. Thus, compounds may require treatment over several days or weeks for identification of effect, making it difficult to distinguish between acute and chronic effects of treatment. On the other hand, upon acquisition of a task, performance is generally quite stable over extended periods of time, thus allowing for repeated drug injection into the same animal over time and comparing the effects to the animal's baseline performance.

We have primarily focused on spatial tasks but it is worth noting that the radial maze is a versatile apparatus that can be easily adapted to evaluate non-spatial memory. As an example, textured inserts placed within the different arms of the radial maze allow for the study of non-spatial cue learning as the animal may utilize intramaze tactile cues as opposed to visuospatial extramaze cues.

B.   Selected Procedures Involving Negative Reinforcement

1.   Water Maze

The Morris water maze usually involves training rodents to escape from water by swimming to a platform.[20] Commonly, the platform is hidden (submerged) and the animals must learn its location relative to visual cues located within the testing room. This learning is referred to as place learning and is usually employed to test complex forms of learning such as spatial reference (learning of a rule) and working (learning new information to successfully complete a task) memory. This task is widely used and is probably one of the most popular cognitive tasks, as it is acquired fairly quickly by rodents (i.e., no more than a couple of days) and it does not require either food or water reinforcement. In addition, rodents are good swimmers and this test is adapted to their level of performance.

a) Apparatus   Testing is conducted in a circular tank (pool) constructed of plastic or fiberglass. We recommend the use of a large and deep pool (i.e., 140 to 180 cm diameters with walls 40 to 70 cm high). Such a pool can be obtained from suppliers of farm equipment. The size of the pool should not be reduced because mice are small animals, as a large pool decreases the probability of escape by chance and also minimizes the use of non-spatial strategies. For example, if the pool is too small, the animal may learn to escape simply by swimming at a fixed distance from the sidewall so that it eventually contacts the escape platform. The walls of the pool should also be high enough to prevent the animals from jumping out. The internal color of the pool will depend on whether the animals tested are dark- or light-colored which will require a white or black pool, respectively. The pool is filled with clear water (24°C) to a depth of 30 to 40 cm based on the height of the walls. If necessary, water can be made opaque by the addition of non-toxic white paint. We do not recommend the use of powdered milk, as it is difficult to clean. The temperature of the water should be accurate to 1 degree and should be colder than the temperature usually recommended for the testing of rats (i.e., 26°C).

The pool should be placed in a fairly large room so that various objects may be located around the pool to serve as visual reference points for the animals. These extra-maze cues must be far enough away to require the mice to use spatial analysis, rather than simple association, to solve the task. Of course, the cues should not be so far away that they cannot be seen or used by the mice. In addition, the cues should be placed at a height in the line of vision of mice swimming in the pool. We use three specific cues (boards with various black patterns, i.e., triangle, squares, or circles) and other general cues such as light and shelves. The target or escape platform is a circular platform 15 cm in diameter made of black, white, or clear Plexiglas to match the internal color of the pool. The surface of the platform should be engraved with a crosshatch of lines to prevent the mice from slipping. The platform should be placed no more than 0.5 cm under the water surface to provide sufficient "escape" from the water, i.e., "the mice should feel that they are out of the water." Four equally spaced points around the edge of the pool designated as north (N), south (S), east (E), and west (W) are used to divide the pool into four imaginary quadrants of equal surface area (NE, NW, SE, and SW). The platform is located in the middle of one of four designated quadrants of the pool (Figure 17.6). The activity of each animal should be monitored using a video-tracking system. Without automated tracking, it is not possible to quantify the behavior with accuracy, with the sole exception of escape latency, a poor and incomplete index of learning. A variety of systems are now available commercially. We have used a tracking system available from San Diego Instruments (California, U.S.A.). Several dependent measures are recorded to assess spatial learning including, but not limited to: swim distance (length of the path that the animal swims to find the platform), latency to climb on the platform, direction of the swim path, and swimming speed. The video tracking system consists of a CDD video camera with a wide-angle lens (4 mm, 90° angle of view) mounted on an adjustable bracket above the center of the pool. The video is connected to a TV monitor and image analyzer. The image analyzer is coupled to a computer. It is highly recommended that the equipment used to record and track the animal's behavior is located in a separate room, as mice are very sensitive to the presence of the experimenter. The use of a video tracking system requires homogenous illumination, which can be difficult to obtain in certain environments. We recommend dimmer-regulated lamps placed just above the pool as well as floodlights placed below the rim of the pool to provide diffuse illumination reflected off the ceiling.

b) Basic procedure: Place navigation task   The simplest and most commonly used procedure involves training to locate a hidden platform in a fixed location over a series of trials, usually conducted over several days (i.e., place navigation). Unlike a simple T-maze in which animals must only make a binary choice between left and right, place navigation requires the mice to learn to swim from any starting position to the escape platform, thereby acquiring a long-term memory of the platform's spatial location (Figure 17.6A).

Our protocol is as follows:

1. Handle mice for at least 1 week prior to the experiment and habituate them to the testing room for 2 days before training.

2. On the first day of the experiment, set up the computer. Fill pool, and adjust the temperature of the water. After the pool is full, check tracking with a non-experimental animal. Be sure that the temperature of the room remains stable across days.

3. Place the first mouse in the water at one of three start positions (center of the SE, NE, or SW quadrant), against the wall with its head pointed toward the pool wall in the center of the "non-platform quadrants." Start the trial with a remote switch as soon as the mouse is released. We use only three start positions that are fairly equidistant to the platform in order to minimize potential bias due to starting proximity to the escape platform.

4. Stand back or go to the adjacent room (equipment room) and stop the trial (remote switch) when the mouse finds the escape platform or when 90 sec have elapsed.

5. If the platform is not found within 90 sec, gently guide the mouse toward it. During training, observe each mouse continuously and remove it from the tank if its head falls below the water.

A. Acquisition Procedure

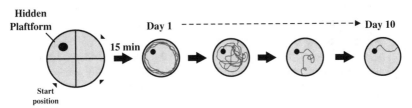

B. Repeated Acquisition Task

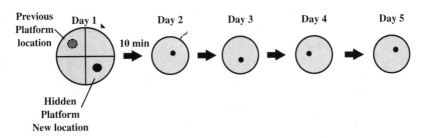

FIGURE 17.6

Schematic representation of the place-navigation task using a water maze. **A.** Acquisition of place-navigation task (Spatial reference memory). **B.** Repeated acquisition task (Spatial working memory).

6. Allow the mouse to remain on the platform for 15 sec.

7. Place the mouse in a holding cage for a period of 15 to 30 min. Intervals between trials enhance learning. Heating pads wrapped in towels are placed in the holding cage to prevent hypothermia.

8. Repeat Steps 3 to 7 from different start locations around the pool until the total number of training trials is reached. We found that mice trained for 2 trials per day learn nearly as quickly as 4 times per day.

9. Following the last trial of the day, towel-dry the mouse and place it under an infrared heat lamp for 5 to 10 min before returning it to its home cage.

10. Record body weight daily to ensure that normal weight gain is maintained. If possible, take the body temperature of each animal prior to the start of training each day. Previous work has shown that hypothermia decreases the level of water maze performance.

11. Repeat Steps 1 to 9 on subsequent days. Training usually requires 10 days. Mice are trained for 5 consecutive days separated by 2 days off (i.e., weekend). All animals undergo the same sequence of starting positions.

12. On the final day or preferably 24 hours after the last training day, conduct a probe trial, i.e., remove the escape platform from the pool and allow the mouse to swim freely for 60 sec. The probe trial provides an index of the animal's tendency to persist around the platform's previous location and is generally considered to be a measure of retention. Place the animal at the start position furthest from the former platform location, and record the following measurements: i) The number of crossings over the former platform location and over regions equivalent to the position of the former platform in the other three quadrants (i.e., annulus crossings); and ii) the distance swum

in each quadrant (measured as a percentage of total distance swum). Animals may be exposed to repeated probe trials during or after the training phase. Keep in mind that certain lesions or drugs can affect the acquisition to escape without affecting performance in a probe trial.

13. A variant to the probe trial task is the **reversal task**. Upon completion of the initial water maze acquisition, each animal receives 2 to 8 trials per day for 2 to 5 days with the platform located in the center of the quadrant diametrically opposed to that during the initial training phase. Training is conducted as described above. This task measures the ability of each animal to adapt to novel conditions (i.e., cognitive flexibility). Inflexibility of learning is an early feature of cognitive deficits related to aging and Alzheimer's disease.

Note: *From our experience, it appears that the primary reward for mice performing the water maze task is the actual rescue by the experimenter after each trial, as opposed to the act of climbing onto the platform and being out of the water. Therefore, the experimenter should be out of sight and inaudible to mice during the 15 sec period on the platform. In an ideal situation, the experimenter should remain outside of the testing room so that each mouse orientates itself to the platform location and not to the experimenter. Time on the platform must be sufficient for the mouse to perceive the platform location, and to orientate itself and assess its surroundings. In practice, mice may not remain on the platform for 10 to 15 sec and may jump back to the pool. If a mouse does not remain on the platform for this time period, wait for the mouse to swim back to the platform before you intervene. If the mouse cannot find the platform, then guide it back. Each mouse must learn that the experimenter removes it from the pool only if it remains on the platform. On the first day of training, most mice will fail to find the platform. Therefore, do not allow them to swim unsuccessfully for more than 90 s, as they may associate swimming with failure and they may panic during the next trial. Mice may also receive a pretraining session consisting of placing the mouse on the platform for 15 sec before starting a trial. This may limit any procedural learning components in subsequent trials by familiarizing the mice with the platform and its spatial location.*

c) Variants   The watermaze is quite flexible, as many paradigms other than place navigation can be explored. Two protocols that we use frequently are the repeated acquisition task (Figure 17.6B) and the visual acuity task.

Repeated acquisition task — Two days after the reference memory task (acquisition and probe trial tests), we usually perform a 5-day spatial working memory (repeated acquisition) test. Each day the platform is placed in a new location, and mice must find it across four successive trials (10-min inter-trial interval).

This test measures the ability of the animals to find a new platform location using the procedural knowledge and spatial cues already learned in the reference memory version of the task. This paradigm involves extinction of the previously learned platform location, and measures adaptation to new spatial problems.

The apparatus, room setup, and lighting are the same as in the reference memory version of the water maze task. Animals are released from the same starting position in all trials on all days. On each of the 5 days, the platform is relocated to a different point. If an animal does not locate the platform (by chance) on the first trial, it is guided to the platform by the experimenter and allowed to remain there for 15 sec. Animals spend an equal amount of time resting on the platform for all trials followed by an intertrial interval of 10 min in a holding cage. If the animal does not find the platform on a subsequent trial (i.e., trial 2 to 4), it will again be guided to it. The difference between distance swum on trials 1 and 2 is averaged over the five days of testing in order to evaluate working memory performance.

Visual acuity task — After the above-described paradigms or prior to the start of training in the pool, the visual acuity of the animal is evaluated. The room setup and lighting are the same as used during the reference and working memory tasks. In contrast to hidden platform training, the platform is raised 1.5 cm above the water surface and is marked with a large visual cue (example: plastic toy positioned on one side of the platform with Velcro). The visible platform is moved randomly to different locations within the four quadrants while the start position remains the same. Animals undergo 4 trials of 60 sec in which swim distance is recorded. This test is used to discern if performance deficiencies are attributable to motivational or sensorimotor defects rather than cognitive deficits. This test can also be used to pre-select animals with good visual acuity and motor coordination prior to hidden platform training.

Although the water maze is used primarily to study spatial learning, various non-spatial protocols have been developed such as the visible platform task mentioned above and the procedural learning. For the procedural learning task, the location of the platform is randomly varied between the four quadrants but is always in the center of each quadrant. Mice trained in this procedural task tend to learn to swim in a circular path at a distance from the wall of the pool that optimizes the chance of bumping into the platform.[21] Thus, this task can be solved without using spatial information and requires different strategies than the ones used to solve the place navigation task.

d) Scope and limitations The water maze is a perfect example of a test that can be used to model hippocampal (i.e., spatial learning) and non-hippocampal (i.e., procedural, non-spatial learning) dependent memory systems. The simplicity of the water maze has made it very popular, especially as the number of various knockout and transgenic mice continuously increases. Advantages of this task include a lack of motivational constraints (i.e., food deprivation), few training trials to acquisition, the absence of local cues within the maze (i.e., scent trails), and a relatively simple experimenter training. However, the water maze task exhibits some strong limitations that are too often ignored. The experimenter should keep in mind that swimming could be stressful or that escape from the water may not be sufficient reward. These two variables may result in an overall lack of performance attempts. The water maze is therefore a delicate balance of motivation provided by stress and reward, and is not as simple as it appears for the test subjects and the experimenters.

Mice also behave differently than rats in this model. As opposed to rats, mice tend to stop swimming and just float or jump. If mice are too stressed, too nervous, or too weak to swim, they may be qualified as cognitively impaired because of poor performance. But is that really the case? Too often experimenters ignore other parameters that could compromise performance independently of spatial learning abilities such as anxiety level, willingness to explore, thermogenesis, and general motor activity. These variables should be evaluated separately to determine the suitability of the water maze task for the test subjects or treatment. Another issue concerns the strategy used by the mice in locating the platform. Can the experimenter guarantee that the animal has learned by reference to spatial cues and not just by luck (i.e., bumping into the submerged platform)? It can be quite difficult for the experimenter to control for the use of non-spatial strategies in this task. Another issue is the use of the probe trial as an index of spatial memory. It is too often forgotten that probe trial and search for hidden platform requires a different level of problem solving and likely the use of different strategy. Persistent swimming in the training quadrant during the probe trial is generally interpreted as an indication of good memory and therefore successful spatial learning. On the other hand, this could also be interpreted as an inability of the animal to adapt to a novel situation and, therefore, a lack of cognitive flexibility. Similarly, an animal that swims to the correct location but does not persevere in the former location of the platform is generally considered to exhibit poor memory, when in fact this animal may have appropriately altered its strategy in response to the lack of an escape platform in its former location. Its strategy may no longer be driven by spatial analysis but survival strategy, and the lack of swim distance in the training quadrant may be indicative of cognitive flexibility. Thus, using probe trial as a measure of good spatial memory requires caution. Our recommendation is to de-emphasize the probe trial, and to thoroughly analyze the acquisition curve and transfer test for group difference. A learning curve that shows progressive improvements in performance over many trials may be more indicative of spatial learning than a probe trial.

2.   Pavlovian Fear Conditioning

The basic procedure of Pavlovian fear conditioning consists in inducing conditioned fear by pairing a neutral stimulus used as the Conditioned Stimulus (CS) (i.e., tone, light, or experimental context) to an aversive Unconditioned Stimulus (US) (i.e., footshock). Autonomic (i.e., heart rate or blood pressure) or behavioral measures (acoustic startle response or amount of freezing behavior) are then used to assess that fear has been acquired. The amount of freezing behavior, an innate defensive response suppression to the CS, is the most widely used parameter. Lesions of the hippocampus abolish learned freezing to context but not to tone, whereas lesions of the amygdala abolish learned freezing to both tone and context.[22] As such, the amygdala is considered to play a fundamental role in the acquisition of unimodal simple associations (i.e., tone-shock) and, *via the hippocampus*, in the acquisition of polymodal associations (i.e., context-shock). The failure of hippocampal-lesioned animals to exhibit conditioned context-shock associations is generally considered as being specifically related to their inability to form an integrative (map-like) representation of the context.

a) **Apparatus** Training is conducted in a conditioning chamber constructed of Plexiglas (30 by 24 by 22-cm high) and placed in a sound-attenuating box. The chamber is equipped with a grid floor (60 stainless steel rods of 2 mm diameter spaced 5 mm apart) connected to a scrambler shock generator to provide foot-shock. A loudspeaker mounted on the ceiling of the Plexiglas chamber is connected to a pure tone generator, allowing presentation of the tone CS. Both the shock and tone generators are coupled to a single control unit. The conditioning chamber is dimly illuminated; a ventilation fan provides a continuous background sound level of 60 dB. The behavior of each mouse is recorded using a video camera mounted 1m above the floor and connected to a monitor recording system.

b) **Basic Procedure**

1. Following a 2-week initial acclimation to collective conditions (20 subjects per cage), house mice individually in a temperature-controlled animal room ($22 \pm 1°C$) on an automatic 12h:12h light/dark cycle (light period: 07.00–19.00) with *ad libitum* access to food and water. Mice (male C57BL/6) are usually 4 to 6 months old at the start of the experiment.

2. Handle (3 to 5 min) mice daily for at least 1 week prior to experiment. The same experimenter should perform handling at the same time each day.

3. Initiate the acquisition paradigm (i.e., conditioning trial). A conditioning trial consists of a 4-min session in the conditioning chamber. Each mouse is placed into the conditioning chamber for 60 sec of acclimation and then exposed to the CS-US paired conditioning procedure. We usually use two paired conditioning procedures, a context conditioning (i.e., paired context-shock condition) or an auditory conditioning (i.e., paired tone-shock condition). The auditory conditioning consists in exposing each mouse to a tone (70 dB, 1 kHz, 20 sec duration) terminated by a foot-shock (0.9 mA, 50 Hz, 3 sec). The tone is presented again 100 sec, followed by a 40 sec delay. Due to ceiling effect, foot-shocks of lower intensity (0.3 to 0.6 mA) are sometimes prefered to demonstrate the effect of a given treatment on conditioned freezing. The context conditioning procedure consists in suppressing the tone CS (i.e., no-tone condition). If the tone needs to be maintained, it should not be paired with the footshock US, and the foot-shock should be delivered before or at a minimum of 30 sec after the tone (i.e., unpaired tone-shock condition). The levels of conditioning to the context depend on whether or not conditioning to the tone (or to other simple phasic CS) may occur. This prediction is based on the assumption that contextual stimuli and simple stimuli compete for association with the US, with the latter stimulus overshadowing the former. Thus, less contextual conditioning to the background context occur when the tone is made predictive for the occurrence of the shock (paired tone-shock condition).

4. After each session, place each mouse back into its home cage and clean the chamber with a 50% alcohol solution between tests.

5. 24 hours after the training session, test mice for freezing behavior, the index of conditioned fear. In the context conditioning paradigm, each mouse is re-exposed to the conditioning chamber for 6 min, during which its behavior is recorded on videotape. In the auditory conditioning paradigm, it is recommended to assess conditioned freezing in an environment which is different from that wherein conditioning took place. We usually keep each mouse in its home cage and place the cage in a room equipped with a speaker and a videotape recording system. Conditioned freezing can be assessed

before or after (i.e,. more than 1 hour) the context test. The 6-min test session is divided into 3 successive periods of 2 min. The first and third periods are used to determine the animal's levels of freezing before and after tone presentation. This represents the baseline which is normally close to zero. The second period is used to re-expose the animal to a tone identical (same frequency and intensity) to the one used for conditioning. The tone is presented for 2 min. In subjects previously conditioned (i.e., paired tone-shock condition) amounts of freezing observed during the second block should be significantly increased with respect to baseline and significantly greater than levels of freezing displayed by controls (i.e., no-tone and unpaired tone-shock groups).

6. After the freezing behavior test, analyze the videotape. The mouse's behavior is scored by one (or preferably two) observer(s) blind to the experimental conditions every 10-sec period of the session. The scoring criteria are either freezing or absence of freezing behavior. Freezing behavior is defined as the absence of visible movement of the body and vibrissae except for respiratory-related movements. The percentage of freezing is calculated per blocks of 2 min of testing by dividing the number of freezing episodes by the total number of observations (i.e., 12) and multiplying by 100. The time the animal spent freezing may also be measured using either a microwave-based detector system[23] or an image analyzer.

Using the procedure described above, the mean percentage of freezing observed during the 6-min period of the context test is about 35 to 40% in the unpaired tone-shock group and is usually about 20 to 25% in the paired group. In the auditory cue test, the mean percentage of freezing in the paired group is about 60 to 70% during tone presentation, and less than 10% prior to and after tone presentation.[24]

c) **Variants**   A variant of the basic task is to repeat the conditioning trials each day for several days. This procedure allows us to assess the rate of acquisition across days and to determine the asymptotic level of conditioning. For this purpose, each daily session is divided into two successive stages, the first one being used to assess freezing to context and to tone, and the second one being devoted to the delivery of context and CS-US associations (conditioning trials). Using the same conditioning procedure as that described above, freezing to contextual cues is assessed during the 60 sec prior to the onset of the first tone presentation and freezing to the tone during the 20 sec of first tone presentation. This procedure allows for the construction of acquisition curves for contextual and elemental conditioning. For the above mentioned reasons, the main weakness of this paradigm is the necessity to assess tone conditioning in the conditioning chamber.

d) **Scope and limitations**   The fear conditioning task is currently widely used to assess hippocampal (i.e., contextual) and non-hippocampal (i.e., cued) dependent forms of learning. This test is simple and rapid but has several limitations.

Firstly, context (as assessed using the fear conditioning task) and spatial processing (as assessed using the water maze task) are different in nature. In both lesioning and transgenic experiments, an impairment of spatial learning may be observed without any concomitant impairment of contextual fear conditioning. Secondly, recent experiments have reported a lack of effect of hippocampal lesions on contextual fear conditioning. This is probably due to the fact that lesioned mice were conditioned to salient features of the context such as electric bars of the conditioning chamber (processed as an elemental stimulus) instead of being conditioned to the context (the true polymodal stimulus). Thus, the experimenter cannot guarantee that changes in amounts of conditioned freezing measured in the context test reflect changes in hippocampal processing.

Another concern is the intensity of the foot shock. US intensity clearly determines the level of learning, i.e., more aversive US produce more conditioning as shown by greater freezing behavior in response to the CS. Thus, a facilitation (or an impairment) of fear conditioning observed after given treatment may simply result from a (nonspecific) change in the perceived intensity of the US. For example, the administration of opiate antagonists produces increased amounts of conditioned freezing due to an increase in the perceived intensity of the foot-shock US.[25] The concurrent use of cued and contextual conditioning generally allows the exclusion of such non-specific effects. Also, several studies indicated that conditioned freezing to the simple CS (i.e., tone or light) may be normal despite severe alterations of conditioned freezing to contextual cues.

IV. General Comments

The main objective of this chapter was not to survey all the procedures that can be used to evaluate learning and memory in mice but to exemplify the basic design and selection of learning and memory tasks for mice. The selection of the learning and memory tasks is one of the most crucial decisions the experimenter faces, especially when there is little or no knowledge of how the treatment or the targeted gene might affect learning and memory. The experimenter should be extremely careful when interpreting results and should always discuss the data within the context of the task used. Our recommendation is to use a broad battery of tests or variant paradigms within the same test that will guarantee the exploration of various memory systems and cognitive processes as mentioned in the above sections. Although there are still debates on the mechanisms underlying certain forms of learning and memory, the selection of learning and memory paradigms should be based on the cognitive processes that the experimenter would like to explore and not the apparatus in which the tasks will be carried out. A sequence of experiments starting from relatively simple tasks to more complex tasks should be included. Non-mnemonic variables and subject selection should also be taken into consideration while interpreting results.

References

1. Tolman, E. C., Cognitive maps in rats and men, *Psychol. Rev.,* 55, 189, 1948.
2. Tolman, E. C., There is more than one kind of learning, *Psychol. Rev.*, 56, 144, 1949.
3. O'Keefe, J., Nadel, L., *The hippocampus as a cognitive map.* Oxford University Press, 1978.
4. Olton, D. S., Becker, J. T., Hendelmann, G. E., Hippocampus, space, and memory, *Brain Behav. Sci.*, 2, 313, 1979.

5. Cohen, N. J., Eichenbaum, H., Memory, amnesia, and the hippocampal system. Cambridge, MIT Press, 1993.
6. Eichenbaum, H., The hippocampal system and declarative memory. In *Memory Systems*, D. L. Schacter, E. Tulving (Eds). Bradford Book, MIT press, pp. 145-201, 1994.
7. Toates, F., The interaction of cognitive and stimulus-response processes in the control of behaviour, *Neurosci. Biobehav. Rev.*, 22, 59, 1998.
8. Jaffard, R., Meunier, M., Role of the hippocampal formation in learning and memory. *Hippocampus*, S3, 203, 1993.
9. Guillou, J. L., Micheau, J., Jaffard, R., The opposite effects of cysteamine on the acquisition of two different tasks in mice are associated with bidirectional testing-induced changes in hippocampal adenylyl cyclase activity, *Behav. Neurosci.*, 112, 900, 1998.
10. Cahill, L., McGaugh, J. L., Modulation of memory storage, *Curr. Opin. Neurobiol.*, 6, 237, 1996.
11. Ammasseuri-Teule, M., Hoffmann, H. J., Rossi-Arnaud, C., Learning in inbred mice: strain-specific abilities across three radial maze problems, *Behav. Genet.*, 23, 405, 1993.
12. Crawley, J. N., Belknap, J. K., Collins, A., Crabbe, J. C., Frankel, W., Henderson, N., Hitzemann, R. J., Maxson, S. C., Miner, L. L., Silva, A. J., Wehner, J. M., Wynshaw-Boris, A., Paylor, R., Behavioral phenotypes of inbred mouse strains: implications and recommendations for molecular studies, *Psychopharmacology*, 132, 107, 1997.
13. Imperato, A., Obinu, M. C., Mascia M. S., Casu, M. A., Zocchi, A., Cabib, S., Puglisi-Allegra, S., Strain-dependent effects of dopamine agonists on acetylcholine release in the hippocampus: an *in vivo* study in mice, *Neuroscience*, 70, 653, 1996.
14. Vekovishcheva, O., Zvartau, E. E., The reactivity of laboratory mice in pharmacological tests as dependent on their zoosocial status, *Eksp. Klin. Farmakol.*, 62, 6, 1999.
15. Destrade, C., Soumireu-Mourat, B., Cardo, B., Effects of postrial hippocampal stimulation on acquisition of operant behavior in the mouse, *Behav. Biol.*, 8, 713, 1973.
16. Messier, C., Mourre, C., Bontempi, B., Sif, J., Lazdunski, M., Destrade, C., Effect of apamin, a toxin that inhibits Ca^{2+}-dependent K^+ channels, on learning and memory processes, *Brain Res.*, 551, 322, 1991.
17. Gauthier, M., Soumireu-Mourat, B., 6-Hydroxydopamine and radio frequency lesions of the lateral entorhinal cortex facilitate an operant appetitive conditioning task in mice, *Neurosc. Lett.*, 24, 193, 1981.
18. Olton, D. S., Samuelson, R. J., *J. Exp. Psychol.: Anim. Behav. Procs.*, 2, 97, 1976.
19. McDonald, R. J., White, N. M., Hippocampal and nonhippocampal contributions to place learning in rats, *Behav. Neurosci.*, 109, 579, 1995.
20. Morris, R. G., Garrud, P., Rawlins, J. N., O'Keefe, J., Place navigation is impaired in rats with hippocampal lesions, *Nature*, 297, 681, 1982.
21. Guillou, J.-L., Rose, G. M., Cooper, D. M., Differential activation of adenylyl cyclases by spatial and procedural learning, *J. Neurosc.*, 19(4), 6183, 1999.
22. Phillips, R. G., Ledoux, J.E., Differential contribution of amygdala and hippocampus to cued and contextual fear conditioning, *Behav. Neurosci.*, 106, 274, 1992.
23. Oler, L. A., Markus, E. J., Age-related deficits on the radial maze and in fear conditioning: hippocampal processing and consolidation, *Hippocampus*, 8, 402, 1998.

24. Desmedt, A., Garcia, R., Jaffard, R. Differential modulation of changes in hippocampal-septal synaptic excitability by the amygdala as a function of either elemental or contextual fear conditioning in mice, *J. Neurosci.*, 18, 480, 1998.

25. Young, S. L., Fanselow, M. S., Associative regulation of pavlovian fear conditioning: unconditional stimulus intensity, incentive shifts, and latent inhibition, *J. Exp. Psychol.: An. Behav. Proc.*, 18, 400, 1992.

26. Marighetto, A., Micheau, J., Jaffard, R., Relationships between testing-induced alterations of hippocampal cholinergic activity and memory performance on two spatial tasks in mice, *Behav. Brain Res.*, 56, 133, 1993.

Index